AF676392

Genetics of Biodegradation and Bioremediation

Genetics of Biodegradation and Bioremediation

Special Issue Editors

Eduardo Santero
Eduardo Díaz

MDPI • Basel • Beijing • Wuhan • Barcelona • Belgrade • Manchester • Tokyo • Cluj • Tianjin

Special Issue Editors
Eduardo Santero
Universidad Pablo de Olavide
Spain

Eduardo Díaz
Centro de Investigaciones Biológicas Margarita Salas (CIB-CSIC)
Spain

Editorial Office
MDPI
St. Alban-Anlage 66
4052 Basel, Switzerland

This is a reprint of articles from the Special Issue published online in the open access journal *Genes* (ISSN 2073-4425) (available at: https://www.mdpi.com/journal/genes/special_issues/biodeg_biorem).

For citation purposes, cite each article independently as indicated on the article page online and as indicated below:

LastName, A.A.; LastName, B.B.; LastName, C.C. Article Title. *Journal Name* **Year**, *Article Number*, Page Range.

ISBN 978-3-03936-236-3 (Hbk)
ISBN 978-3-03936-237-0 (PDF)

© 2020 by the authors. Articles in this book are Open Access and distributed under the Creative Commons Attribution (CC BY) license, which allows users to download, copy and build upon published articles, as long as the author and publisher are properly credited, which ensures maximum dissemination and a wider impact of our publications.
The book as a whole is distributed by MDPI under the terms and conditions of the Creative Commons license CC BY-NC-ND.

Contents

About the Special Issue Editors

Eduardo Santero received a Ph.D. in biology from the University of Seville, Spain, in 1986. Full professor at the University Pablo de Olavide since 2003, he has been vice-chancellor for research at the same university (2003–2007), and deputy director (2007–2011) and director (2011–2015) of the Andalusian Center for Developmental Biology. This is a joint research center funded by the Spanish Research Council (CSIC), the Andalusian Regional Government Junta de Andalucía, and the Universidad Pablo de Olavide (UPO) in Seville, where he is currently principal investigator. His research group focuses on transcriptional and posttranscriptional gene regulation mechanisms in bacteria of environmental interest, mainly those regulating biodegradation genes, and the design and construction of regulatory circuits for environmental and biomedical applications. He has more than 80 publications, six patents, and countless communications in congresses. He is a member of the Editorial Board of Genes, a frequent reviewer for many journals, and a member of grant evaluation panels for national and international institutions.

Eduardo Díaz received a Ph.D. in biology from the Complutense University of Madrid (Spain) in 1991. He is currently a research scientist at the Department of Microbial and Plant Biotechnology of the Biological Research Center Margarita Salas (CIB) of the Spanish National Research Council (CSIC); he heads its environmental microbiology group, which studies biochemical and genetic aspects of the bacterial metabolism of pollutants and their biotechnological applications (bioremediation, biosensors, bioconversions). He has more than 90 publications published in international peer-reviewed journals, one book, fifteen book chapters, eight patents, and more than 170 communications at congresses and workshops. He is a member of the editorial board of five highly reputed microbiology journals (*J. Bacteriol.*, *Curr. Opin. Biotechnol.*, *Environ. Microbiol.*, *Microb. Biotechnol.*, *Appl. Environ. Microbiol.*), a collaborator at the Bioscience and Biotechnology Area of the Spanish State Research Agency (AEI), and the cofounder of a biotechnology start-up.

genes

MDPI

Editorial

Special Issue: Genetics of Biodegradation and Bioremediation

Eduardo Santero [1,*] and Eduardo Díaz [2]

1 Department of Molecular Biology and Biochemical Engineering. Centro Andaluz de Biología del Desarrollo, Universidad Pablo de Olavide/Consejo Superior de Investigaciones Científicas/Junta de Andalucía, 41013 Seville, Spain

2 Department of Microbial and Plant Biotechnology, Centro de Investigaciones Biológicas Margarita Salas, Agencia Estatal Consejo Superior de Investigaciones Científicas, 28040 Madrid, Spain; ediaz@cib.csic.es

* Correspondence: esansan@upo.es

Received: 14 April 2020; Accepted: 15 April 2020; Published: 17 April 2020

Abstract: Many different biodegradation pathways, both aerobic and anaerobic, have already been characterised, and the phylogenetic relationships among catabolic genes within the different types of pathways have been studied. However, new biodegradation activities and their coding genes are continuously being reported, including those involved in the catabolism of emerging contaminants or those generally regarded as non-biodegradable. Gene regulation is also an important issue for the efficient biodegradation of contaminants. Specific induction by the substrate and over-imposed global regulatory networks adjust the expression of the biodegradation genes to the bacterial physiological needs. New biodegradation pathways can be assembled in a particular strain or in a bacterial consortium by recruiting biodegradation genes from different origins through horizontal gene transfer. The abundance and diversity of biodegradation genes, analysed by either genomic or metagenomic approaches, constitute valuable indicators of the biodegradation potential of a particular environmental niche. This knowledge paves the way to systems metabolic engineering approaches to valorise biowaste for the production of value-added products.

Keywords: biodegradation; bioremediation; valorisation; catabolic pathway; mobile DNA; anaerobic biodegradation; gene regulation

---

Industrial activity has resulted in the continuous production of huge amounts of contaminant molecules, many of which are very persistent in nature because of their stable chemical structure, low biodegradability, or toxicity, thus leading to their accumulation in the environment to concentrations that may affect natural populations of living beings. However, some microorganisms, mainly bacteria, have evolved to metabolise these contaminants, in many cases using them as nutrient and/or energy sources. Besides the importance of these biodegrading strains to mitigate contamination of natural areas, the study of their biodegradation pathways has revealed very interesting features of the catabolic activities involved, how they evolve in really short periods of time, and how they can be spread among different bacteria.

This book reprints 12 contributions that show recent advances in our knowledge of different aspects of biodegradation of particular contaminants in different bacteria at biochemical, genetic, and genomic levels.

One of the most abundant contaminants on Earth is the xenobiotic plastic polyethylene terephthalate (PET). It is chemically very stable and highly refractory to microbial biodegradation. However, a number of bacteria have been recently reported that have PET hydrolase activities and further ability to use the resulting monomers (ethylene glycol and terephthalate) as carbon and energy sources. These biodegradation activities that allow PET to be used as a feedstock and the possibility of biosynthesising PET from sustainable substrates in a circular bio-PET economy not dependent on fossil fuels are reviewed by Salvador et al. [1].

Two additional reviews deal with related aromatic contaminants that can be found in crude oils. The review by Phale et al. [2] focuses on the mechanisms of assembly of biodegradation pathways of the well-known biaromatic naphthalene and its derivative, carbaryl, that have apparently evolved by recruiting different catabolic activities via horizontal gene transfer through mobile genetic elements. Tetralin, found in oil but also industrially produced from naphthalene or anthracene, is similar to naphthalene, but one of the rings is alicyclic. The biodegradation pathway by which a conventional enzymatic team involved in the biodegradation of one aromatic ring is able to metabolise both the aromatic and the alicyclic ring of tetralin is thoroughly reviewed at biochemical, genetics, genomics, and gene regulation levels [3].

Steroids are complex molecules composed of four alicyclic rings, which are produced by eukaryotic cells and also synthesised by the pharma industry. Although they are very recalcitrant due to their hydrophobic nature and the absence of functional groups, several steroid-metabolising bacteria have been isolated, and their biodegradation pathways characterised. The different strategies to metabolise steroids by different bacteria, including a comparative analysis of the genes responsible for these activities, are reviewed by Olivera and Luengo [4]. Steroid biodegradation is also the focus of an original research paper by Ibero et al. [5], where the testosterone biodegradation pathway of a *Novosphingobium* strain is inferred from in silico genomic analysis and subsequently confirmed experimentally.

Besides bearing the genes that code for biodegradation capability, it is crucial for an efficient catabolic process that these genes are expressed at sufficiently high levels and only when the substrate to be metabolised is present in order to prevent the wasteful production of catabolic enzymes and reactions when they are not needed or advantageous. Regulatory systems, which may evolve independently of the catabolic genes, then have to adjust the range of molecules to which they respond to the range of molecules that can be metabolised by the regulated catabolic pathway in order to prevent gratuitous induction by a molecule that cannot be metabolised. Besides this specific regulation, catabolic pathways may be controlled by an over-imposed regulatory circuit, resulting in a catabolite repression phenomenon that prevents their expression when the bacterium can use other more favourable nutrients. The importance of a regulatory system leading to sufficient expression of the catabolic genes is shown by Igeño et al. [6], who demonstrate that a strain of *Pseudomonas pseudoalcaligenes* with the potential to use furfuryl alcohol, furfural, and furoic acid as carbon and energy sources can only grow on these substrates when it gains a particular mutation in a regulatory gene coding for an AraC-type transcriptional activator. Durante-Rodríguez et al. [7] documented the regulation of an anaerobic benzoate biodegradation pathway by identifying the *bzdR* repressor gene in different *Azoarcus*/*Aromatoleum* strains and characterising the repressor binding and repressor complex formation at the target promoter-regulatory region. The relevance of the specificity of the regulatory protein for the response to a particular inducer and how this specificity may be tuned to create variants that respond synergistically to several effectors is shown by Tumen-Velasquez et al. [8] by comparing the effector specificity of two very similar LysR-type activators, the CatM and BenM paralogs in *Acinetobacter baylyi* ADP1. Finally, the sophisticated regulatory system consisting of three regulatory circuits that control the expression of the tetralin biodegradations genes is reviewed by Floriano et al. [3].

Genes coding for a catabolic pathway are always in a genomic context, either chromosomal or extra-chromosomal (e.g., plasmid, transposon, integrative-conjugative element), and are expressed in bacteria with different metabolic capabilities and physiologies. Exploring the genomic contexts of catabolic genes becomes essential to know the potential for mobility of such genes, as well as to understand the physiology and metabolism of the host strain, including the biodegradation potential of other contaminants, and the similarities and differences among different biodegradation strains. Genomic analysis of a strain able to degrade the highly toxic cyanide has revealed its potential to also degrade furfural [5]. Genomic analysis by Imperato et al. [9] of two naphthalene-degrading

Pseudomonas strains isolated from an oil-contaminated field revealed that their genomes also bear genes for the degradation of other contaminants such, as alkanes, BTEX (acronym for benzene, toluene, ethyl benzene and xylenes), anthranilate, or terephthalate. Whereas both strains contain a plasmid, only the large plasmid of one of the strains bear the whole set of *nah* genes required to use naphthalene as the carbon and energy source. The other strain apparently has the *nah* genes associated with mobile DNA sequences but integrated into the chromosome. Naphthalene biodegradation genes associated with mobile DNA elements is also reviewed by Phale et al [2]. Genomic analysis of *Sphingopyxis lindanitolerans*, a strain able to degrade the γ-hexachlorocyclohexane pesticide, by Kaminski et al. [10] revealed that it contains two plasmids that bear different *lin* genes associated with insertion sequences, which together encode the entire lindane biodegradation pathway. Genomic analysis of ten strains isolated from biphenyl-contaminated sites by Hirose et al. [11] showed the very high incidence of *bph* genes associated with integrative conjugative elements and revealed the molecular basis of its dissemination potential by lateral transfer among different bacteria.

Bacteria are not alone in nature, and they interact with each other. In many instances, biodegradation is achieved by a consortium of microorganisms that may have similar or even complementary activities, rather than by a single population. This cooperation is even more evident when the pollution is caused by a mixture of contaminants, such as that produced by spills or leakage of petroleum hydrocarbons. In these cases, bioremediation of a polluted site requires the concerted action of different populations of bacteria with different catabolic capacities. The characterisation of these bacterial consortia is crucial for the understanding of bioremediation processes of complex mixtures of pollutants. The paper by Garrido-Sanz et al. [12] analyses the population composition of a consortium isolated from a diesel-contaminated site, showing how this composition changes when growing in the presence of a specific hydrocarbon, and assigning functions to particular members of the consortium.

In summary, all the contributions presented in this book highlight the importance of unravelling the genetic determinants responsible for the astonishing genomic plasticity of biodegradation strains to rapidly adapt to the presence of new molecules that are continuously released to the environment, some of which are emerging priority pollutants. Moreover, these studies pave the way to systems metabolic engineering approaches to valorise some abundant contaminants as feedstock for the production of value-added products in a circular bioeconomy strategy.

Author Contributions: Both authors contributed equally to original draft preparation, as well as the reviewing and editing process. All authors have read and agree to the published version of the manuscript.

Funding: This editorial was funded by grants BIO2014–57545-R and SAF2017-85785-R, co-funded by the Spanish Ministerio de Educación y Ciencia and the European Regional Development Fund, to E.S.; and by grants BIO2016-79736-R, CSIC 2019 20E005, and H2020 Grant 760994 from the Spanish Ministerio de Economía y Competitividad, CSIC and European Union, to E.D.

Acknowledgments: We are very grateful to all authors and reviewers for their contributions to this issue.

Conflicts of Interest: The authors declare no conflict of interest. The funders had no role in the writing of the manuscript.

References

1. Salvador, M.; Abdulmutalib, U.; Gonzalez, J.; Kim, J.; Smith, A.A.; Faulon, J.-P.; Wei, R.; Zimmermann, W.; Jiménez, J.-I. Microbial Genes for a Circular and Sustainable Bio-PET Economy. *Genes* **2019**, *10*, 373. [CrossRef] [PubMed]
2. Phale, P.S.; Shah, B.A.; Malhotra, H. Variability in Assembly of Degradation Operons for Naphthalene and its derivative, Carbaryl, Suggests Mobilization through Horizontal Gene Transfer. *Genes* **2019**, *10*, 569. [CrossRef] [PubMed]
3. Floriano, B.; Santero, E.; Reyes-Ramírez, F. Biodegradation of Tetralin: Genomics, Gene Function and Regulation. *Genes* **2019**, *10*, 339. [CrossRef] [PubMed]

4. Olivera, E.R.; Luengo, J.M. Steroids as Environmental Compounds Recalcitrant to Degradation: Genetic Mechanisms of Bacterial Biodegradation Pathways. *Genes* **2019**, *10*, 512. [CrossRef] [PubMed]
5. Ibero, J.; Galán, B.; Díaz, E.; García, J.L. Testosterone Degradative Pathway of *Novosphingobium tardaugens*. *Genes* **2019**, *10*, 871. [CrossRef]
6. Igeño, M.I.; Macias, D.; Blasco, R. Case of Adaptive Laboratory Evolution (ALE): Biodegradation of Furfural by *Pseudomonas pseudoalcaligenes* CECT 5344. *Genes* **2019**, *10*, 499. [CrossRef]
7. Durante-Rodríguez, G.; Gutiérrez-del-Arroyo, P.; Vélez, M.; Díaz, E.; Carmona, M. Further Insights into the Architecture of the P_N Promoter That Controls the Expression of the *bzd* Genes in *Azoarcus*. *Genes* **2019**, *10*, 489. [CrossRef] [PubMed]
8. Tumen-Velasquez, M.P.; Laniohan, N.S.; Momany, C.; Neidle, E.L. Engineering CatM, a LysR-Type Transcriptional Regulator, to Respond Synergistically to Two Effectors. *Genes* **2019**, *10*, 421. [CrossRef] [PubMed]
9. Imperato, V.; Portillo-Estrada, M.; McAmmond, B.M.; Douwen, Y.; Van Hamme, J.D.; Gawronski, S.W.; Vangronsveld, J.; Thijs, S. Genomic Diversity of Two Hydrocarbon-Degrading and Plant Growth-Promoting *Pseudomonas* Species Isolated from the Oil Field of Bóbrka (Poland). *Genes* **2019**, *10*, 443. [CrossRef] [PubMed]
10. Kaminski, M.A.; Sobczak, A.; Dziembowski, A.; Lipinski, L. Genomic Analysis of γ-Hexachlorocyclohexane-Degrading *Sphingopyxis lindanitolerans* WS5A3p Strain in the Context of the Pangenome of *Sphingopyxis*. *Genes* **2019**, *10*, 688. [CrossRef] [PubMed]
11. Hirose, J.; Fujihara, H.; Watanabe, T.; Kimura, N.; Suenaga, H.; Futagami, T.; Goto, M.; Suyama, A.; Furukawa, K. Biphenyl/PCB Degrading *bph* Genes of Ten Bacterial Strains Isolated from Biphenyl-Contaminated Soil in Kitakyushu, Japan: Comparative and Dynamic Features as Integrative Conjugative Elements (ICEs). *Genes* **2019**, *10*, 404. [CrossRef] [PubMed]
12. Garrido-Sanz, D.; Redondo-Nieto, M.; Guirado, M.; Pindado Jiménez, O.; Millán, R.; Martín, M.; Rivilla, R. Metagenomic Insights into the Bacterial Functions of a Diesel-Degrading Consortium for the Rhizoremediation of Diesel-Polluted Soil. *Genes* **2019**, *10*, 456. [CrossRef] [PubMed]

© 2020 by the authors. Licensee MDPI, Basel, Switzerland. This article is an open access article distributed under the terms and conditions of the Creative Commons Attribution (CC BY) license (http://creativecommons.org/licenses/by/4.0/).

genes

Review

Microbial Genes for a Circular and Sustainable Bio-PET Economy

Manuel Salvador [1], Umar Abdulmutalib [1], Jaime Gonzalez [1], Juhyun Kim [1], Alex A. Smith [1], Jean-Loup Faulon [2,3,4], Ren Wei [5], Wolfgang Zimmermann [5] and Jose I. Jimenez [1,*]

1 Faculty of Health and Medical Sciences, University of Surrey, Guildford GU2 7XH, UK; m.salvadordelara@surrey.ac.uk (M.S.); u.abdulmutalib@surrey.ac.uk (U.A.); j.gonzalezgutierrezdelaconch@surrey.ac.uk (J.G.); juhyun.kim@surrey.ac.uk (J.K.); alex.smith@surrey.ac.uk (A.A.S.)
2 Micalis Institute, INRA, AgroParisTech, Université Paris-Saclay, 78350 Jouy-en-Josas, France; jean-loup.faulon@inra.fr
3 SYNBIOCHEM Centre, Manchester Institute of Biotechnology, University of Manchester, Manchester M1 7DN, UK
4 CNRS-UMR8030/Laboratoire iSSB, Université Paris-Saclay, 91000 Évry, France
5 Department of Microbiology and Bioprocess Technology, Institute of Biochemistry, Leipzig University, 04103 Leipzig, Germany; wei@uni-leipzig.de (R.W.); wolfgang.zimmermann@uni-leipzig.de (W.Z.)
* Correspondence: j.jimenez@surrey.ac.uk; Tel.: +44-14-8368-4557

Received: 23 April 2019; Accepted: 13 May 2019; Published: 16 May 2019

Abstract: Plastics have become an important environmental concern due to their durability and resistance to degradation. Out of all plastic materials, polyesters such as polyethylene terephthalate (PET) are amenable to biological degradation due to the action of microbial polyester hydrolases. The hydrolysis products obtained from PET can thereby be used for the synthesis of novel PET as well as become a potential carbon source for microorganisms. In addition, microorganisms and biomass can be used for the synthesis of the constituent monomers of PET from renewable sources. The combination of both biodegradation and biosynthesis would enable a completely circular bio-PET economy beyond the conventional recycling processes. Circular strategies like this could contribute to significantly decreasing the environmental impact of our dependence on this polymer. Here we review the efforts made towards turning PET into a viable feedstock for microbial transformations. We highlight current bottlenecks in degradation of the polymer and metabolism of the monomers, and we showcase fully biological or semisynthetic processes leading to the synthesis of PET from sustainable substrates.

Keywords: plastics; biodegradation; sustainability; upcycling; biotransformations; polyethylene terephthalate; terephthalate; ethylene glycol

1. Introduction

Thermoplastic polymers, some of which constitute the majority of the commonly known plastics, are extremely useful materials endowed with properties that make them ideal for applications such as insulation and packaging [1,2]. They are durable, water-proof and versatile materials that have become almost essential in our lives. In fact, in 2017 the contribution of plastics to the European economy reached a market size of EUR 355 billion while employing 1.5 million people [3]. Plastics are light-weight and have significantly contributed to decreasing transportation costs and extending the shelf life of food [4]. Their success as a material is only comparable to their detrimental environmental impact. The accumulation of plastic waste in the environment has become an extremely serious concern [5,6]. Plastic pollution is present in every single niche of the planet, with dramatic effects on ecosystems, especially in marine environments, affecting equally large and small fauna and flora [6,7].

Plastics possess two key features: they are barely degradable by environmental physical, chemical and especially by biological processes [8], and they have low production costs, which make their reuse not economically competitive. While these individual properties are desirable, when combined they lead to the current problem we are facing: the accumulation of recalcitrant and polymers in the environment that can degrade into microplastics with potential toxic effects [9]. The story of plastic pollution is a story of mismanagement of an otherwise valuable resource. Numerous recent studies have highlighted the poor recycling rates of plastics compared to other materials. For instance, a recent report estimates the amount of virgin plastics produced from oil of over 8 billion metric tons, out of which only 9% have been recycled [10]. This reflects a saturated traditional recycling industry and emphasises the need for novel approaches to plastic management, including the possibility of harnessing microbial activities to use plastic waste as a feedstock for biotransformations [11–14].

Out of all plastics, polyesters such as polyethylene terephthalate (PET) are in a good position for becoming a sustainable polymer compared to other oil-derived counterparts. PET is obtained from the polymerisation of the constituent monomers terephthalic acid (TPA) and ethylene glycol (EG) (Figure 1). It is durable, relatively easy to mould by blowing, which results in an almost inert, hard and stiff polymer that has been adopted by the beverage industry as the main material for the production of bottles [15,16]. PET has, in addition, the highest collection rates of all plastics even though reused PET is only a small fraction of the total PET consumed: The US National Association for PET Container Resources (NAPCOR) reported that out of the approximately 3 million tons of new PET bottles reaching the market in 2017, only 29% of them were made from collected and recycled PET, a nearly 5% decrease compared to recycling rates of the previous year [17].

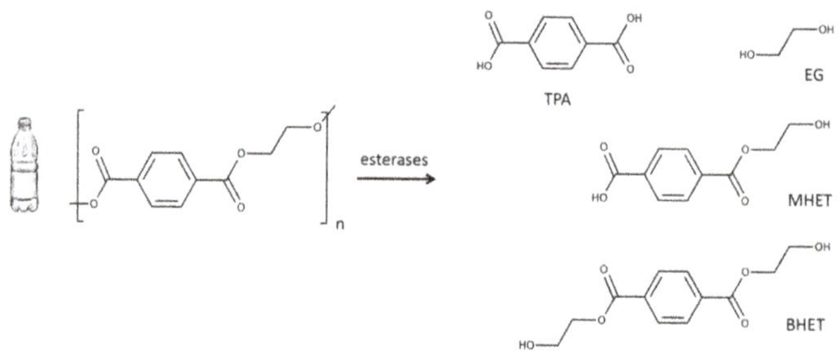

Figure 1. Enzymatic hydrolysis of polyethylene terephthalate (PET) results in a mixture of terephthalic acid (TPA) and ethylene glycol (EG) and, to a lesser extent, the incomplete hydrolysis products bis-(2-hydroxyethyl) terephthalate (BHET) and mono-(2-hydroxyethyl) terephthalate (MHET).

As a polyester, PET can be depolymerized as a more effective alternative to mechanical recycling [18]. Methods of depolymerization include glycolysis, methanolysis, hydrolysis, aminolysis and ammonolysis [19]. Among them, glycolysis has recently emerged as a key technology for recycling PET waste. Glycolysis is the process of PET degradation by glycols at high temperatures and in the presence of catalysts such as metal acetates [20]. Compared to other methods, glycolysis has the great advantage of enabling the recycling of coloured and opaque PET that cannot be otherwise recycled due to the presence of the pigments. The resulting monomers TPA and EG can be re-used to produce PET, as well as other polymers of interest [21]. Glycolysis and related methods contribute to a more sustainable PET economy, although they also have drawbacks such as the energy cost of the high temperatures required and the long reaction times needed for effective depolymerization [21].

Biological activities capable of catalysing PET hydrolysis under mild reaction conditions are emerging as an alternative to chemical PET depolymerization methods [22]. As a result, a number of enzymes from different microorganisms have been characterised [23–25], facilitating the implementation of PET as a biotechnological feedstock [11,26]. We argue that this strategy is more versatile than chemical methods because, if funnelled to the central microbial metabolism, the monomers obtained can be transformed into a plethora of molecules by harnessing advances in synthetic biology and metabolic engineering. This would contribute to creating a path for revenue from PET waste beyond current recycling activities. It could thereby help to mitigate the impact of PET environmental release and promote the competitive development of a next generation of environmentally friendly materials.

Given the interesting physicochemical properties of PET and its potential use as a substrate in biotechnology, in this article we review the genes that are required for a sustainable and circular PET economy. In our view, to accomplish this goal it is required to (i) improve the kinetics of PET enzymatic hydrolysis; (ii) link the metabolism of the resulting monomers to relevant biosynthetic pathways and (iii) engineer biological systems for the production of PET monomers TPA and EG from renewable sources.

2. PET Metabolism

The enzymatic hydrolysis of PET involves the release of constituent monomers TPA and EG due to the action of esterases. The resulting monomers can be degraded by microorganisms endowed with the appropriate metabolic pathways for these compounds. TPA is converted into protocatechuate (PCA) that will undergo dioxygenolytic cleavage and degradation through different routes prior to reaching the central metabolism [27–30]. Similarly, EG is assimilated through different pathways depending on the microorganism. For instance, it can be transformed into acetate via acetyl-CoA in *Acetobacterium woodii* [31], whereas in some strains in *Pseudomonas putida* it is funnelled directly to the Krebs cycle via isocitrate [32]. In this section we will focus on the genes responsible for these activities and their (co)occurrence in different bacterial taxa.

2.1. Enzymatic Hydrolysis of PET

Different types of hydrolases have shown to be active against the PET polymer. These enzymes are lipases, esterases, cutinases and carboxylesterases isolated from fungi and bacteria (see [14,25] for recent reviews on this topic). They belong to the α/β hydrolase superfamily and have evolved in a different context and for a different function [33]. For instance, the original role of the cutinases from the genus *Thermobifida* is to hydrolyse the plant polyester cutin. Among the different variants of these enzymes, the ones endowed with certain properties (e.g., a more accessible active site) display the highest activity against PET [34]. A recent bioinformatic analysis has investigated the distribution of genes encoding for homologs of these esterases in terrestrial and marine metagenomes and has allowed to identify 504 new hydrolases [35]. The two main conclusions of this study are: (i) genes potentially encoding polyester hydrolases are rare, and (ii) their taxonomic distribution seems to be related to the niche studied, with *Actinobacteria* or *Proteobacteria* being more prominent hosts in terrestrial environments, whereas *Bacteroidetes* are the most frequent hosts in marine metagenomes [35].

As a new-to-nature polymer, PET constitutes a challenge for any of the hydrolases that are active against it. In this sense, it is worth highlighting that not all types of PET are equally susceptible to microbial degradation. Depending on processing and thermal treatments, PET can occur in an amorphous form or a semi-crystalline form [36]. It has been shown that the extent of enzymatic polyester hydrolysis depends on the degree of its crystallinity and chain orientation [37]. In the amorphous regions, the polymer chains are less densely packed and are more susceptible to hydrolytic attack compared to the crystalline regions. The enzymatic degradation rate of the polyester correlates with the temperature difference between the melting temperature of the polymer and the hydrolysis temperature. The polymer chain can be considered to be more mobile and accessible to enzymatic attack when close to the glass transition temperature (Tg) of amorphous PET [38]. Therefore, increased enzymatic hydrolysis rates of PET are expected when performing the reaction at temperatures near the Tg of the amorphous polyester (67–71 °C). This suggests that efficient PET hydrolysis needs to be conducted by thermostable polyester hydrolases such as the cutinases TfCut2 and HiC isolated, respectively, from the thermophilic actinomycete *Thermobifida fusca* [23] or the fungus *Thermomyces insolens* [37], both of which, especially the latter, have been reported to be active for long periods of time at temperatures of up to 70 °C. Engineered post-translational modifications (e.g., glycosylation) can then be used on these polyester hydrolases to improve thermal properties of the enzymes further [39]. Hydrolysis at those temperatures is obviously not compatible with most bioprocesses using whole-cell catalysts, especially those involving engineered mesophilic organisms such as *Escherichia coli* which can grow up to a maximum temperature of 48–50 °C only after evolutionary adaptation and at a fitness cost [40,41]. The bacterium *Ideonella sakaiensis* has been reported to be capable of growing on PET as a sole carbon source due to the secretion of a PET hydrolase [24]. When tested in vitro and in mesophilic conditions (below the Tg of PET), this enzyme shows very low degradation rates of PET and, even though this activity could be increased somewhat by directed evolution [42], potential hydrolysis yields are far from being able to sustain industrial bioprocesses.

Another important factor affecting the performance of the enzymes hydrolysing PET is their inhibition mediated by mono-(2-hydroxyethyl) terephthalate (MHET) and bis-(2-hydroxyethyl) terephthalate (BHET), by-products of an incomplete hydrolysis [43]. These molecules are oligomers of TPA and EG that act as competitive inhibitors of the enzymes [44]. Even though it is possible to design reactors that allow for a continuous removal of MHET and BHET [45], this is likely to pose a challenge for the biodegradation of PET using whole cells. Other solutions have been tested such as the use of mixtures of hydrolases that act synergistically [46], or the selective modification of amino acid residues of the polyester hydrolase involved in the interaction with the inhibitors [47]. These factors emphasise the need for obtaining enzymes, either by direct screening or by modification of existing ones, which are not susceptible to inhibition by MHET and BHET and can therefore be used to develop efficient bioprocesses using PET as the substrate.

2.2. Metabolism of TPA

TPA is transformed into PCA by the pathway encoded by the *tph* genes. These genes encode two sequential catabolic steps: the addition of two hydroxyl groups in positions 4 and 5 of TPA by the activity of the TPA dioxygenase TphA1A2A3 producing 1,6-dihydroxycyclohexa-2, 4-diene-dicarboxylate (DCD), and the removal of the carboxyl group in position 6 by the action of the 1,2-dihydroxy-3,5-cyclohexadiene-1,4-dicarboxylate dehydrogenase TphB (Figure 2A). The genes responsible for those activities have been characterised in the actinomycete *Rhodococcus* sp. strain DK17 [48], in the β-proteobacteria *Comamonas testosteroni* YZW-D [49], and in *Comamonas* sp. strain E6 [50]. In addition to the catabolic *tph* genes, both organisms encode within this cluster the transcriptional regulator TphR (Figure 2B). TphR has been described as an IclR-type activator that responds to the inducer TPA [51]. *Comamonas* sp. strain E6 also contains the extra gene *tphC*, which encodes a permease involved in the uptake of TPA using the tripartite aromatic acid transporter [52].

We conducted a systematic analysis of the presence of the *tph* genes in the genomes available in public databases. As a result, we identified genes sharing a significant identity (greater than 35% for all the genes in the cluster with *tphA2* greater than 65%) and similar genetic organisation in only a limited number of organisms, which are representative of β-proteobacteria (*Comamonas*, *Ideonella* and *Ramlibacter*) and γ-proteobacteria (*Pseudomonas*), as well as of actinomycetes (*Rhodococcus*). In the genus *Rhodococcus* the *tph* genes are associated with plasmids with the exception of *Rhodococcus opacus* 1CP in which the cluster of genes was identified in the chromosome. In all the genomes investigated, the four catabolic genes were conserved in the same order. All clusters contain a regulatory gene encoding an IclR-type transcriptional regulator upstream the catabolic genes and in a divergent orientation. More diversity was observed in the putative transport of TPA inside the cell: all the β-proteobacteria utilized the transporter *tphC*, whereas the rest of the organisms contained a previously unidentified MFS transporter of the AAHS family (aromatic acid:H^+ symporter; named *tphK*) homologous to the *p*-hydroxybenzoate transporter *pcaK* [53] (Figure 2B).

2.3. Metabolism of PCA

The PCA resulting from the activity of the Tph enzymes follows different pathways depending on the organism. This suggests that the *tph* genes can act as an independent metabolic module regardless of the type of PCA metabolism present in the TPA degrading strain. In fact, two copies of this cluster of genes are harboured by two different plasmids in *Rhodococcus* sp. strain DK17, indicating that this pathway can be mobilised by horizontal gene transfer into species containing one of the widespread PCA degradation pathways [48]. All PCA pathways share an initial dioxygenolytic step in which the aromatic ring is cleaved. Until now, three different pathways have been reported depending on the cleavage position in the aromatic ring. They are known as the *ortho-*, *meta-* and *para-*cleavage pathways and their initial reaction is catalysed by a PCA-3,4-, 4,5- and 2,3-dioxygenase, respectively (Figure 3) [27,29,54]. For simplicity, we will refer from now on to the nomenclature of the enzymes to discriminate between the pathways.

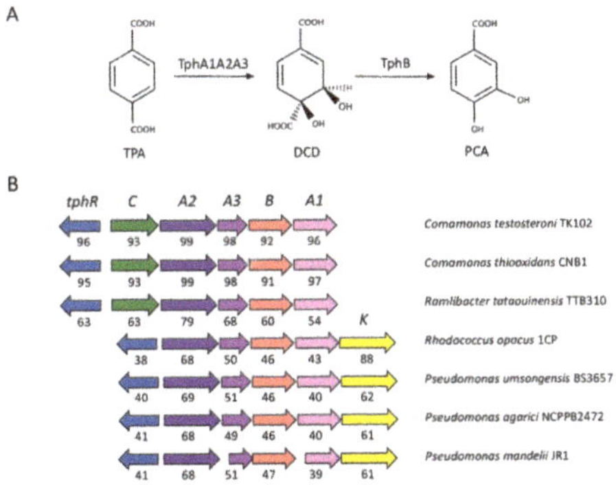

Figure 2. (**a**) TPA metabolism reported in bacteria. The names of the molecules and abbreviations are: terephthalic acid, TPA; 1,6-dihydroxycyclohexa-2,4-diene-dicarboxylate, DCD; protocatechuate, PCA. (**b**) Genetic organisation of the *tph* genes identified in several genomes available in databases. Numbers below arrows indicate the percentage of identity compared to the orthologous genes present in *Comamonas* sp. E6 (accession: AB238679; [50]) with the exception of the *tphK* genes that were compared to the ortholog present in the plasmid pDK3 of *Rhodococcus* sp. DK17 (accession: AY502076; [48]). Plots were produced with SyntTax (http://archaea.u-psud.fr/SyntTax; [55]).

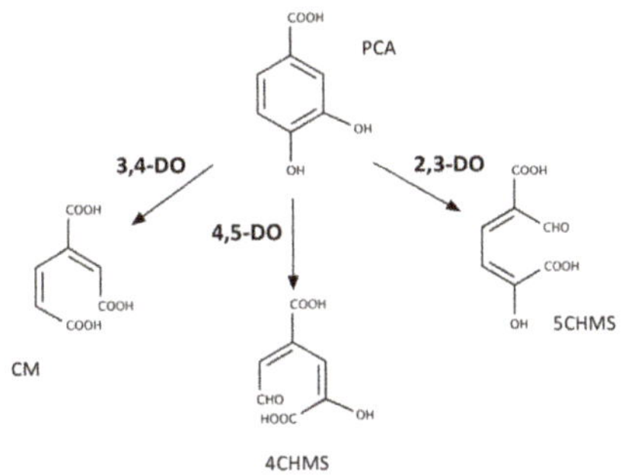

Figure 3. Types of dioxygenase-mediated reactions involved in PCA cleavage by bacteria. DO: dioxygenase; CM: 3-carboxy-cis,cis-muconate; 4CHMS: 4-carboxy-2-hydroxymuconate semialdehyde; 5CHMS: 5-carboxy-2-hydroxymuconate-6-semialdehyde.

Using the sequences of characterised PCA dioxygenases, we conducted a bioinformatics search of the pathways likely involved in the metabolism of PCA that are present in the genomes in which we had previously identified the genes responsible for the conversion of TPA into PCA. Out of the three pathways, the PCA-2,3-dioxygenase was not present in any of them. Among the β-proteobacteria, *C. testosteroni*, *C. thiooxydans* and *R. tataouinensis* have homologs of the PCA-4,5-dioxygenase in their genomes, whereas *I. sakaiensis*, the different species of *Pseudomonas* and *R. opacus* contain the PCA-3,4-dioxygenase pathway. These results are consistent with previous observations showing that a PCA-3,4-dioxygenase activity is present in cells of *Rhodococcus* sp. strain DK17 growing on TPA [48], whereas a PCA-4,5-dioxygenase activity was identified in *Comamonas* sp. strain E6 [50]. Likewise, *I. sakaiensis* has been reported to contain a *tph* cluster and PCA-3,4-pathway [24].

The diversity of PCA metabolic pathways is an important factor when considering developing bioprocesses based on PET. Depending on the pathway used, a range of metabolites can be produced with different applications in mind. Out of them, the PCA-3,4-dioxygenolytic pathway has been thoroughly studied. This route is one of the branches of the β-ketoadipate pathway that connects the metabolism of aromatics converging on either catechol (e.g., benzoate) or PCA (e.g., 4-hydrozybenzoate) with the central metabolism of certain bacterial species [30]. The β-ketoadipate pathway has traditionally been used as a way of incorporating toxic and recalcitrant aromatic molecules in the central metabolism of bacteria, including nitrophenols and polychlorinated arenes. It is also an important path for funnelling the degradation products of lignocellulosic waste that could be used for the synthesis of other molecules of interest [56]. Strikingly, despite the metabolic diversity of the pathways involved which could allow for the production of molecules with interesting properties (e.g., functionalised lactones), complete mineralization of PCA continues to be the main application of the PCA metabolism. Only recently, PCA obtained from lignin-derived aromatics has been used for synthesis of the industrially relevant metabolite adipic acid [57]. This has not been achieved by the action of any of the described PCA pathways, but by the conversion of PCA into catechol catalysed by a PCA decarboxylase. Catechol is then transformed into *cis,cis*-muconate by the action of a catechol-1,2-dioxygenase, and the latter is hydrogenated abiotically to adipic acid in the presence of a catalyst [58].

2.4. Metabolism of EG

The metabolism of EG is more diverse compared to TPA. In acetogens, EG is oxidised to ethanol and acetaldehyde that is eventually converted to acetate via acetyl-CoA [31]. In other bacterial species, however, EG is degraded via the formation of glyoxylate (GLA) (Figure 4A) [59,60]. Activities responsible for the conversion of EG into GLA have been identified in multiple organisms. These initial steps are catalysed by dehydrogenases with broad specificity involved in the metabolism of short-chain alcohols and aldehydes such as the propanediol oxidoreductase of *E. coli* (also known as lactaldehyde reductase AldA) [61]. In *Pseudomonas aeruginosa* and *P. putida*, the initial reaction is carried out by periplasmic alcohol dehydrogenases that depend on pyrroloquinoline quinone for their activity [32,62]. Once GLA is produced, the pathway proceeds to intermediates of the central metabolism through different routes depending on the organism. For instance, whereas in *Escherichia coli* the pathway continues to acetyl-CoA via 3-phosphoglycerate—this is called the "canonical" pathway [63]—it has been proposed that some strains of *P. putida* make use of the shunt that funnels GLA to the Krebs cycle via isocitrate or malate [32,64]. The genetic determinants of the canonical GLA pathway have been identified in different microorganisms. The reactions are catalysed by the enzymes GLA carboligase (Gcl), tartronate semialdehyde reductase (GlxR) and glycerate-2-kinase (GlxK), all of which are encoded in the same cluster of genes in *E. coli* K12 and *Pseudonocardia dioxanivorans* strain CB1190 [65,66].

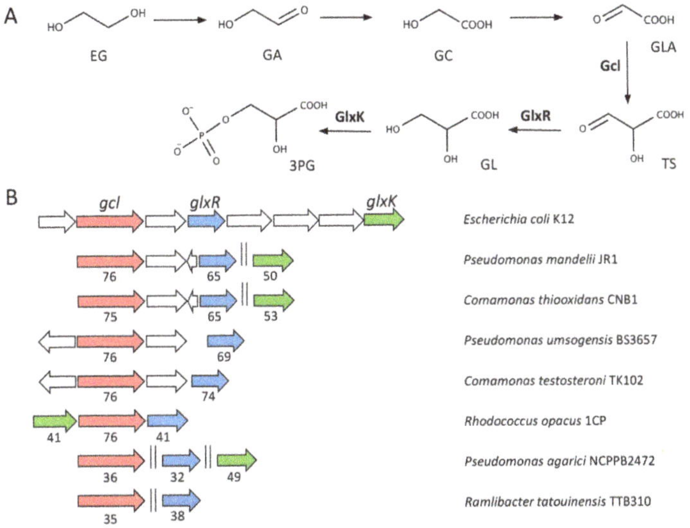

Figure 4. (**a**) EG metabolism via GLA. The GLA canonical pathway described in the text is shown. The 3-phosphoglycerate (3PG) produced is later funnelled into the central metabolism via acetyl-CoA. The names of the molecules and abbreviations are: ethylene glycol, EG; glycoladehyde, GA; glycolate, GC; glyoxylate, GLA; tartronate semialdehyde, TS; glycerate, GL. (**b**) Genetic organisation of the genes involved in GLA metabolism identified in several genomes available in databases. Numbers below arrows indicate the percentage of identity compared to the orthologous genes present in *E. coli* K12 (accession: AP009048; [67]). Plots were produced with SyntTax (http://archaea.u-psud.fr/SyntTax; [55]).

Using the sequences of FucO and Gcl from *E. coli* as probes, we conducted an analysis of the likelihood of the occurrence of activities for EG degradation in different bacteria. Homologs to *fucO* are widespread and present in all organisms investigated (not shown). Added to the broad substrate specificity of the enzymes active against EG, this suggests that EG degradation is a relatively common feature in bacteria. Likewise, the canonical pathway for GLA degradation seems ubiquitous as *gcl* is conserved in a very large number of bacterial species (not shown). As TPA degradation genes are not as frequently present in bacterial genomes, next we investigated the presence of activities for EG degradation in the strains that we had previously identified as carriers of the *tph* genes for TPA mineralisation (Figure 4B). All of them contain homologs to *fucO* or alcohol dehydrogenases similar to *pedE* described in *Pseudomonas* species. Moreover, all of them contain homologs to *gcl*, *glxR* and *glxK*, although only the genetic organization of these genes in *R. opacus* resembles that of *E. coli*. Contrary to the case of TPA, our synteny search did not identify conserved transporters involved in the uptake of EG or GLA. Likewise, no regulatory elements controlling the expression of the genes responsible for GLA degradation could be found.

Taken all together, these results indicate that most organisms capable of degrading TPA are also likely able to degrade EG, thereby enabling a more efficient usage of the products resulting from PET hydrolysis. In this sense, it has been recently demonstrated that EG can be readily transformed into the bioplastic polyhydroxyalkanoate in an engineered strain of *P. putida* KT2440 [64], underlining the usability of microorganisms for the conversion of oil-derived plastics into bioplastics.

3. Anabolism of Monomers Used for Bio-PET Synthesis

Bio-based PET, also known as bio-PET, is the common term used to refer to a PET polymer in which at least a fraction of the constituent monomers is obtained from biological—and therefore renewable—sources. In this section we will review recent efforts to produce TPA and EG involving microorganisms at any step (Figure 5). These methods can be fully or at least partially biotic and may involve abiotic physico-chemical steps. Even if not completely green, these synthetic processes promise to decrease the dependence on virgin PET derived from fossil feedstocks and may certainly contribute to a fully circular and sustainable PET economy.

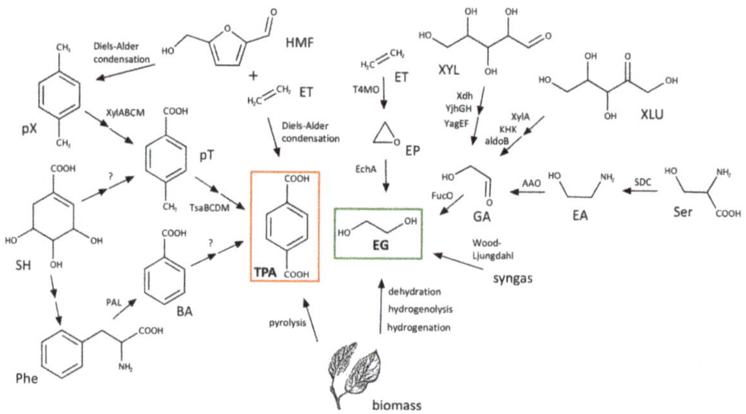

Figure 5. Selected pathways and processes used to produce TPA and EG from renewable sources. The names of the molecules and abbreviations are: *p*-xylene, pX; *p*-toluate, pT; shikimate, SH; phenylalanine, Phe; benzoate, BA; 5-(hydromethyl)furfural, HMF; ethylene, ET; ethylene oxide, EP; xylose, XYL; xylulose, XLU; glycoaldehyde, GA; serine, Ser; ethanolamine, EA. If known, the names of enzymes/processes responsible for the different conversions are shown next to the arrows.

3.1. Biosynthesis of TPA

The microbial biosynthesis of aromatic compounds has not been characterised with the same level of detail as their degradation. Despite this, there are a number of pathways that render aromatic compounds and generally involve the metabolism of aromatic amino acids and the shikimate pathway, or the condensation of molecules such as *cis,cis*-muconate [68]. Unfortunately, none of the currently known pathways are likely to allow for the direct production of TPA from central intermediates. It has been proposed, however, that the shikimate pathway could be used to produce *p*-toluate that could later be transformed into TPA, although the activities required for this pathway have not been identified [69]. Inspired by this, we have conducted a retrosynthesis analysis of plausible biochemical reactions that could render TPA using as substrates molecules present in the metabolism of *E. coli*. This allows for the formulations of reactions that are chemically plausible (e.g., because the mechanism involves reactive groups following known mechanistic rules), even though this might be in the absence of any biochemical evidence [70]. This method is particularly useful for guiding the screening of genomic or metagenomic libraries in search of genes coding for enzymes capable of catalysing a proposed reaction, as well as for the lab-directed evolution of known enzymes for the efficient catalysis of novel reactions. Our analysis resulted in a number of pathways leading to benzoate that can be obtained from phenylalanine, which itself is produced from shikimic acid [71]. The last step, however, will be more difficult to take place biotically as it would involve the conversion of benzoate into TPA by direct incorporation in the aromatic ring of a carboxylic group coming from bicarbonate, a step that is typically conducted at high temperatures and in the presence of metal catalysts [72,73].

Another possibility for the sustainable production of TPA is to use aromatics obtained from renewable sources such as lignin [74]. A recent work shows that TPA biosynthesis can be achieved from *p*-xylene [75]. This process was successfully implemented in *E. coli* by the heterologous expression of segments of two different pathways. In this process, *p*-xylene is first converted into toluic acid by the action of the xylene monooxygenase (XylMA), benzyl alcohol dehydrogenase (XylB) and the benzaldehyde dehydrogenase (XylC) of the TOL pathway for the degradation of toluene and xylene encoded in the pWW0 plasmid of *P. putida* mt-2 [76]. These enzymes oxidise, respectively, one of the methyl groups of xylene to a carboxylic group via formation of the corresponding alcohol and aldehyde [77]. Toluic acid is later transformed into TPA by the action of a toluene sulfate monooxygenase (TsaMB), a 4-carboxybenzaldehyde dehydrogenase (TsaC) and a 4-carboxybenzalcohol dehydrogenase (TsaD) present in *C. testosteroni* T2 [78].

This biosynthetic pathway poses a significant improvement in terms of sustainability compared to conventional chemical methods [75], but obtaining *p*-xylene from renewable sources also poses a considerable challenge. This has been solved by using isobutanol [79,80] or biomass as substrates for different chemical transformations. Pyrolysis of biomass [81], as well as the Diels-Alders condensation of ethylene with different types of biomass-derived molecules (e.g., furans) can be used to produce *p*-xylene or TPA [82–85]. Ethylene itself can be produced by different biosynthetic pathways, some of which have been harnessed to produce high levels of this molecule in engineered bacteria [86,87].

3.2. Biosynthesis of EG

Given the difficulties of obtaining TPA from sources other than fossil feedstocks, bio-PET typically refers to a PET polymer in which only EG is obtained from renewable sources [88]. EG accounts for 30% of the mass of the polymer and, therefore, this is usually the maximum percentage of bio components encountered in bio-PET. As recently reviewed in [89], there are a number of artificial pathways that have been engineered to obtain EG from renewable plant feedstocks using microorganisms. Among them, biosynthesis of EG in bacteria can be achieved in high yields by a pentose pathway that uses xylose as a substrate (Figure 5). Xylose is first transformed into xylonate by the action of a dehydrogenase. After the subsequent action of a dehydratase and an aldolase, glycoaldehyde is obtained, which is finally reduced to EG by a reductase [90–92]. This pathway has been extensively engineered to increase production yields that currently reach a 98% of the theoretical maximum and constitute a promising alternative for the synthesis of EG [93].

The engineered xylose pathway is not the only way of obtaining EG. It can also be produced from glucose in *Saccharomyces cerevisiae* using glycolytic enzymes [94] and via the synthesis of serine in an engineered pathway in *E. coli* [95]. Serine is transformed into ethanolamine by a plant serine decarboxylase. Ethanolamine is later transformed into glycolaldehyde by an oxidase and the latter reduced to EG by a reductase (Figure 5). The pathway has been artificially reconstituted in *E. coli* and is also amenable to metabolic engineering efforts to improve production yields. More recent work has shown the feasibility of using synthesis gas (syngas) for the production of EG harnessing the Wood-Ljungdahl pathway of carbon fixation present in acetogenic bacterial species such as *Moorella thermoacetica* and *Clostridium ljungdahlii* [96]. In another approach, EG was obtained from gaseous alkenes by a strain of *E. coli* that expresses recombinantly a monooxygenase and an epoxide hydrolase [97].

Similarly to TPA, EG can also be directly obtained from biomass. This can be achieved through the dehydration of cellulosic ethanol [98], the hydrogenolysis of xylitol [99] and the hydrogenation of corn stalk [100]. This reflects a wide diversity of options for the production of EG that could be used to replace the chemical procedures relying on fossil feedstocks.

4. Future Prospects and Concluding Remarks

Here we have reviewed the potential use of PET as a feedstock for microbial biotransformations. We have identified the challenges of large-scale PET enzymatic hydrolysis and proposed strategies for the enhancement of this process using enzymes—and possibly organisms—capable of being stable and active near the Tg of the polymer.

Similarly to the case of glycolytic procedures, TPA and EG resulting from hydrolysis could be used for the synthesis of fresh PET, but we also advocate for their biotransformation into molecules or processes with added value. An example of this could be their use in microbial fuel cells for the production of electricity that has been achieved using TPA as a carbon source [101]. TPA metabolism is neither widespread nor diverse in the genomes currently available. This could constitute a bottleneck for the development of future applications that currently have adipic acid as the main target of molecules funnelled through the PCA pathways. EG metabolic genes, on the contrary, are found in numerous organisms and encode a more diverse metabolism, likely enabling a variety of applications.

Hand in hand with an efficient degradation of PET, a circular economy of this polymer requires a sustainable large-scale synthesis of TPA and EG. We have reviewed a number of efforts made for the biosynthesis of bio-PET using renewable sources. On this front a milestone seems to have been reached recently with the production of the first bottle that is completely made of monomers obtained from biological sources [102]. Any method, including those reaching a maximum of 30% bio-PET, have a significantly lower carbon footprint compared with the synthesis of oil-derived plastics and are worth pursuing. By using plant biomass, it is possible to contribute to CO_2 fixation, although a major breakthrough would be to obtain TPA or EG with engineered microorganisms directly from CO_2.

Overall, the prospects for a circular bio-based economy of PET are encouraging, and most of the technological hurdles for either biodegradation or biosynthesis have already been overcome, or there are alternatives or clear strategies to overcome them. Although bio-approaches to the PET economy might not be as profitable as the current *status quo* in the short term, there is an undeniable pressure from the general public to manage PET differently, and this is already producing changes in policies and regulations. In our view, this will at least partially bridge the strict financial gap compared to chemical processes, which will enable itself the diversification of applications of PET including its upcycling in other molecules. In the long run, this will have a positive impact on recycling rates and will also lower the environmental release of PET waste, therefore contributing to solving an imperative environmental concern.

Funding: J.I.J., W.Z., M.S., A.A.S., J.K. and R.W. would like to acknowledge the support received from the European Union's Horizon 2020 research and innovation programme under grant agreement no. 633962 for the project P4SB. J.I.J., A.A.S. and J.K. acknowledge the support from the Biotechnology and Biological Sciences Research Council (BBSRC) (grant BB/M009769/1). J.L.F. acknowledges the support from the Engineering and Physical Sciences Research Council (EPSRC) (grant EP/N025504/1). U.A. is the recipient of a PhD studentship from the Petroleum Technology Development Fund of Nigeria.

Conflicts of Interest: The authors declare no conflict of interest. The funders had no role in the design of the study; in the collection, analyses, or interpretation of data; in the writing of the manuscript, or in the decision to publish the results.

References

1. Andrady, A.L.; Neal, M.A. Applications and societal benefits of plastics. *Philos. Trans. R. Soc. B Biol. Sci.* **2009**, *364*, 1977–1984. [CrossRef] [PubMed]
2. Scalenghe, R. Resource or waste? A perspective of plastics degradation in soil with a focus on end-of-life options. *Heliyon* **2018**, *4*, e00941. [CrossRef]
3. PlasticsEurope: Plastics—The Facts 2018. Available online: https://www.plasticseurope.org/en/resources/publications/619-plastics-facts-2018 (accessed on 16 May 2019).
4. Upasen, S.; Wattanachai, P. Packaging to prolong shelf life of preservative-free white bread. *Heliyon* **2018**, *4*, e00802. [CrossRef] [PubMed]
5. Ryan, P.G.; Moore, C.J.; van Franeker, J.A.; Moloney, C.L. Monitoring the abundance of plastic debris in the marine environment. *Philos. Trans. R. Soc. B Biol. Sci.* **2009**, *364*, 1999–2012. [CrossRef] [PubMed]
6. Barnes, D.K.A.; Galgani, F.; Thompson, R.C.; Barlaz, M. Accumulation and fragmentation of plastic debris in global environments. *Philos. Trans. R. Soc. B Biol. Sci.* **2009**, *364*, 1985–1998. [CrossRef] [PubMed]
7. Derraik, J.G. The pollution of the marine environment by plastic debris: A review. *Mar. Pollut. Bull.* **2002**, *44*, 842–852. [CrossRef]
8. Kubowicz, S.; Booth, A.M. Biodegradability of plastics: Challenges and misconceptions. *Environ. Sci. Technol.* **2017**, *51*, 12058–12060. [CrossRef]
9. Alimba, C.G.; Faggio, C. Microplastics in the marine environment: Current trends in environmental pollution and mechanisms of toxicological profile. *Environ. Toxicol. Pharmacol.* **2019**, *68*, 61–74. [CrossRef]

10. Geyer, R.; Jambeck, J.R.; Law, K.L. Production, use, and fate of all plastics ever made. *Sci. Adv.* **2017**, *3*, e1700782. [CrossRef]
11. Wierckx, N.; Prieto, M.A.; Pomposiello, P.; de Lorenzo, V.; O'Connor, K.; Blank, L.M. Plastic waste as a novel substrate for industrial biotechnology. *Microb. Biotechnol.* **2015**, *8*, 900–903. [CrossRef] [PubMed]
12. Wierckx, N.; Narancic, T.; Eberlein, C.; Wei, R.; Drzyzga, O.; Magnin, A.; Ballerstedt, H.; Kenny, S.T.; Pollet, E.; Avérous, L.; et al. Plastic biodegradation: Challenges and opportunities. In *Consequences of Microbial Interactions with Hydrocarbons, Oils, and Lipids: Biodegradation and Bioremediation*; Springer: Cham, Switzerland, 2018; pp. 1–29.
13. Narancic, T.; O'Connor, K.E. Microbial biotechnology addressing the plastic waste disaster. *Microb. Biotechnol.* **2017**, *10*, 1232–1235. [CrossRef] [PubMed]
14. Koshti, R.; Mehta, L.; Samarth, N. Biological recycling of polyethylene terephthalate: A mini-review. *J. Polym. Environ.* **2018**, *26*, 3520–3529. [CrossRef]
15. Arroyo, M. Thermoplastic polyesters. In *Handbook of Thermoplastics*; Olabisi, O., Ed.; Marcel Dekker: New York, NY, USA, 1997; pp. 417–448.
16. Ji, L.N. Study on preparation process and properties of polyethylene terephthalate (PET). *Appl. Mech. Mater.* **2013**, *312*, 406–410. [CrossRef]
17. NAPCOR. Report on Postconsumer PET Container Recycling Activity in 2017. Available online: https://napcor.com/wp-content/uploads/2018/11/NAPCOR_2017RateReport_FINAL.pdf (accessed on 16 May 2019).
18. Nikles, D.E.; Farahat, M.S. New motivation for the depolymerization products derived from poly(ethylene terephthalate) (PET) waste: A review. *Macromol. Mater. Eng.* **2005**, *290*, 13–30. [CrossRef]
19. Al-Sabagh, A.M.; Yehia, F.Z.; Eshaq, G.; Rabie, A.M.; ElMetwally, A.E. Greener routes for recycling of polyethylene terephthalate. *Egypt. J. Pet.* **2016**, *25*, 53–64. [CrossRef]
20. Sinha, V.; Patel, M.; Patel, J. Pet Waste Management by Chemical Recycling: A Review. *J. Polym. Environ.* **2010**, *18*, 8–25. [CrossRef]
21. Furtwengler, P.; Avérous, L. Renewable polyols for advanced polyurethane foams from diverse biomass resources. *Polym. Chem.* **2018**, *9*, 4258–4287. [CrossRef]
22. Webb, H.K.; Arnott, J.; Crawford, R.J.; Ivanova, E.P. Plastic degradation and its environmental implications with special reference to poly(ethylene terephthalate). *Polymers* **2013**, *5*, 1–18. [CrossRef]
23. Roth, C.; Wei, R.; Oeser, T.; Then, J.; Föllner, C.; Zimmermann, W.; Sträter, N. Structural and functional studies on a thermostable polyethylene terephthalate degrading hydrolase from *Thermobifida fusca*. *Appl. Microbiol. Biotechnol.* **2014**, *98*, 7815–7823. [CrossRef]
24. Yoshida, S.; Hiraga, K.; Takehana, T.; Taniguchi, I.; Yamaji, H.; Maeda, Y.; Toyohara, K.; Miyamoto, K.; Kimura, Y.; Oda, K. A bacterium that degrades and assimilates poly(ethylene terephthalate). *Science* **2016**, *351*, 1196–1199. [CrossRef] [PubMed]
25. Wei, R.; Oeser, T.; Zimmermann, W. Synthetic polyester-hydrolyzing enzymes from thermophilic actinomycetes. *Adv. Appl. Microbiol.* **2014**, *89*, 267–305.
26. Wei, R.; Zimmermann, W. Biocatalysis as a green route for recycling the recalcitrant plastic polyethylene terephthalate. *Microb. Biotechnol.* **2017**, *10*, 1302–1307. [CrossRef] [PubMed]
27. Kasai, D.; Fujinami, T.; Abe, T.; Mase, K.; Katayama, Y.; Fukuda, M.; Masai, E. Uncovering the protocatechuate 2,3-cleavage pathway genes. *J. Bacteriol.* **2009**, *191*, 6758–6768. [CrossRef] [PubMed]
28. Maruyama, K.; Shibayama, T.; Ichikawa, A.; Sakou, Y.; Yamada, S.; Sugisaki, H. Cloning and characterization of the genes encoding enzymes for the protocatechuate meta-degradation pathway of *Pseudomonas ochraceae* NGJ1. *Biosci. Biotechnol. Biochem.* **2004**, *68*, 1434–1441. [CrossRef]
29. Frazee, R.W.; Livingston, D.M.; LaPorte, D.C.; Lipscomb, J.D. Cloning, sequencing, and expression of the *Pseudomonas putida* protocatechuate 3,4-dioxygenase genes. *J. Bacteriol.* **1993**, *175*, 6194–6202. [CrossRef]
30. Harwood, C.S.; Parales, R.E. The β-ketoadipate pathway and the biology of self-identity. *Annu. Rev. Microbiol.* **1996**, *50*, 553–590. [CrossRef]
31. Trifunović, D.; Schuchmann, K.; Müller, V. Ethylene glycol metabolism in the acetogen *Acetobacterium woodii*. *J. Bacteriol.* **2016**, *198*, 1058–1065. [CrossRef]
32. Mückschel, B.; Simon, O.; Klebensberger, J.; Graf, N.; Rosche, B.; Altenbuchner, J.; Pfannstiel, J.; Huber, A.; Hauer, B. Ethylene glycol metabolism by *Pseudomonas putida*. *Appl. Environ. Microbiol.* **2012**, *78*, 8531–8539. [CrossRef] [PubMed]

33. Zimmermann, W.; Billig, S. Enzymes for the biofunctionalization of poly(ethylene terephthalate). In *Biofunctionalization of Polymers and Their Applications*; Nyanhongo, G.S., Steiner, W., Gübitz, G., Eds.; Springer: Berlin/Heidelberg, Germany, 2011; pp. 97–120.
34. Herrero Acero, E.; Ribitsch, D.; Steinkellner, G.; Gruber, K.; Greimel, K.; Eiteljoerg, I.; Trotscha, E.; Wei, R.; Zimmermann, W.; Zinn, M.; et al. Enzymatic surface hydrolysis of PET: Effect of structural diversity on kinetic properties of cutinases from *Thermobifida*. *Macromolecules* **2011**, *44*, 4632–4640. [CrossRef]
35. Danso, D.; Schmeisser, C.; Chow, J.; Zimmermann, W.; Wei, R.; Leggewie, C.; Li, X.; Hazen, T.; Streit, W.R. New insights into the function and global distribution of polyethylene terephthalate (PET)-Degrading bacteria and enzymes in marine and terrestrial metagenomes. *Appl. Environ. Microbiol.* **2018**, *84*, e02773-17. [CrossRef] [PubMed]
36. Wang, Z.; Ma, Z.; Li, L. Flow-induced crystallization of polymers: Molecular and thermodynamic considerations. *Macromolecules* **2016**, *49*, 1505–1517. [CrossRef]
37. Ronkvist, Å.M.; Xie, W.; Lu, W.; Gross, R.A. Cutinase-catalyzed hydrolysis of poly(ethylene terephthalate). *Macromolecules* **2009**, *42*, 5128–5138. [CrossRef]
38. Marten, E.; Müller, R.-J.; Deckwer, W.-D. Studies on the enzymatic hydrolysis of polyesters. II. Aliphatic–aromatic copolyesters. *Polym. Degrad. Stab.* **2005**, *88*, 371–381. [CrossRef]
39. Shirke, A.N.; White, C.; Englaender, J.A.; Zwarycz, A.; Butterfoss, G.L.; Linhardt, R.J.; Gross, R.A. Stabilizing Leaf and branch Compost Cutinase (LCC) with glycosylation: Mechanism and effect on PET hydrolysis. *Biochemistry* **2018**, *57*, 1190–1200. [CrossRef]
40. Guyot, S.; Pottier, L.; Hartmann, A.; Ragon, M.; Hauck Tiburski, J.; Molin, P.; Ferret, E.; Gervais, P. Extremely rapid acclimation of *Escherichia coli* to high temperature over a few generations of a fed-batch culture during slow warming. *Microbiologyopen* **2014**, *3*, 52–63. [CrossRef]
41. Rudolph, B.; Gebendorfer, K.M.; Buchner, J.; Winter, J. Evolution of *Escherichia coli* for growth at high temperatures. *J. Biol. Chem.* **2010**, *285*, 19029–19034. [CrossRef]
42. Austin, H.P.; Allen, M.D.; Donohoe, B.S.; Rorrer, N.A.; Kearns, F.L.; Silveira, R.L.; Pollard, B.C.; Dominick, G.; Duman, R.; El Omari, K.; et al. Characterization and engineering of a plastic-degrading aromatic polyesterase. *Proc. Natl. Acad. Sci. USA* **2018**, *115*, E4350–E4357. [CrossRef]
43. Wei, R.; Zimmermann, W. Microbial enzymes for the recycling of recalcitrant petroleum-based plastics: How far are we? *Microb. Biotechnol.* **2017**, *10*, 1308–1322. [CrossRef] [PubMed]
44. Barth, M.; Oeser, T.; Wei, R.; Then, J.; Schmidt, J.; Zimmermann, W. Effect of hydrolysis products on the enzymatic degradation of polyethylene terephthalate nanoparticles by a polyester hydrolase from *Thermobifida fusca*. *Biochem. Eng. J.* **2015**, *93*, 222–228. [CrossRef]
45. Barth, M.; Wei, R.; Oeser, T.; Then, J.; Schmidt, J.; Wohlgemuth, F.; Zimmermann, W. Enzymatic hydrolysis of polyethylene terephthalate films in an ultrafiltration membrane reactor. *J. Membr. Sci.* **2015**, *494*, 182–187. [CrossRef]
46. Carniel, A.; Valoni, É.; Nicomedes, J.; Gomes, A.d.C.; Castro, A.M.d. Lipase from *Candida antarctica* (CALB) and cutinase from *Humicola insolens* act synergistically for PET hydrolysis to terephthalic acid. *Process Biochem.* **2017**, *59*, 84–90. [CrossRef]
47. Wei, R.; Oeser, T.; Schmidt, J.; Meier, R.; Barth, M.; Then, J.; Zimmermann, W. Engineered bacterial polyester hydrolases efficiently degrade polyethylene terephthalate due to relieved product inhibition. *Biotechnol. Bioeng.* **2016**, *113*, 1658–1665. [CrossRef]
48. Choi, K.Y.; Sul, W.J.; Kim, Y.M.; Kim, E.; Kim, D.; Zylstra, G.J.; Chae, J.-C. Molecular and biochemical analysis of phthalate and terephthalate degradation by *Rhodococcus* sp. strain DK17. *FEMS Microbiol. Lett.* **2005**, *252*, 207–213. [CrossRef] [PubMed]
49. Wang, Y.Z.; Zhou, Y.; Zylstra, G.J. Molecular analysis of isophthalate and terephthalate degradation by Comamonas testosteroni YZW-D. *Environ. Health Perspect.* **1995**, *103*, 9–12. [PubMed]
50. Sasoh, M.; Masai, E.; Ishibashi, S.; Hara, H.; Kamimura, N.; Miyauchi, K.; Fukuda, M. Characterization of the terephthalate degradation genes of *Comamonas* sp. strain E6. *Appl. Environ. Microbiol.* **2006**, *72*, 1825–1832. [CrossRef] [PubMed]
51. Kasai, D.; Kitajima, M.; Fukuda, M.; Masai, E. Transcriptional regulation of the terephthalate catabolism operon in Comamonas sp. strain E6. *Appl. Environ. Microbiol.* **2010**, *76*, 6047–6055. [CrossRef] [PubMed]

52. Hosaka, M.; Kamimura, N.; Toribami, S.; Mori, K.; Kasai, D.; Fukuda, M.; Masai, E. Novel tripartite Aromatic Acid Transporter Essential for Terephthalate Uptake in *Comamonas* sp. Strain E6. *Appl. Environ. Microbiol.* **2013**, *79*, 6148–6155. [CrossRef] [PubMed]
53. Nichols, N.N.; Harwood, C.S. PcaK, a high-affinity permease for the aromatic compounds 4-hydroxybenzoate and protocatechuate from *Pseudomonas putida*. *J. Bacteriol.* **1997**, *179*, 5056–5061. [CrossRef] [PubMed]
54. Noda, Y.; Nishikawa, S.; Shiozuka, K.; Kadokura, H.; Nakajima, H.; Yoda, K.; Katayama, Y.; Morohoshi, N.; Haraguchi, T.; Yamasaki, M. Molecular cloning of the protocatechuate 4,5-dioxygenase genes of *Pseudomonas paucimobilis*. *J. Bacteriol.* **1990**, *172*, 2704–2709. [CrossRef]
55. Oberto, J. SyntTax: A web server linking synteny to prokaryotic taxonomy. *BMC Bioinf.* **2013**, *14*, 4. [CrossRef] [PubMed]
56. Wells Jr, T.; Ragauskas, A.J. Biotechnological opportunities with the b-ketoadipate pathway. *Trends Biotechnol.* **2012**, *30*, 627–637. [CrossRef]
57. Johnson, C.W.; Salvachúa, D.; Khanna, P.; Smith, H.; Peterson, D.J.; Beckham, G.T. Enhancing muconic acid production from glucose and lignin-derived aromatic compounds via increased protocatechuate decarboxylase activity. *Metab. Eng. Commun.* **2016**, *3*, 111–119. [CrossRef] [PubMed]
58. Vardon, D.R.; Franden, M.A.; Johnson, C.W.; Karp, E.M.; Guarnieri, M.T.; Linger, J.G.; Salm, M.J.; Strathmann, T.J.; Beckham, G.T. Adipic acid production from lignin. *Energy Environ. Sci.* **2015**, *8*, 617–628. [CrossRef]
59. Child, J.; Willetts, A. Microbial metabolism of aliphatic glycols bacterial metabolism of ethylene glycol. *Biochim. Biophys. Acta Gen. Subj.* **1978**, *538*, 316–327. [CrossRef]
60. Kataoka, M.; Sasaki, M.; Hidalgo, A.-R.G.D.; Nakano, M.; Shimizu, S. Glycolic acid production using ethylene glycol-oxidizing microorganisms. *Biosci. Biotechnol. Biochem.* **2001**, *65*, 2265–2270. [CrossRef] [PubMed]
61. Boronat, A.; Aguilar, J. Rhamnose-induced propanediol oxidoreductase in *Escherichia coli*: Purification, properties, and comparison with the fucose-induced enzyme. *J. Bacteriol.* **1979**, *140*, 320–326.
62. Wehrmann, M.; Billard, P.; Martin-Meriadec, A.; Zegeye, A.; Klebensberger, J. Functional role of lanthanides in enzymatic activity and transcriptional regulation of pyrroloquinoline quinone-dependent alcohol dehydrogenases in *Pseudomonas putida* KT2440. *MBio* **2017**, *8*, e00570-17. [CrossRef]
63. Boronat, A.; Caballero, E.; Aguilar, J. Experimental evolution of a metabolic pathway for ethylene glycol utilization by *Escherichia coli*. *J. Bacteriol.* **1983**, *153*, 134–139.
64. Franden, M.A.; Jayakody, L.N.; Li, W.-J.; Wagner, N.J.; Cleveland, N.S.; Michener, W.E.; Hauer, B.; Blank, L.M.; Wierckx, N.; Klebensberger, J.; et al. Engineering *Pseudomonas putida* KT2440 for efficient ethylene glycol utilization. *Metab. Eng.* **2018**, *48*, 197–207. [CrossRef]
65. Cusa, E.; Obradors, N.; Baldomà, L.; Badía, J.; Aguilar, J. Genetic analysis of a chromosomal region containing genes required for assimilation of allantoin nitrogen and linked glyoxylate metabolism in *Escherichia coli*. *J. Bacteriol.* **1999**, *181*, 7479–7484. [PubMed]
66. Grostern, A.; Sales, C.M.; Zhuang, W.-Q.; Erbilgin, O.; Alvarez-Cohen, L. Glyoxylate metabolism Is a key feature of the metabolic degradation of 1,4-dioxane by *Pseudonocardia dioxanivorans* strain CB1190. *Appl. Environ. Microbiol.* **2012**, *78*, 3298–3308. [CrossRef] [PubMed]
67. Wanner, B.L.; Wishart, D.; Blattner, F.R.; Thomas, G.H.; Plunkett Guy, I.I.I.; Mori, H.; Keseler, I.M.; Glasner, J.D.; Rudd, K.E.; Serres, M.H.; et al. *Escherichia coli* K-12: A cooperatively developed annotation snapshot—2005. *Nucleic Acids Res.* **2006**, *34*, 1–9.
68. Huccetogullari, D.; Luo, Z.W.; Lee, S.Y. Metabolic engineering of microorganisms for production of aromatic compounds. *Microb. Cell Fact.* **2019**, *18*, 41. [CrossRef]
69. Osterhout, R.E.; Burgard, A.P.; Pharkya, P.; Burk, P. Microorganisms and Methods for the Biosynthesis of Aromatics, 2,4-Pentadienoate and 1,3-Butadiene. U.S. Patent 8715957 B2, 26 July 2011.
70. Delépine, B.; Duigou, T.; Carbonell, P.; Faulon, J.-L. RetroPath2.0: A retrosynthesis workflow for metabolic engineers. *Metab. Eng.* **2018**, *45*, 158–170. [CrossRef] [PubMed]
71. Moore, B.S.; Hertweck, C.; Hopke, J.N.; Izumikawa, M.; Kalaitzis, J.A.; Nilsen, G.; O'Hare, T.; Piel, J.; Shipley, P.R.; Xiang, L.; et al. Plant-like biosynthetic pathways in bacteria: From benzoic acid to chalcone. *J. Nat. Prod.* **2002**, *65*, 1956–1962. [CrossRef]
72. Lind, W.; Campbell, R. Preparation of Potassium Terephthalate. U.S. Patent 3761515 A, 21 October 1971.
73. Bernhard, R. Production of Terephthalic Acid. U.S. Patent 2823229 A, 20 June 1956.

74. Graglia, M.; Kanna, N.; Esposito, D. Lignin refinery: Towards the preparation of renewable aromatic building blocks. *ChemBioEng Rev.* **2015**, *2*, 377–392. [CrossRef]
75. Luo, Z.W.; Lee, S.Y. Biotransformation of *p*-xylene into terephthalic acid by engineered *Escherichia coli*. *Nat. Commun.* **2017**, *8*, 15689. [CrossRef] [PubMed]
76. Franklin, F.C.; Bagdasarian, M.; Bagdasarian, M.M.; Timmis, K.N. Molecular and functional analysis of the TOL plasmid pWWO from *Pseudomonas putida* and cloning of genes for the entire regulated aromatic ring meta cleavage pathway. *Proc. Natl. Acad. Sci. USA* **1981**, *78*, 7458–7462. [CrossRef] [PubMed]
77. Harayama, S.; Rekik, M.; Wubbolts, M.; Rose, K.; Leppik, R.A.; Timmis, K.N. Characterization of five genes in the upper-pathway operon of TOL plasmid pWW0 from *Pseudomonas putida* and identification of the gene products. *J. Bacteriol.* **1989**, *171*, 5048–5055. [CrossRef] [PubMed]
78. Junker, F.; Kiewitz, R.; Cook, A.M. Characterization of the p-toluenesulfonate operon tsaMBCD and tsaR in *Comamonas testosteroni* T-2. *J. Bacteriol.* **1997**, *179*, 919–927. [CrossRef]
79. Dedov, A.G.; Loktev, A.S.; Karavaev, A.A.; Moiseev, I.I. A novel direct catalytic production of *p*-xylene from isobutanol. *Mendeleev Commun.* **2018**, *28*, 352–353. [CrossRef]
80. Peters, M.; Taylor, J.; Jenni, M.; Manzer, L.; Henton, D. Integrated process to selectively convert renewable isobutanol to p-xylene. U.S. Patent US 20110087000 A1, 6 October 2010.
81. Chang, R.; Zhu, L.; Jin, F.; Fan, M.; Liu, J.; Jia, Q.; Tang, C.; Li, Q. Production of bio-based *p*-xylene via catalytic pyrolysis of biomass over metal oxide-modified HZSM-5 zeolites. *J. Chem. Technol. Biotechnol.* **2018**, *93*, 3292–3301. [CrossRef]
82. Pacheco, J.J.; Davis, M.E. Synthesis of terephthalic acid via Diels-Alder reactions with ethylene and oxidized variants of 5-hydroxymethylfurfural. *Proc. Natl. Acad. Sci. USA* **2014**, *111*, 8363–8367. [CrossRef] [PubMed]
83. Shiramizu, M.; Toste, F.D. On the Diels–Alder approach to solely biomass-derived polyethylene terephthalate (PET): Conversion of 2,5-dimethylfuran and acrolein into *p*-xylene. *Chem. A Eur. J.* **2011**, *17*, 12452–12457. [CrossRef] [PubMed]
84. Maneffa, A.; Priecel, P.; Lopez-Sanchez, J.A. Biomass-derived renewable aromatics: Selective routes and outlook for *p*-xylene commercialisation. *ChemSusChem* **2016**, *9*, 2736–2748. [CrossRef] [PubMed]
85. Williams, C.L.; Chang, C.-C.; Do, P.; Nikbin, N.; Caratzoulas, S.; Vlachos, D.G.; Lobo, R.F.; Fan, W.; Dauenhauer, P.J. Cycloaddition of biomass-derived furans for catalytic production of renewable *p*-xylene. *ACS Catal.* **2012**, *2*, 935–939. [CrossRef]
86. Van de Poel, B.; Cooper, E.D.; Delwiche, C.F.; Chang, C. An evolutionary perspective on the plant hormone ethylene. In *Ethylene in Plants;* Springer: Dordrecht, The Netherlands; pp. 109–134.
87. Digiacomo, F.; Girelli, G.; Aor, B.; Marchioretti, C.; Pedrotti, M.; Perli, T.; Tonon, E.; Valentini, V.; Avi, D.; Ferrentino, G.; et al. Ethylene-producing bacteria that ripen fruit. *ACS Synth. Biol.* **2014**, *3*, 935–938. [CrossRef] [PubMed]
88. Steeman, A. PET is PET—Petro-PET or Bio-PET. Available online: https://bestinpackaging.com/2011/07/13/pet-is-pet-petro-pet-or-bio-pet/ (accessed on 3 March 2019).
89. Salusjärvi, L.; Havukainen, S.; Koivistoinen, O.; Toivari, M. Biotechnological production of glycolic acid and ethylene glycol: Current state and perspectives. *Appl. Microbiol. Biotechnol.* **2019**, *103*, 2525–2535. [CrossRef] [PubMed]
90. Liu, H.; Ramos, K.R.M.; Valdehuesa, K.N.G.; Nisola, G.M.; Lee, W.-K.; Chung, W.-J. Biosynthesis of ethylene glycol in *Escherichia coli*. *Appl. Microbiol. Biotechnol.* **2013**, *97*, 3409–3417. [CrossRef] [PubMed]
91. Alkim, C.; Cam, Y.; Trichez, D.; Auriol, C.; Spina, L.; Vax, A.; Bartolo, F.; Besse, P.; François, J.M.; Walther, T. Optimization of ethylene glycol production from (d)-xylose via a synthetic pathway implemented in *Escherichia coli*. *Microb. Cell Fact.* **2015**, *14*, 127. [CrossRef]
92. Cam, Y.; Alkim, C.; Trichez, D.; Trebosc, V.; Vax, A.; Bartolo, F.; Besse, P.; François, J.M.; Walther, T. Engineering of a synthetic metabolic pathway for the assimilation of (d)-xylose into value-added chemicals. *ACS Synth. Biol.* **2016**, *5*, 607–618. [CrossRef] [PubMed]
93. Cabulong, R.B.; Valdehuesa, K.N.G.; Ramos, K.R.M.; Nisola, G.M.; Lee, W.-K.; Lee, C.R.; Chung, W.-J. Enhanced yield of ethylene glycol production from d-xylose by pathway optimization in *Escherichia coli*. *Enzyme Microb. Technol.* **2017**, *97*, 11–20. [CrossRef] [PubMed]
94. Uranukul, B.; Woolston, B.M.; Fink, G.R.; Stephanopoulos, G. Biosynthesis of monoethylene glycol in *Saccharomyces cerevisiae* utilizing native glycolytic enzymes. *Metab. Eng.* **2019**, *51*, 20–31. [CrossRef] [PubMed]

95. Pereira, B.; Zhang, H.; De Mey, M.; Lim, C.G.; Li, Z.-J.; Stephanopoulos, G. Engineering a novel biosynthetic pathway in *Escherichia coli* for production of renewable ethylene glycol. *Biotechnol. Bioeng.* **2016**, *113*, 376–383. [CrossRef]
96. Islam, M.A.; Hadadi, N.; Ataman, M.; Hatzimanikatis, V.; Stephanopoulos, G. Exploring biochemical pathways for mono-ethylene glycol (MEG) synthesis from synthesis gas. *Metab. Eng.* **2017**, *41*, 173–181. [CrossRef] [PubMed]
97. Desai, S.H.; Koryakina, I.; Case, A.E.; Toney, M.D.; Atsumi, S. Biological conversion of gaseous alkenes to liquid chemicals. *Metab. Eng.* **2016**, *38*, 98–104. [CrossRef] [PubMed]
98. Ji, N.; Zhang, T.; Zheng, M.; Wang, A.; Wang, H.; Wang, X.; Shu, Y.; Stottlemyer, A.L.; Chen, J.G. Catalytic conversion of cellulose into ethylene glycol over supported carbide catalysts. *Catal. Today* **2009**, *147*, 77–85. [CrossRef]
99. Sun, J.; Liu, H. Selective hydrogenolysis of biomass-derived xylitol to ethylene glycol and propylene glycol on supported Ru catalysts. *Green Chem.* **2011**, *13*, 135–142. [CrossRef]
100. Pang, J.; Zheng, M.; Wang, A.; Zhang, T. Catalytic hydrogenation of corn stalk to ethylene glycol and 1,2-propylene glycol. *Ind. Eng. Chem. Res.* **2011**, *50*, 6601–6608. [CrossRef]
101. Song, T.; Xu, Y.; Ye, Y.; Chen, Y.; Shen, S. Electricity generation from terephthalic acid using a microbial fuel cell. *J. Chem. Technol. Biotechnol.* **2009**, *84*, 356–360. [CrossRef]
102. The Coca Cola Company Coca-Cola Produces World's First PET Bottle Made Entirely From Plants. Available online: https://www.coca-colacompany.com/press-center/press-releases/coca-cola-produces-worlds-first-pet-bottle-made-entirely-from-plants (accessed on 5 March 2019).

© 2019 by the authors. Licensee MDPI, Basel, Switzerland. This article is an open access article distributed under the terms and conditions of the Creative Commons Attribution (CC BY) license (http://creativecommons.org/licenses/by/4.0/).

genes

MDPI

Review

Variability in Assembly of Degradation Operons for Naphthalene and Its Derivative, Carbaryl, Suggests Mobilization through Horizontal Gene Transfer

Prashant S. Phale *, Bhavik A. Shah and Harshit Malhotra

Department of Biosciences and Bioengineering, Indian Institute of Technology-Bombay, Powai, Mumbai 400 076, India

* Correspondence: pphale@iitb.ac.in

Received: 30 May 2019; Accepted: 11 July 2019; Published: 27 July 2019

Abstract: In the biosphere, the largest biological laboratory, increased anthropogenic activities have led microbes to evolve and adapt to the changes occurring in the environment. Compounds, specifically xenobiotics, released due to such activities persist in nature and undergo bio-magnification in the food web. Some of these compounds act as potent endocrine disrupters, mutagens or carcinogens, and therefore their removal from the environment is essential. Due to their persistence, microbial communities have evolved to metabolize them partially or completely. Diverse biochemical pathways have evolved or been assembled by exchange of genetic material (horizontal gene transfer) through various mobile genetic elements like conjugative and non-conjugative plasmids, transposons, phages and prophages, genomic islands and integrative conjugative elements. These elements provide an unlimited opportunity for genetic material to be exchanged across various genera, thus accelerating the evolution of a new xenobiotic degrading phenotype. In this article, we illustrate examples of the assembly of metabolic pathways involved in the degradation of naphthalene and its derivative, Carbaryl, which are speculated to have evolved or adapted through the above-mentioned processes.

Keywords: xenobiotics; naphthalene; Carbaryl; horizontal gene transfer; mobile genetic elements; transposons; integrative conjugative elements; pathway assembly; evolution

1. Introduction

Anthropogenic activities like manufacturing, industrialization and combustion of fossil fuels have led to the release of a large number of compounds in the environment, which were previously unknown to the biosphere. These compounds, known as xenobiotics, are toxic to various life forms and persist in the environment for a long period of time. Monocyclic and polycyclic aromatic hydrocarbons (PAHs) form a major group of xenobiotics. Due to the highly reduced nature, resonance stabilized structure, greater hydrophobicity, and lower aqueous solubility, they are resistant to biodegradation (recalcitrant). They act as metabolic inhibitors, endocrine disrupters [1] and further reactive epoxide formed due to metabolic activation of these compounds by microsomal enzymes interacts with DNA, causing mutation(s) or chromosomal damage [2–4] which may lead to cytotoxicity, genotoxicity and/or carcinogenicity [5]. Repeated application and release of these compounds in the environment has exerted a selection pressure onto the microbes, resulting in the evolution of novel/unique degradative pathways. Although processes such as volatilization, chemical oxidation and photo-oxidation contribute to degradation, bioremediation is the most efficient and cost-effective method for complete removal of these compounds [6,7]. In this review, we focus on the metabolism and various genetic elements involved in the assembly of the degradative pathways for naphthalene and its derivative, Carbaryl.

Naphthalene is ubiquitous in the environment and used extensively in the manufacture of various compounds including pesticides and plastics, as well as for domestic consumption (mothballs, insect repellent). Being hydrophobic, it partitions into the membranes [8] and gets accumulated in tissues of aquatic organisms, leading to toxicity [9,10]. In humans, exposure to naphthalene occurs via skin, oral ingestion or inhalation of fumes. Acute toxicity causes hemolytic anemia [11], whereas exposure to higher doses causes cerebral oedema and chronic renal failure [12]. Naphthalene has been classified as a potential human carcinogen based on the studies performed on experimental animals [13].

Carbaryl (1-naphthyl *N*-methylcarbamate), a derivative of naphthalene, is a broad-spectrum carbamate family insecticide that has been manufactured and used since 1960. It acts as a competitive inhibitor of the enzyme acetylcholine esterase of the central nervous system leading to paralysis [14]. It is highly toxic to aquatic invertebrates, amphibians, bees, earthworms and humans [15–21]. It has also been classified as a likely human carcinogen based on vascular tumour formation in mice [22]. In nature, various bacterial species belonging to genus *Pseudomonas, Rhodococcus, Mycobacterium, Nocardia, Bacillus*, Vibrio, Marinobacter, *Micrococcus* and *Sphingomonas* have been reported to degrade naphthalene and Carbaryl.

2. Microbial Adaptation to Aromatics and Xenobiotics

Aromatics including PAHs and xenobiotics like pesticides, being toxic and persistent, act as a selection force on the microbial community to evolve degradative pathways and possibly remove sensitive microorganisms from the population [23–25]. Microbes adapt to these environmental challenges and acquire the ability to degrade these compounds using the following strategies:

a) Quite a few bacterial isolates have the ability to completely degrade and utilize mono as well as polycyclic aromatics like benzoate, phthalate isomers and their esters, chlorobenzoate, naphthalene, biphenyls, etc. [23,26–41]. Such organisms harbor genes encoding all enzymes responsible for the complete metabolism of these compounds as a sole source of carbon and energy. Further, these genes could be arranged as operon(s) so as to finetune their expression to achieve optimum degradation efficiency. In a few organisms, both the carbon source as well as detoxification pathways are functional. For example, in *Pseudomonas putida* CSV86, 1-methylnaphthalene, which is highly toxic and used as insect repellent, is metabolized by two pathways. It is transformed and detoxified by side chain hydroxylation pathway to 1-naphthoic acid, a less toxic and more water-soluble metabolite, which is released into the medium as a dead-end product [29]. Whereas in other pathway, 1-methylnaphthalene is ring-hydroxylated and further metabolized to central carbon pathway intermediates, thus acting as sole source of carbon and energy [42].

b) In biosphere, microbial communities degrade a diverse range of pollutants. Also, specific consortia are constructed in the laboratory based on their metabolic properties and compatibility of organisms to degrade the pollutants. Members of such communities/consortia achieve complete mineralization by working together, which is mutually beneficial. Initial transformation of the parent compound by one member yields metabolite(s) which may be a dead-end product(s) for its metabolic machinery, which can then be subsequently used by other member(s). To list a few examples: consortia of *Pseudomonas* spp. 50552 and 50581 involved in complete degradation of Carbaryl [43]; EC20 consortia of *Pseudomonas, Mesorhizobium, Achromobacter, Stenotrophomonas,* and *Halomonas* involved in BTEX (benzene, toluene, ethylbenzene and xylene) degradation [44]; consortia of five fungal and eight bacterial isolates to degrade phenanthrene, pyrene, and benzo(a)pyrene [45]; ASDC consortia of *Rhodococcus* sp., *Bacillus* sp., and *Burkholderia* sp. degrading chrysene [46]. Several bacteria, algae and fungi have been reported to transform methylnaphthalene into naphthoic acid [47,48] which can then be metabolized by other microbes like *Stenotrophomonas maltophilia* CSV89 [49].

c) Efficient degradation of certain xenobiotics can be achieved by supplementing with specific nutrient(s). For example, isophthalate, which is used in the plastic and textile industry, acts as a competitive inhibitor for the enzyme glutamate dehydrogenase (GDH), involved in the C/N metabolism. When the growth medium is supplemented with glutamate, GDH is relieved from the competitive inhibition by isophthalate, thereby enhancing the degradation efficiency of isophthalate in *Pseudomonas* sp. [50]. On the other hand, soil isolate *Acinetobacter lwofii* ISP4 degrades isophthalate faster and more efficiently without any supplementation of glutamate by expressing GDH, which has less sensitivity to inhibition by isophthalate and is synthesized in more quantity [51]. These examples indicate the adaptation by the microbes at the metabolic level for their effective survival.

3. Horizontal Gene Transfer Elements Involved in Catabolism of Aromatics

Genetic variation in microorganisms is the major driving factor in adaptation to environmental conditions. The metabolic diversity observed in the organism(s) can be attributed to the plasticity of the genome which is acquired through various strategies, resulting in the generation of genetic variants and ultimately adaptation, as described by Arber [52]. These include: (i) small local change(s) in the nucleotide sequence, (ii) reshuffling of gene fragments within the genome, and/or iii) horizontal gene transfer (HGT). The third strategy allows for evolution in quantum leaps [53] and is mediated through mobile genetic elements (MGEs) like plasmids (conjugative and non-conjugative), transposons, integrative conjugative elements (ICEs) and genomic islands (GEIs), as well as phages and prophages. The association of MGEs with catabolic gene clusters, as well as high levels of similarity in gene organization and nucleotide sequences among phylogenetically and geographically distant microbial species suggest that HGT is the major player in acquisition and assembly of the degradative pathways in microorganisms [25,54–57]. The presence of PAH-degradation genes on MGEs is an indication of ease of mobilization of catabolic genes and hence better adaptability to the polluted environment [58].

Catabolic plasmids involved in the degradation of camphor, octane, naphthalene, salicylate, etc. were first discovered in the early 70s [59–63]. These plasmids can either be conjugative or non-conjugative in nature. Plasmid isolation and characterization, and plasmid curing by chemicals as well as by nutritional conditions, southern hybridization, conjugation and transformation experiments established the association of the degradative property to the plasmid(s) [64–67]. Advances in genomic techniques and approaches have further provided insights into the evolutionary aspects of degradation properties [68–70]. Presence of insertion elements harboring functional or non-functional transposase and integrase like features have been seen to be associated with degradative genes in large number of catabolic plasmids like pWW0 and NAH7 amongst others [71,72]. The ability of some of these plasmids to move across various genera and strains make them carriers of the degradative property. Plasmids play an important role in the evolution of metabolic diversity in Pseudomonads [73].

Catabolic transposons form the major group of MGEs involved in the transfer of xenobiotic degradation property [74–76]. Transposons can be grouped into three classes based on the nucleotide sequence homology, genetic organization and mechanistic properties [77]. Class-I elements include composite transposons and simple insertion sequences (ISs). Simple ISs only contain the elements essential for transposition and are flanked by inverted or direct repeats (DRs). Composite transposons have genes encoding features other than transposition and are flanked by very similar ISs in direct or inverted position. For example, catabolic transposon Tn*5280*, harboring genes encoding chlorobenzene dioxygenase and dehydrogenase, was found to be part of the plasmid pP51 in *Pseudomonas* sp. strain P51 [78]. Class II transposons, known to transpose by the formation of co-integrate, are known to carry gene fragments greater than 50 kb. Transposon Tn*4651* which harbors the *xyl* operon, is a part of Tn*4653* and is located on the plasmid pWW0 in *P. putida* mt-2 [71,79]. The third class of transposons is called as conjugative transposons.

Elements like ICEs refer to a group of MGEs which had earlier been classified under different groups like conjugative transposons, integrative plasmids and GEIs [25,80]. These elements are excised by site-specific recombination yielding circular intermediates which are then transferred by conjugation to other bacteria and can possibly re-integrate into the genome [25,80]. Elements like GEIs have been described as MGEs residing on the genome in close proximity to a gene for a tRNA and harbor a gene for integrase and/along with a short duplication of the insertion element at the other end [53,54,81].

Pathogenicity islands (PAIs), a class of GEIs associated with the virulence phenotype, were first discovered in *E. coli* [82] and since then a wide plethora has been uncovered. Various disease associated toxins [83,84], antibiotic resistance genes [85] and even superantigens [86] are harbored on these MGEs. The similarity between the insertion sites of PAIs and phages at tRNA loci suggests that these islands are acquired through phage mediated HGT [87]. Also, certain PAIs harbor open reading frames that show high similarity to integrase of bacteriophages [88]. In comparison, conjugative transposons contain multiple integration sites and are not confined to tRNA genes. A few ICEs/GEIs are also reported to harbor genes involved in the degradation of various aromatic compounds; for example: *clc* element, for the metabolism of chlorobenzoate in *Pseudomonas* sp. strain B13 [89]; ICE$_{XTD}$ for *m*-xylene, toluene and cumene degradation in *Azoarcus* sp. CIB [90]; ICE$_{CSV86}$ for naphthalene degradation in *P. putida* CSV86 [91]; ICEclc JB2 for metabolism of *o*-halobenzoate and *o*-hydroxybenzoate in *P. aeroginosa* JB2 [92]; ICE$_{KKS102}$4677 for the metabolism of polychlorinated biphenyl/biphenyl in *Acidovorax* sp. strain KKS102 [93]; proposed ICE harboring Tn*4371*, for the metabolism of biphenyl and 4-chlorobiphenyl [94]; and *phn* island for the metabolism of phenanthrene in *Delftia* sp. Cs1-4 [95]. The features of ICEs involved in aromatic compound degradation are depicted in Figure 1.

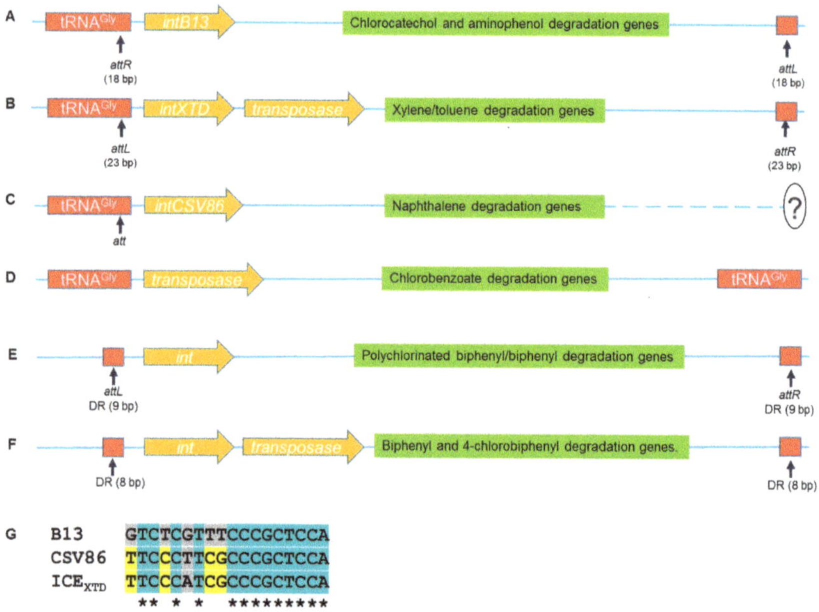

Figure 1. The structural features of various integrative conjugative elements (ICEs) involved in aromatic compound metabolism from: (**A**) *clc* element of *Pseudomonas* sp. B13; (**B**) ICE$_{XTD}$ of *Azoarcus* sp. CIB; (**C**) ICE$_{CSV86}$ element of *Pseudomonas putida* CSV86; (**D**) *clc* element of *Pseudomonas aeruginosa* JB2; (**E**) ICE$_{KKS102}$4677 from *Acidovorax* sp. strain KKS102; and (**F**) Tn*4371* transposon most likely to be an ICE from *Cupriavidus oxalaticus*. Panel (**G**) represents multiple sequence alignment of attachment site ('*att*') region of the 3′ end of tRNA^{Gly} of *Pseudomonas* sp.B13, as B13, *Pseudomonas putida* CSV86 as CSV86 and *Azoarcus* sp. CIB as ICE$_{XTD}$. Question mark in panel C depicts the incomplete/unidentified downstream region and *attR* site. Figure is not to the scale. tRNA: Transfer-RNA.

The *clc* element is the first GEI, reported and characterized in detail from *Pseudomonas* sp. strain B13 and involved in the degradation of 3-chlorobenzoate [89]. ICE*clc*$_{B13}$, ICE$_{XTD}$ and ICE$_{CSV86}$ are associated with tRNA^{Gly} having *att* site at the 3′ end of tRNA (72% identity among all three *att* and 94% among ICE$_{XTD}$ and ICE$_{CSV86}$, Figure 1G). In *P. aeruginosa* JB2, the *att* site was found to be absent. The *phn* island involved in phenanthrene degradation from *Delftia* sp. Cs1-4 is structurally distinct from the *clc* family ICEs as it is not associated with the tRNA gene and has an SXT/R391 type mobilization system [95]. $ICE_{KKS102}4677$ harbors *bph* operon and shows similar arrangement as observed in *clc* of B13 and ICE_{CSV86}, consisting of direct repeats (9 bp, *attL* and *attR* region in $ICE_{KKS102}4677$) with *attL* at the start, followed by integrase. Also, it has been found to be integrated with the chromosome and shows low conjugal transfer frequency as observed in the case of CSV86 [93,96]. Transposon Tn*4371* consists of transposase (*tnpA*) and phage-like integrase (*int*) genes [94]. Tn*4371* and $ICE_{KKS102}4677$ are flanked by 8 bp and 9 bp direct repeats (55.6% identity), respectively, which are not associated with/in proximity to $tRNA^{Gly}$.

4. Assembly of Naphthalene Degradation Pathway

Naphthalene, the simplest PAH, is used as a model compound to understand aromatic degradation pathways, enzymes and genetics. The number of microbes has been reported to metabolize naphthalene (Table 1A). The degradation is initiated by hydroxylation of one of the aromatic rings to yield 1,2-dihydroxynaphthalene, which is subsequently metabolized to salicylic acid. Generated salicylic acid is metabolized either via catechol (*meta* or *ortho* ring-cleavage) or gentisic acid to yield central carbon pathway metabolites like organic acids (Figure 2, Table 1A). Apart from these well studied routes, isolate *Bacillus thermoleovorans* was reported to degrade naphthalene via phthalic acid [97]. Based on biochemical, enzyme induction and regulation studies, naphthalene degradation pathway is segmented into *upper pathway* (naphthalene to salicylate, *nah* operon) and *lower pathway* (salicylate to central carbon pathway either via catechol, *sal* operon or gentisate, *gen*/*sgp* operon).

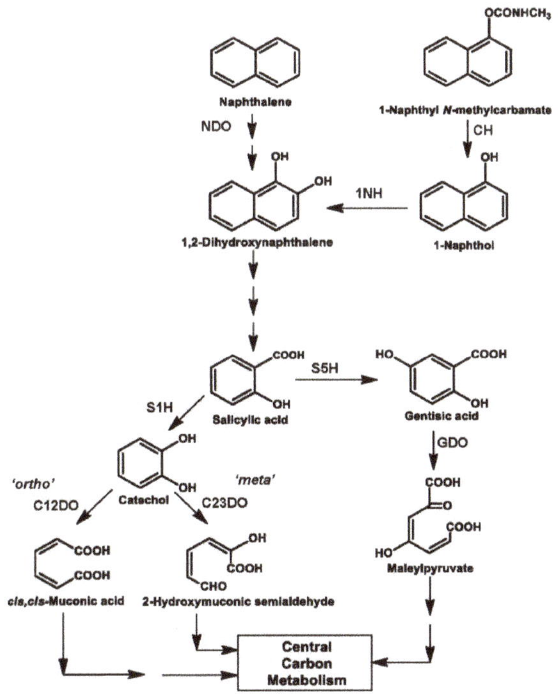

Figure 2. Metabolic diversity observed in naphthalene and Carbaryl degradation pathways from various soil bacterial isolates. Abbreviations are: NDO, naphthalene 1,2-dioxygenase; CH, Carbaryl hydrolase; 1NH, 1-naphthol 2-hydroxylase; S1H, salicylate 1-monoocxygenase; S5H, salicylate 5-hydroxylase; C12DO, catechol 1,2-dioxygenase; C23DO, catechol 2,3-dioxygenase; GDO, gentisate 1,2-dioxygenase. *ortho* and *meta* pathway indicates the mode of aromatic ring-cleavage by these enzymes.

Metabolic diversity observed in naphthalene degradation pathways (Figure 2) is possible due to acquisition or exchange of genetic material. Genes for naphthalene degradation were found to be present on plasmid, chromosome, transposon or ICE (Table 1). The arrangement of genes encoding enzymes involved in naphthalene degradation is depicted in Figure 3. The degradation genes are arranged as two inducible operons: upper pathway (*nah*) and lower pathway (*sal* or *sgp/gen*) operon [98–100]. These operons are induced by salicylic acid and its analogue like 2-aminobenzoate and 2-hydroxybenzyl alcohol [100]. Among naphthalene degrading strains, the function of *nahR* was found to be highly conserved [101].

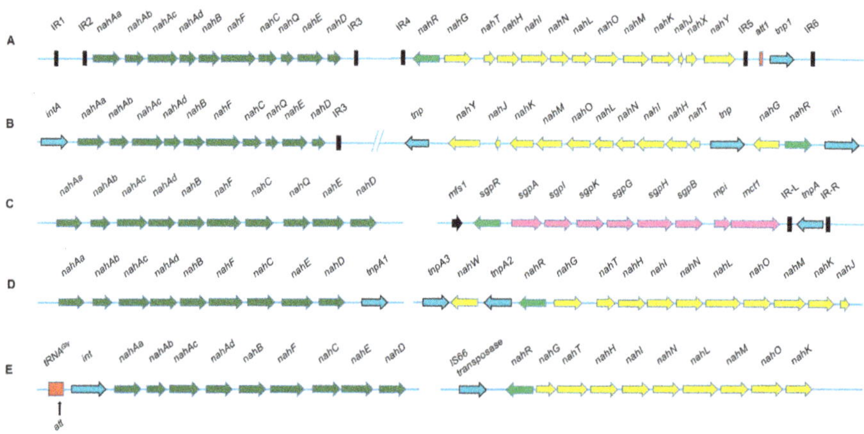

Figure 3. The diversity observed with respect to location, transcription and arrangement of genes involved in naphthalene degradation from: (**A**) *Pseudomonas putida* G7 as Tn*4655* transposon on plasmid NAH7 flanked by IR1 and IR6; (**B**) Pseudomonas *putida* strain NCIB9816-4 on conjugative plasmid pDTG1; (**C**) *Pseudomonas putida* AK5 on conjugative plasmid pAK5 via gentisate pathway; (**D**) *Pseudomonas stutzeri* AN10 on chromosome; (**E**) *Pseudomonas putida* CSV86 on ICE showing $tRNA^{Gly}$ and integrase while the other end is still not known. Arrow indicates the direction of transcription. Genes responsible for upper pathway, '*nah*' operon (naphthalene to salicylic acid metabolism) are depicted in green; lower pathway, '*sal*' operon (salicylic acid to central carbon metabolites via catechol) are depicted in yellow; while those in pink are responsible for salicylate metabolism via gentisic acid, '*sgp/gen*' operon. Genes depicted in cyan arrows indicates transposase or integrase, black boxes depict IR elements, while red box indicates tRNA gene with '*att*' site, light green arrow represents the regulators of the operons and the black arrow represents the putative integral membrane transport protein. Figure is not to the scale. Genes and proteins encoded are: *nahAa*, 2Fe-2S iron-sulfur cluster binding domain-containing protein, reductase; *nahAb*, non-heme iron oxygenase ferredoxin subunit; *nahAc*, naphthalene 1,2-dioxygenase; *nahAd*, naphthalene 1,2-dioxygenase subunit beta; *nahB*, *cis*-naphthalene dihydrodiol dehydrogenase; *nahF*, salicylaldehyde dehydrogenase; *nahC*, 1,2-dihydroxynaphthalene dioxygenase (oxygenase); *nahE*, 2-hydroxybenzalpyruvate aldolase; *nahD*, 2-hydroxychromene-2-carboxylate isomerase; *nahG*, salicylate 1-monooxygenase; *nahT*, 2Fe-2S iron-sulfur cluster binding domain-containing protein (chloroplast-type ferredoxin); *nahH*, catechol 2,3-dioxygenase; *nahI*, 2-hydroxymuconic semialdehyde dehydrogenase; *nahN*, 2-hydroxymuconic semialdehyde dehydrogenase; *nahL*, 2-oxopent 4-enoate hydratase; *nahO*, 4-hydoxy-2-oxovalerate aldolase; *nahM*, acetaldehyde dehydrogenase; *nahK*, 4-oxalocrotonate decarboxylase/2-oxo 3-hexendioate decarboxylase; *nahJ*, 4-oxalocrotonate tautomerase family protein; *nahY*, methyl accepting chemotaxis protein; *nahX*, unknown; *sgpA*, reductase component of salicylate 5-hydroxylase; *sgpI*, gentisate 1,2-dioxygenase; *sgpK*, fumarylacetoacetate hydrolase/fumarylpyruvate hydrolase; *sgpG*, salicylate 5-hydroxylase large oxygenase subunit; *sgpH*, salicylate 5-hydroxylase small oxygenase subunit; *sgpB*, ferredoxin component of salicylate 5-hydroxylase; *mpi*, maleylacetoacetate isomerase; *mct1*, methyl-accepting chemotaxis sensory transducer; *nahR* and *sgpR*, LysR family transcriptional regulator.

In *P. putida* G7, the plasmid NAH7 is known to harbor naphthalene degradation genes which are proposed to be part of a defective transposon which requires Tn*4653* transposase for mobilization [99]. In NAH7, the transcription for *nah* and *sal* operons is in the same direction. In *P. putida* strain NCIB9816-4, genes were found to be present on a conjugative plasmid pDTG1 as two operons (~15 kb apart) which are transcribed in opposite direction [102]. As compared to NAH7 and pDTG1, a non-conjugative pAK5 (IncP-7 group) from *P. putida* strain AK5 encodes naphthalene degradation via gentisate pathway as *nah* and *gen/sgp* operon [103]. In *P. putida* strain PMD-1, upper pathway genes (*nah* operon) were found to be located on the chromosome, while salicylate degradation genes (*sal* operon) were present on the conjugative plasmid pMWD-1 [104]. However, in *P. stutzeri* AN10, the naphthalene degradation genes (*nah* and *sal* operons) were found to be located on the chromosome and hypothesized to be recruited through transposition, recombination and rearrangement events [105,106]. Besides plasmids and chromosomes, naphthalene degradation genes were hypothesized to be present on ICE in *P. putida* CSV86 [91]. The draft genome analysis revealed that the *nah* operon is present next to $tRNA^{Gly}$ (along with *att* region) and phage-like integrase [91], suggesting the involvement of a GEI, referred to as ICE_{CSV86} (Figure 3). This ICE is located on the chromosome and the degradation genes are arranged as *nah* and *sal* operons, which were transferred by conjugation at a very low frequency [96].

Interestingly, irrespective of the location of genes for naphthalene degradation either on plasmid, chromosome + plasmid, chromosome or ICE, the gene arrangement in *nah* and *sal* operon is almost conserved among various naphthalene degraders (Figure 3). Further, metabolic variations in isolates are generated through the regulation of operons and specificity of the enzyme(s), thus granting an advantage to the strain to metabolize naphthalene and other PAHs effectively [107,108].

Table 1. Naphthalene and Carbaryl degradation pathways and location of genes from various isolates.

Organism	Pathway, ring-cleavage mode	Chromosome/plasmid (kb); Operon size (kb)	References
A. Naphthalene degradation			
Pseudomonas putida G7	Catechol, *meta*	Plasmid, NAH7 (82); *nah* (10) and *sal* (8)	[99]
Pseudomonas putida strain NCIB 9816-4	Catechol, *meta*	Plasmid, pDTG1 (88); *nah* (9.5) and *sal* (13.4)	[102]
Pseudomonas sp. strain ND6	Catechol, *meta*	Plasmid, pND6-1 (102); *nah* (10) and *sal* (18)	[109]
Pseudomonas fluorescens strain PC20	Catechol, *meta*	Plasmid, pNAH20 (88); *nah* (6) and *sal* (13)	[110]
Pseudomonas putida strain AK5	Gentisic acid	Plasmid, pAK5 (-); *nah* (6.7) and *sal* (12.2)	[103]
Pseudomonas sp. AS1	Catechol, *ortho*	Plasmid, pAS1 (82)	[111]
Pseudomonas putida strain PMD-1	Catechol, *meta*	Chromosome and Plasmid, pMWD-1	[104]
Pseudomonas stutzeri AN10	Catechol, *meta*	Chromosome, *nah* (11.5) and *sal* (16)	[105,106]
Pseudomonas putida CSV86	Catechol, *meta*	ICE, *nah* (8.2) and *sal* (9.8)	[91]
B. Carbaryl degradation			
Achromobacter sp.	Hydroquinone, Catechol,	- *	[112]
Pseudomonas sp. NCIB 12043	Gentisic acid	-	[113]
Pseudomonas sp. NCIB 12042	Catechol, *meta*	-	[113]
Rhodococcus sp. NCIB 12038	Gentisic acid	-	[113]

Table 1. *Cont.*

Organism	Pathway, ring-cleavage mode	Chromosome/plasmid (kb); Operon size (kb)	References
Consortia *Pseudomonas* spp. isolate 50581 and 50552	Catechol	Plasmid, pCD1 (50) in isolate 50581 encodes Carbaryl hydrolase; Chromosome encodes degradative enzymes for 1-naphthol in isolate 50552	[43]
Blastobacter sp. strain M501	Hydrolysis to 1-naphthol	-	[114]
Sphingomonas sp. strain CF06	Gentisic acid	Plasmids pCF01, pCF02, pCF03, pCF04, and pCF05, role of each plasmid is not clear	[67]
Arthrobacter sp. RC100	Gentisic acid	Plasmid, pRC1 (110) encodes Carbaryl hydrolase; Plasmid, pRC2 (120) encodes enzymes for 1-naphthol to gentisic acid; chromosome encodes enzymes for utilization of gentisic acid	[115]
Rhizobium sp. strain AC100	Partial hydrolysis to 1-naphthol	Plasmid, pAC200 encodes Carbaryl hydrolase	[116]
Micrococcus sp.	Gentisic acid	-	[117]
Pseudomonas sp. strain C4	Gentisic acid	Chromosome	[30]
Pseudomonas sp. strain C5	Gentisic acid	Chromosome	[30]
Pseudomonas sp. strain C6	Gentisic acid	Chromosome	[30]
Burkholderia sp. C3	Catechol and Gentisic acid	-	[118]
Pseudomonas sp. strain C7	Gentisic acid	-	[119]
Pseudomonas putida XWY-1	Gentisic acid	Plasmid, pXWY (400) encoding all enzymes of Carbaryl degradation	[120]

Note: *: -, Not known/reported

5. Assembly of Carbaryl Degradation Pathway

Carbaryl has been used extensively in the agriculture sector since 1960. Repeated application has led microbes to adapt and utilize Carbaryl as the sole source of carbon and energy (Table 1B). The degradation pathway is initiated by hydrolyzing the ester bond by the enzyme Carbaryl hydrolase (CH) to yield 1-napthol. Generated 1-naphthol is ring-hydroxylated by the enzyme 1-naphthol hydroxylase (1NH) yielding 1,2-dihydroxynaphthalene, which is further metabolized via salicylic acid and gentisic acid. Few Carbaryl degraders also metabolize salicylic acid via catechol route [44] (Table 1B and Figure 2). The metabolic steps involved in the conversion of 1,2-dihydroxynaphthalene to salicylic acid are common for naphthalene and Carbaryl degradation (Figure 2). Interestingly, in naphthalene degraders salicylic acid is predominantly metabolized via catechol, whereas in Carbaryl degraders, it is metabolized mainly through the gentisic acid route (Table 1B and Figure 2).

A large number of genes involved in Carbaryl degradation have been proposed to be plasmid borne [121]. *Arthrobacter* sp. RC100 harbors three plasmids (pRC1, pRC2, and pRC300) of which conjugative plasmids, pRC1 and pRC2, encode enzymes for Carbaryl degradation up to gentisic acid. Enzymes involved in the conversion of gentisic acid to central carbon metabolites are encoded by genes located on the chromosome [115]. A *Rhizobium* sp. strain AC100, which transforms Carbaryl to 1-naphthol, harbors a plasmid pAC200 carrying gene *cehA* encoding CH as a part of Tn*ceh* transposon flanked by insertion element-like sequence (*ist*A and *istB*, Figure 4). *ist*A and *istB*, which comprise the ISR*sp3* element, showed homology to IS*1600* element from *A. eutrophus* NH9 and of IS*1326* from *P. aeruginosa* of the IS*21* family. Treating cells with plasmid curing agents like mitomycin-C resulted in deletion of *cehA* from the Tn*ceh* leaving behind the insertion elements as a part of plasmid pAC200d due to homologous recombination among ISRsp3 [116]. This depicts the instability of the system and loss of catabolic genes by the activation of IS elements. *Sphingomonas* sp. strain CF06, which degrades carbofuran and its derivatives (including Carbaryl), harbors five plasmids- pCF01, pCF02, pCF03, pCF04, and pCF05. Plasmid curing and conjugation experiments revealed that the degradation property was associated with plasmid(s). Southern hybridization studies revealed a high degree of sequence similarity among pCF01 and pCF02, and pCF03 and pCF04, indicating gene duplication, which could be due to lack of positive regulatory system. Various IS elements were found to be present on pCF01, pCF02 and pCF03. The presence of active IS elements as well as gene duplications indicate that the system is in an early stage of evolution [67]. In Carbaryl degrading consortia of two *Pseudomonas* spp., the strain 50581 harbors a conjugative plasmid, pCD1 (50 kb) which encodes for CH, while chromosomally located genes from strain 50552 encode enzymes for 1-naphthol degradation [43].

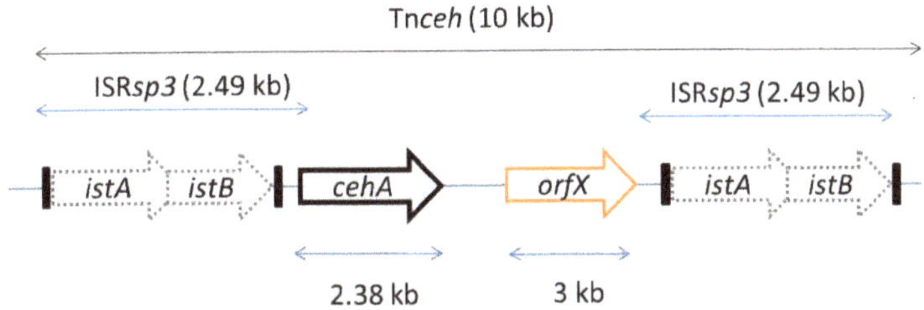

Figure 4. Organization of insertion elements observed in Carbaryl degrading *Rhizobium* sp. strain AC101 harboring gene *cehA* encoding Carbaryl hydrolase. Figure is not to scale.

Soil isolates *Pseudomonas* spp. strains C4, C5pp and C6 utilizes Carbaryl via salicylic acid and gentisic acid (Figure 2, [30]). Based on metabolic studies like enzyme induction, cell respiration and biotransformation, the Carbaryl degradation pathway was hypothesized to be divided into *upper*, *middle* and *lower* segments; and genes are probably arranged as three distinct operons induced by respective carbon source [122]. The degradation property was found to be very stable and could not be cured by nutritional conditions or chemical agents. Southern hybridization analyses indicate genes encoding degradative enzymes are located on the host genome [122]. The draft genome sequence of *Pseudomonas* sp. C5pp, primer walking, and gap filling PCR studies yielded a 76.3 kb sequence, referred to as Supercontig-A, harboring all genes required for Carbaryl degradation (Figure 5), [123,124]. Genes encoding enzymes responsible for the conversion of Carbaryl to salicylic acid (*upper* operon), salicylic acid to gentisic acid (*middle* operon), and gentisic acid to central carbon metabolites (*lower* operon) are probably under the regulation of transcription regulators encoded by *mcbG*, *mcbH* and *mcbN*, respectively (Figure 5).

The draft genome (6.15 Mb) analysis revealed the presence of 42 MGEs and 36 GEIs, out of which 17 MGEs were located in Supercontig-A (76.3 kb), indicating this region could be a hotspot for genome alterations [123,124]. The *upper* and *middle* operon region showed lower G+C content (54%) compared to the genome (62.65%) of strain C5pp. Whereas, the *lower* operon region showed comparable (60%) G+C content. This skewing of G+C content suggests that genes of the *upper* and *middle* pathways probably have a different ancestral origin and have been acquired through HGT events and assembled in proximity to each other. This is evident from the fact that 40% of MGEs are located in the Supercontig A (which is 1% of the total genome) [123,124]. Of the 36 GEIs, 3 were found to be a part of Supercontig-A. The *upper* pathway genes, except the regulator *mcbG* constitute a single GEI (13.8 kb), whereas the *middle* (6.5 kb) and the *lower* (9.6 kb) pathway genes, including their respective regulators, were established to be parts of two distinct GEIs (Figure 5). Interestingly, *P. putida* XWY-1, isolated from waste water in Shangong, China, 4750 km away from the site of isolation of the strain C5pp (Mumbai, India), showed a similar gene arrangement and nucleotide sequence for the genes involved in Carbaryl mineralization; however, these genes were reported to be plasmid borne [120].

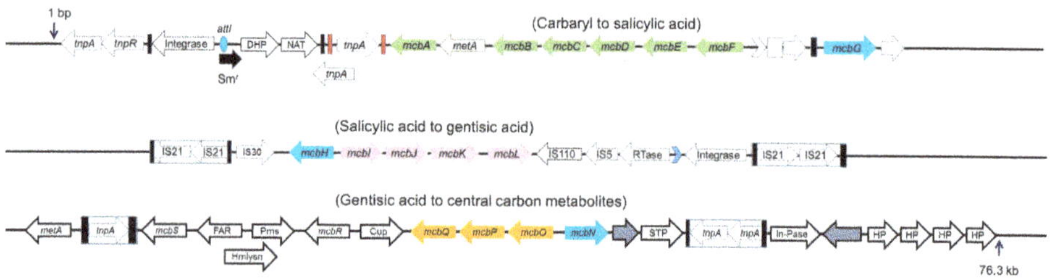

Figure 5. The organization of genes encoding enzymes responsible for Carbaryl degradation as three distinct transposons in *Pseudomonas* sp. strain C5pp. Genes are located on 76.3 kb long Supercontig-A (Accession number: KU522233.1). Inverted repeats (IRs) are depicted in red and black boxes. Arrow direction indicates the probable transcription direction. Blue color arrows depict genes encoding the probable regulators. Figure is not to the scale. Genes and protein encoded are transposase (*tnpA* and *tnpR*); integrase; *attI* site; streptomycin resistance (Sm^r); dihydropterate synthetase (DHP); N-acetyl transferase (NAT); *mcbA*, Carbaryl hydrolase; *metA*, conserved protein; *mcbB*, 1,2-dihydroxynaphthalene dioxygenase; *mcbC*, 1-naphthol 2-hydoxylase; *mcbD*, 2-hydroxychromene 2-carboxylate isomerase; *mcbE*, *trans-o*-hydroxybenzylidene pyruvate hydratase-aldolase; *mcbF*, salicylaldehyde dehydrogenase; *mcbG*, LysR family regulator flanked by truncated transposons (depicted as dotted lines). IS21 insertion elements with inverted repeats; IS30 insertion element; *mcbH*, transcriptional regulator NahR; *mcbI*, ferredoxin reductase; *mcbJ*, salicylate 5-hydroxylase large oxygenase component; *mcbK*, salicylate 5-hydroxylase small oxygenase component; *mcbL*, ferredoxin; other transposase and insertion elements present downstream of the transposon. *metA*, conserved protein; *tnpA*, transposase; *mcbS*, TetR family transcriptional regulator; FAR, fusaric acid resistance; Hmlysn, hemolysin; Pms, permease; *mcbR*, LysR family transcriptional regulator; Cup, pirin and cupin2 superfamily related protein; *mcbQ*, maleylpyruvate isomerase; *mcbP*, fumaryl pyruvate hydrolase; *mcbO*, gentisate dioxygenase; *mcbN*, LysR family transcription regulator; transposase; STP, serine/threonine phosphatase; *tnpA*, transposase; In-Pase, inositol phosphatase; recombinase; HP, hypothetical protein.

In *Pseudomonas* sp. strain C5pp, the upstream region of the *upper* pathway genes showed class-I integron features, which showed high identity (95–99%) with Tn*6217* [124]. Other features present were transposase, 25 bp left-end repeat (92% homology to IR_i), *attI* site, 5′ and 3′ conserved segment, resistance to streptomycin and two additional genes. The regulator *mcbG* is flanked by truncated transposases, which showed a similarity to ISP*a20* of IS3 family at the left end and ISP*st7* of IS5 family at the right end. The presence of partial transposase sequences indicate that their function has probably been lost due to decay linked recombination events and these might be the leftover sequences of previously functional ICE. The region consisting of *middle* pathway genes, *mcbIJKL*, flanked by IS21 family insertion repeats, exhibits class-I composite transposon like features and is referred to as the catabolic transposon Tn*C5ppsal* [123]. Members belonging to class-I composite transposon have been known to harbor gene clusters involved in degradation of xenobiotics [25]. The transposases present at the 3′ end of the cluster show a high degree of similarity to the IS110 and IS5 family, reverse transcriptase, group-II intron D1-D4-2, and leucine-zipper class of integrase. As depicted in Figure 5, the insertion elements of the IS21 family are present in inverted orientation, which could be the reason for the stability of the catabolic transposon in the genome of strain C5pp. Also, it has been reported that the increase in size of the intervening sequence of the composite transposon lowers the transposition frequency [123,124]. The lower pathway genes encoding for catabolism of gentisic acid to central carbon metabolites, are hypothesized to be a part of class-I composite transposon, which is bordered by non-identical insertion elements showing sequence similarity with ISP*a1635* and IS*481* [123,124]. In addition to this, the prediction of the probable ancestral origin of the genes using the IMG-JGI server indicates that *mcbBC, mcbEF*, and *mcbIJKL* were probably derived from beta-proteobacteria; *mcbA* and *mcbM* from alpha-proteobacteria; and *mcbD, mcbGH* and *mcbNOPQ* from gamma-proteobacteria.

The steps involved in the conversion of 1,2-dihydroxynaphthalene to salicylate appears to be common in both naphthalene and Carbaryl degradation pathways. Pair-wise sequence alignment (nucleotide as well as amino acid) of genes *nahC, nahD, nahE* and *nahF* (encoding these metabolic steps) from various naphthalene degrading Pseudomonads (Figure 3) and genes *mcbB, mcbD, mcbE*, and *mcbF* from Carbaryl degrading *Pseudomonas* sp. C5pp (Figure 5) revealed higher (~90–100%) identity amongst 2-hydroxybenzylidene pyruvate hydratase-aldolase as well as salicylaldehyde dehydrogenase encoding genes while 2-hydroxychromene 2-carboxylate isomerase showed moderate (~44–52%) identity. The phylogenetic analysis of Carbaryl degradation genes has been described in detail by Trivedi et al. [123,124]. Interestingly, 1,2-Dihydroxynaphthalene dioxygenase (12DHNDO) from strain C5pp, which belongs to class-II extradiol dioxygenases displayed low (4–10%) identity with 12DHNDOs from naphthalene degraders (class-I) as compared to protocatechuate 3,4-dioxygenase from lignin degraders *Pseudomonas humi* (74%) and *Burkholderia* sp. LIG30 (73%) [123,124]. Among various naphthalene degraders, irrespective of location, these genes displayed high (86–98%) identity. In strain C5pp, genes encoding CH and 1NH were hypothesized to be acquired from other degraders. Phylogenetic analysis revealed that CH from C5pp belongs to a new family of esterase with new conserved motif [124]. Enzyme 1NH from strain C5pp, which showed 47% activity on 2,4-dichlorophenol as substrate, yielded 55% identity with 2,4-dichlorophenol 6-monoxygenase from *Paraburkholderia zhejiangensis* [123,124]. This suggests that the strain has probably acquired the relevant gene, followed by mutations to evolve the enzyme, which will now accept 1-naphthol as its primary substrate. All of these observations indicate that the Carbaryl degradation genes must have been acquired through HGT followed by multiple transposition events leading to integration into the genome, giving a stable degradation phenotype in *Pseudomonas* sp. strain C5pp even in the absence of selection pressure.

Besides HGT events and assembly of genes in the form of operons, *Pseudomonas* sp. C5pp shows cellular level adaptation by compartmentalization of enzymes to tackle the toxicity of 1-naphthol [125,126]. This adaptation helps in degradation of higher concentration of Carbaryl (1% tested so far). The ability to metabolize Carbaryl at high concentration can be attributed to the presence of a low affinity CH (K_m, 100 μM) in the periplasm of the bacterium [126]. This enzyme catalyses the conversion of Carbaryl to 1-naphthol (more toxic and recalcitrant than Carbaryl) in the periplasm, minimizing the interaction of 1-naphthol with cytoplasmic components. 1-Naphthol, transported across the membrane through diffusion and partition processes into the cytoplasm, is ring-hydroxylated by high affinity cytosolic 1NH (K_m, 10μM) to 1,2-dihydroxynaphthalene and subsequently metabolized to central carbon pathway intermediates. The compartmentalization of enzymes involved in the degradation suggests a successful strategy evolved by *Pseudomonas* sp. strain C5pp for efficient degradation of Carbaryl at high concentration.

6. Conclusions

Bacteria, which are present ubiquitously in nature, with a simple unicellular structure and ability to grow faster, have made their niches very dynamic, metabolically as well as genotypically plastic, and hence hot spots for the evolution of new trait(s). During the process of adaptation to strenuous conditions, microbes might acquire new genes and functions to enhance survivability. In addition, they might mutate already existing and/or acquired genes. The former approach is mediated through MGEs, which are the major players responsible for genome plasticity. Microbes evolve genes for novel pathways encoding xenobiotic degradation by the above mechanisms, which are further assembled as transcription unit(s)/operon(s), so as to finetune their regulation. Thus, only the required enzymes of the degradative pathway or segment of the pathway are activated by the respective carbon source. Further, the mutations in the acquired gene(s), shuffling/rearrangement of genes and compartmentalization of metabolic steps grant an additional advantage to the organism to adapt and utilize the compound more efficiently, as observed in the case of Carbaryl metabolism by *Pseudomonas* sp. C5pp. Apart from plasmids and transposons, the genes for aromatic degradation have also been found to be present as a part of GEIs/ICEs in the genome of various organisms including *Pseudomonas*. Integration of the degradation property into the host genome may impart additional stability to the phenotype, as generally the plasmids are lost in the absence of selection pressure. Therefore, by employing various strategies, microbes display an ability to adapt to challenging environments.

Author Contributions: All three authors contributed equally towards conceptualization, writing original draft, review and editing. P.S.P. acquired funds through various government agencies to support the research activities.

Funding: PP acknowledges research grants from BRNS, DBT and DST, Govt. of India.

Acknowledgments: Thanks to Sravanti Kasarlawar for suggestions and discussion. BAS thanks DST for project Junior Research Fellowship and HM thanks CSIR, Govt. of India for Junior Research Fellowship.

Conflicts of Interest: The authors declare no conflict of interest.

References

1. Danzo, B.J. Environmental xenobiotics may disrupt normal endocrine function by interfering with the binding of physiological ligands to steroid receptors and binding proteins. *Environ. Health Perspect.* **1997**, *105*, 294–301. [CrossRef] [PubMed]
2. Ohnishi, S.; Kawanishi, S. Double base lesions of DNA by a metabolite of carcinogenic benzo[a]pyrene. *Biochem. Biophys. Res. Commun.* **2002**, *290*, 778–782. [CrossRef] [PubMed]
3. Conney, A.H. Induction of microsomal enzymes by foreign chemicals and carcinogenesis by polycyclic aromatic hydrocarbons: GHA Clowes memorial lecture. *Cancer Res.* **1982**, *42*, 4875–4917.
4. Stegeman, J.J.; Lech, J.J. Cytochrome P-450 monooxygenase systems in aquatic species: Carcinogen metabolism and biomarkers for carcinogen and pollutant exposure. *Environ. Health Perspect.* **1991**, *90*, 101–109. [PubMed]
5. Pashin, Y.V.; Bakhitova, L.M. Mutagenic and carcinogenic properties of polycyclic aromatic hydrocarbons. *Environ. Health Perspect.* **1979**, *30*, 185–189. [CrossRef] [PubMed]
6. Wild, S.R.; Jones, K.C. Polynuclear aromatic hydrocarbons in the United Kingdom environment: A preliminary source inventory and budget. *Environ. Pollut.* **1995**, *88*, 91–108. [CrossRef]
7. Haritash, A.K.; Kaushik, C.P. Biodegradation aspects of polycyclic aromatic hydrocarbons (PAHs): A review. *J. Hazard. Mater.* **2009**, *169*, 1–15. [CrossRef] [PubMed]
8. Sikkema, J.; de Bont, J.A.; Poolman, B. Interactions of cyclic hydrocarbons with biological membranes. *J. Biol. Chem.* **1994**, *269*, 8022–8028. [PubMed]
9. Hansen, B.H.; Altin, D.; Vang, S.H.; Nordtug, T.; Olsen, A.J. Effects of naphthalene on gene transcription in *Calanus finmarchicus* (Crustacea: Copepoda). *Aquat. Toxicol.* **2008**, *86*, 157–165. [CrossRef] [PubMed]
10. Elumalai, M.; Balasubramanian, M.P. Influence of naphthalene on esterase activity during vitellogenesis of marine edible crab, *Scylla serrata*. *Bull. Environ. Contam. Toxicol.* **1999**, *62*, 743–748. [CrossRef]
11. Valaes, T.; Doxiadis, S.A.; Fessas, P. Acute hemolysis due to naphthalene inhalation. *J. Pediatr.* **1963**, *63*, 904–915. [CrossRef]
12. Gupta, R.; Singhal, P.C.; Muthusethupathy, M.A.; Malik, A.K.; Chugh, K.S. Cerebral oedema and renal failure following naphthalene poisoning. *J. Assoc. Phys. India* **1979**, *27*, 347–348.
13. National Toxicology Program. Toxicology and carcinogenesis studies of naphthalene (CAS No. 91-20-3) in F344/N rats (inhalation studies). *Natl. Toxicol. Program Tech. Rep. Ser.* **2000**, *500*, 1–173.
14. Smulders, C.J.; Bueters, T.J.; Van Kleef, R.G.; Vijverberg, H.P. Selective effects of carbamate pesticides on rat neuronal nicotinic acetylcholine receptors and rat brain acetylcholinesterase. *Toxicol. Appl. Pharmacol.* **2003**, *193*, 139–146. [CrossRef] [PubMed]
15. Moeller, F.E. Effect of pollen availability on poisoning of honey bees by Carbaryl applied to sweet corn. *J. Econ. Entomol.* **1971**, *64*, 1314–1315. [CrossRef] [PubMed]
16. Kaur, K.; Dhawan, A. Effect of Carbaryl on tissue composition, maturation, and breeding potential of *Cirrhina mrigala* (Ham.). *Bull. Environ. Contam. Toxicol.* **1996**, *57*, 480–486. [CrossRef] [PubMed]

17. Boone, M.D.; Semlitsch, R.D.; Little, E.E.; Doyle, M.C. Multiple stressors in amphibian communities: Effects of chemical contamination, bullfrogs, and fish. *Ecol. Appl.* **2007**, *17*, 291–301. [CrossRef]
18. Eraslan, G.; Kanbur, M.; Silici, S. Effect of Carbaryl on some biochemical changes in rats: The ameliorative effect of bee pollen. *Food Chem. Toxicol.* **2009**, *47*, 86–91. [CrossRef]
19. Bridges, C.M. Tadpole swimming performance and activity affected by acute exposure to sub lethal levels of Carbaryl. *Environ. Toxicol. Chem.* **1997**, *16*, 1935–1939. [CrossRef]
20. Bulen, B.J.; Distel, C.A. Carbaryl concentration gradients in realistic environments and their influence on our understanding of the tadpole food web. *Arch. Environ. Contam. Toxicol.* **2011**, *60*, 343–350. [CrossRef]
21. Lima, M.P.; Cardoso, D.N.; Soares, A.M.; Loureiro, S. Carbaryl toxicity prediction to soil organisms under high and low temperature regimes. *Ecotoxicol. Environ. Saf.* **2015**, *114*, 263–272. [CrossRef] [PubMed]
22. USEPA/Office of Pesticides Programs. Interim Reregistration Eligibility Decision for Carbaryl. p.2. Available online: https://www3.epa.gov/pesticides/chemsearch/regactions/reregistration/iredPC-056801_22-Oct-04.pdf (accessed on 15 April 2019).
23. Gibson, D.T.; Subramanian, V. Microbial degradation of aromatic hydrocarbons. In *Microbial Degradation of Organic Compounds*; Gibson, D.T., Ed.; Microbiology Series; Marcel Dekker Inc.: New York, NY, USA; Basel, Switzerland, 1984; Volume 13, pp. 181–252.
24. Singleton, I. Microbial metabolism of xenobiotics: Fundamental and applied research. *J. Chem. Technol. Biotech. Int. Res. Process Environ. Clean Technol.* **1994**, *59*, 9–23. [CrossRef]
25. Nojiri, H.; Shintani, M.; Omori, T. Divergence of mobile genetic elements involved in the distribution of xenobiotic-catabolic capacity. *Appl. Microbiol. Biotechnol.* **2004**, *64*, 154–174. [CrossRef]
26. Furukawa, K.; Matsumura, F. Microbial metabolism of polychlorinated biphenyls. Relative degradability of polychlorinated biphenyl components by *Alkaligenes* species. *J. Agric. Food Chem.* **1976**, *24*, 251–256. [CrossRef]
27. Chatterjee, D.K.; Kellogg, S.T.; Hamada, S.; Chakrabarty, A.M. Plasmid specifying total degradation of 3-chlorobenzoate by a modified ortho pathway. *J. Bacteriol.* **1981**, *146*, 639–646.
28. Baggi, G.; Barbieri, P.; Galli, E.; Tollari, S. Isolation of a *Pseudomonas stutzeri* strain that degrades o-xylene. *Appl. Environ. Microbiol.* **1987**, *53*, 2129–2132.
29. Mahajan, M.C.; Phale, P.S.; Vaidyanathan, C.S. Evidence for the involvement of multiple pathways in the biodegradation of 1-and 2-methylnaphthalene by *Pseudomonas putida* CSV86. *Arch. Microbiol.* **1994**, *161*, 425–433. [CrossRef] [PubMed]
30. Swetha, V.P.; Phale, P.S. Metabolism of Carbaryl via 1,2-dihydroxynaphthalene by soil isolates *Pseudomonas* sp. strains C4, C5, and C6. *Appl. Environ. Microbiol.* **2005**, *71*, 5951–5956. [CrossRef] [PubMed]
31. Vamsee-Krishna, C.; Mohan, Y.; Phale, P.S. Biodegradation of phthalate isomers by *Pseudomonas aeruginosa* PP4, *Pseudomonas* sp. PPD and *Acinetobacter lwoffii* ISP4. *Appl. Microbiol. Biotechnol.* **2006**, *72*, 1263–1269. [CrossRef] [PubMed]
32. John, R.C.; Essien, J.P.; Akpan, S.B.; Okpokwasili, G.C. Polycyclic aromatic hydrocarbon-degrading bacteria from aviation fuel spill site at Ibeno, Nigeria. *Bull. Environ. Contam. Toxicol.* **2012**, *88*, 1014–1019. [CrossRef] [PubMed]
33. Zhang, H.; Jiang, X.; Lu, L.; Xiao, W. Biodegradation of polychlorinated biphenyls (PCBs) by the novel identified cyanobacterium *Anabaena* PD-1. *PLoS ONE* **2015**, *10*, e0131450. [CrossRef] [PubMed]
34. Kanaly, R.A.; Harayama, S. Biodegradation of high-molecular-weight polycyclic aromatic hydrocarbons by bacteria. *J. Bacteriol.* **2000**, *182*, 2059–2067. [CrossRef] [PubMed]
35. Smith, M.R. The biodegradation of aromatic hydrocarbons by bacteria. *Biodegradation* **1990**, *1*, 191–206. [CrossRef] [PubMed]
36. Samanta, S.K.; Singh, O.V.; Jain, R.K. Polycyclic aromatic hydrocarbons: Environmental pollution and bioremediation. *Trends Biotechnol.* **2002**, *20*, 243–248. [CrossRef]
37. González-Gaya, B.; Martínez-Varela, A.; Vila-Costa, M.; Casal, P.; Cerro-Gálvez, E.; Berrojalbiz, N.; Jiménez, B. Biodegradation as an important sink of aromatic hydrocarbons in the oceans. *Nat. Geosci.* **2019**, *12*, 119. [CrossRef]
38. Srivastava, S.; Kumar, M. Biodegradation of polycyclic aromatic hydrocarbons (PAHs): A sustainable approach. In *Sustainable Green Technologies for Environmental Management*; Shah, S., Ramanan, V., Prasad, R., Eds.; Springer: Singapore, 2019; pp. 111–139.

39. Ladino-Orjuela, G.; Gomes, E.; da Silva, R.; Salt, C.; Parsons, J.R. Metabolic pathways for degradation of aromatic hydrocarbons by bacteria. In *Reviews of Environmental Contamination and Toxicology*; de Voogt, P., Ed.; Springer: Cham, Switzerland, 2016; Volume 237, pp. 105–121.
40. Liang, D.W.; Zhang, T.; Fang, H.H.; He, J. Phthalates biodegradation in the environment. *Appl. Microbiol. Biotechnol.* **2008**, *80*, 183. [CrossRef]
41. Kumar, V.; Sharma, N.; Maitra, S.S. Comparative study on the degradation of dibutyl phthalate by two newly isolated *Pseudomonas* sp. V21b and *Comamonas* sp. 51F. *Biotechnol. Rep.* **2017**, *15*, 1–10. [CrossRef]
42. Basu, A.; Dixit, S.S.; Phale, P.S. Metabolism of benzyl alcohol via catechol ortho-pathway in methylnaphthalene-degrading *Pseudomonas putida* CSV86. *Appl. Microbiol. Biotechnol.* **2003**, *62*, 579–585. [CrossRef]
43. Chapalamadugu, S.; Chaudhry, G.R. Hydrolysis of Carbaryl by a *Pseudomonas* sp. and construction of a microbial consortium that completely metabolizes Carbaryl. *Appl. Environ. Microbiol.* **1991**, *57*, 744–750.
44. Deng, Y.; Yang, F.; Deng, C.; Yang, J.; Jia, J.; Yuan, H. Biodegradation of BTEX aromatics by a haloduric microbial consortium enriched from a sediment of Bohai Sea, China. *Appl. Biochem. Biotechnol.* **2017**, *183*, 893–905. [CrossRef]
45. Zafra, G.; Absalón, Á.E.; Anducho-Reyes, M.Á.; Fernandez, F.J.; Cortés-Espinosa, D.V. Construction of PAH-degrading mixed microbial consortia by induced selection in soil. *Chemosphere* **2017**, *172*, 120–126. [CrossRef]
46. Vaidya, S.; Devpura, N.; Jain, K.; Madamwar, D. Degradation of chrysene by enriched bacterial consortium. *Front. Microbiol.* **2018**, *9*, 1333. [CrossRef]
47. Cerniglia, C.E.; Freeman, J.P.; Althaus, J.R.; van Baalen, C. Metabolism and toxicity of 1-and 2-methylnaphthalene and their derivatives in cyanobacteria. *Arch. Microbiol.* **1983**, *136*, 177–183. [CrossRef]
48. Cerniglia, C.E. Microbial metabolism of polycyclic aromatic hydrocarbons. *Adv. Appl. Microbiol.* **1984**, *30*, 31–71.
49. Phale, P.S.; Mahajan, M.C.; Vaidyanathan, C.S. A pathway for biodegradation of 1-naphthoic acid by *Pseudomonas maltophilia* CSV89. *Arch. Microbiol.* **1995**, *163*, 42–47. [CrossRef]
50. Vamsee-Krishna, C.; Phale, P.S. Bacterial degradation of phthalate isomers and their esters. *Indian J. Microbiol.* **2008**, *48*, 19–34. [CrossRef]
51. Vamsee-Krishna, C.; Phale, P.S. Bypassing isophthalate inhibition by modulating glutamate dehydrogenase (GDH): Purification and kinetic characterization of NADP-GDHs from isophthalate-degrading *Pseudomonas aeruginosa* strain PP4 and *Acinetobacter lwoffii* strain ISP4. *J. Bacteriol.* **2010**, *192*, 801–806. [CrossRef]
52. Arber, W. Genetic variation: Molecular mechanisms and impact on microbial evolution. *FEMS Microbiol. Rev.* **2000**, *24*, 1–7. [CrossRef]
53. Hacker, J.; Carniel, E. Ecological fitness, genomic islands and bacterial pathogenicity: A Darwinian view of the evolution of microbes. *EMBO Rep.* **2001**, *2*, 376–381. [CrossRef]
54. van der Meer, J.R.; Sentchilo, V. Genomic islands and the evolution of catabolic pathways in bacteria. *Curr. Opin. Biotechnol.* **2003**, *14*, 248–254. [CrossRef]
55. Springael, D.; Top, E.M. Horizontal gene transfer and microbial adaptation to xenobiotics: New types of mobile genetic elements and lessons from ecological studies. *Trends Microbiol.* **2004**, *12*, 53–58. [CrossRef]
56. Van Der Meer, J.R.; De Vos, W.M.; Harayama, S.; Zehnder, A.J. Molecular mechanisms of genetic adaptation to xenobiotic compounds. *Microbiol. Mol. Biol. Rev.* **1992**, *56*, 677–694.
57. Nagata, Y.; Kato, H.; Ohtsubo, Y.; Tsuda, M. Mobile genetic elements involved in the evolution of bacteria that degrade recalcitrant xenobiotic compounds. In *DNA Traffic in the Environment*; Hiromi, N., Taku, O., Eds.; Springer: Singapore, 2019; pp. 215–244.
58. Johnsen, A.R.; Wick, L.Y.; Harms, H. Principles of microbial PAH-degradation in soil. *Environ. Pollut.* **2005**, *133*, 71–84. [CrossRef]
59. Chakrabarty, A.M. Genetic basis of the biodegradation of salicylate in *Pseudomonas*. *J. Bacteriol.* **1972**, *112*, 815–823.
60. Chakrabarty, A.M.; Chou, G.; Gunsalus, I.C. Genetic regulation and extra chromosomal nature of octane degradative pathway in *Pseudomonas*. *Proc. Natl. Acad. Sci. USA* **1973**, *70*, 1137–1140. [CrossRef]

61. Dunn, N.W.; Gunsalus, I.C. Transmissible plasmid coding early enzymes of naphthalene oxidation in *Pseudomonas putida*. *J. Bacteriol.* **1973**, *114*, 974–979.
62. Rheinwald, J.G.; Chakrabarty, A.M.; Gunsalus, I.C. A transmissible plasmid controlling camphor oxidation in *Pseudomonas putida*. *Proc. Natl. Acad. Sci. USA* **1973**, *70*, 885–889. [CrossRef]
63. Worsey, M.J.; Williams, P.A. Metabolism of toluene and xylenes by *Pseudomonas putida* (*arvilla*) mt-2: Evidence for a new function of the TOL plasmid. *J. Bacteriol.* **1975**, *124*, 7–13.
64. Mulbry, W.W.; Karns, J.S.; Kearney, P.C.; Nelson, J.O.; McDaniel, C.S.; Wild, J.R. Identification of a plasmid-borne parathion hydrolase gene from *Flavobacterium* sp. by southern hybridization with *opd* from *Pseudomonas diminuta*. *Appl. Environ. Microbiol.* **1986**, *51*, 926–930.
65. Hayatsu, M.; Hirano, M.; Tokuda, S. Involvement of two plasmids in fenitrothion degradation by *Burkholderia* sp. strain NF100. *Appl. Environ. Microbiol.* **2000**, *66*, 1737–1740. [CrossRef]
66. Wu, S.J.; Hu, Z.H.; Zhang, L.L.; Yu, X.; Chen, J.M. A novel dichloromethane-degrading *Lysinibacillus sphaericus* strain wh22 and its degradative plasmid. *Appl. Microbiol. Biotechnol.* **2009**, *82*, 731–740. [CrossRef]
67. Feng, X.; Ou, L.T.; Ogram, A. Plasmid-mediated mineralization of carbofuran by *Sphingomonas* sp. strain CF06. *Appl. Environ. Microbiol.* **1997**, *63*, 1332–1337.
68. Maeda, K.; Nojiri, H.; Shinatani, M.; Yoshida, T.; Habe, H.; Omori, T. Complete nucleotide sequence of carbazole/dioxin degrading plasmid pCAR1 in *Pseudomonas resinovorans* strain CA10 indicates its mosaicity and the presence of large catabolic transposon Tn4676. *J. Mol. Biol.* **2003**, *326*, 21–33. [CrossRef]
69. Martinez, B.; Tomkins, J.; Wackett, L.P.; Wing, R.; Sadowsky, M.J. Complete nucleotide sequence and organization of the atrazine catabolic plasmid pADP-1 from *Pseudomonas* sp. strain ADP. *J. Bacteriol.* **2001**, *183*, 5684–5697. [CrossRef]
70. Sota, M.; Kawasaki, H.; Tsuda, M. Structure of haloacetate catabolic IncP-1β plasmid pUO1 and genetic mobility of its residing haloacetate-catabolic transposon. *J. Bacteriol.* **2003**, *185*, 6741–6745. [CrossRef]
71. Tsuda, M.; Iino, T. Identification and characterization of Tn4653, a transposon covering the toluene transposon Tn4651 on TOL plasmid pWW0. *Mol. Gen. Genet.* **1988**, *213*, 72–77. [CrossRef]
72. Tsuda, M.; Iino, T. Naphthalene degrading genes on plasmid NAH7 are on a defective transposon. *Mol. Gen. Genet.* **1990**, *223*, 33–39. [CrossRef]
73. Obayori, O.S.; Salam, L.B. Degradation of polycyclic aromatic hydrocarbons: Role of plasmids. *Sci. Res. Essays* **2010**, *5*, 4093–4106.
74. Weightman, A.J.; Topping, A.W.; Hill, K.E.; Lee, L.L.; Sakai, K.; Slater, J.H.; Thomas, A.W. Transposition of DEH, a broad-host-range transposon flanked by ISPpu12, in *Pseudomonas putida* is associated with genomic rearrangements and dehalogenase gene silencing. *J. Bacteriol.* **2002**, *184*, 6581–6591. [CrossRef]
75. Williams, P.A.; Jones, R.M.; Shaw, L.E. A third transposable element, ISPpu12, from the toluene-xylene catabolic plasmid pWW0 of *Pseudomonas putida* mt-2. *J. Bacteriol.* **2002**, *184*, 6572–6580. [CrossRef]
76. Tsuda, M.; Genka, H. Identification and characterization of Tn4656, a novel class II transposon carrying a set of toluene degrading genes from TOL plasmid pWW53. *J. Bacteriol.* **2001**, *183*, 6215–6224. [CrossRef]
77. Grindley, N.D.; Reed, R.R. Transpositional recombination in prokaryotes. *Annu. Rev. Biochem.* **1985**, *54*, 863–896. [CrossRef]
78. van der Meer, J.R.; Zehnder, A.J.; de Vos, W.M. Identification of a novel composite transposable element, Tn5280, carrying chlorobenzene dioxygenase genes of *Pseudomonas* sp. strain P51. *J. Bacteriol.* **1991**, *173*, 7077–7083. [CrossRef]
79. Tsuda, M.; Minegishi, K.I.; Iino, T. Toluene transposons Tn4651 and Tn4653 are class II transposons. *J. Bacteriol.* **1989**, *171*, 1386–1393. [CrossRef]
80. Burrus, V.; Pavlovic, G.; Decaris, B.; Guédon, G. Conjugative transposons: The tip of the iceberg. *Mol. Microbiol.* **2002**, *46*, 601–610. [CrossRef]
81. Hentschel, U.; Hacker, J. Pathogenicity islands: The tip of the iceberg. *Microb. Infect.* **2001**, *3*, 545–548. [CrossRef]
82. Blum, G.; Ott, M.; Lischewski, A.; Ritter, A.; Imrich, H.; Tschäpe, H.; Hacker, J. Excision of large DNA regions termed pathogenicity islands from tRNA-specific loci in the chromosome of an *Escherichia coli* wild-type pathogen. *Infect. Immun.* **1994**, *62*, 606–614.

83. Yamaguchi, T.; Nishifuji, K.; Sasaki, M.; Fudaba, Y.; Aepfelbacher, M.; Takata, T.; Ohara, M.; Komatsuzawa, H.; Amagai, M.; Sugai, M. Identification of the *Staphylococcus aureus etd* pathogenicity island which encodes a novel exfoliative toxin, ETD, and EDIN-B. *Infect. Immun.* **2002**, *70*, 5835–5845. [CrossRef]
84. Parreira, V.R.; Gyles, C.L. A novel pathogenicity island integrated adjacent to the thrW tRNA gene of avian pathogenic *Escherichia coli* encodes a vacuolating autotransporter toxin. *Infect. Immun.* **2003**, *71*, 5087–5096. [CrossRef]
85. Luck, S.N.; Turner, S.A.; Rajakumar, K.; Sakellaris, H.; Adler, B. Ferric dicitrate transport system (Fec) of *Shigella flexneri* 2a YSH6000 is encoded on a novel pathogenicity island carrying multiple antibiotic resistance genes. *Infect. Immun.* **2001**, *69*, 6012–6021. [CrossRef]
86. Novick, R.P. Mobile genetic elements and bacterial toxinoses: The superantigen-encoding pathogenicity islands of *Staphylococcus aureus*. *Plasmid* **2003**, *49*, 93–105. [CrossRef]
87. Inouye, S.; Sunshine, M.G.; Six, E.W.; Inouye, M. Retronphage phi R73: An *E. coli* phage that contains a retroelement and integrates into a tRNA gene. *Science* **1991**, *252*, 969–971. [CrossRef]
88. Cheetham, B.F.; Katz, M.E. A role for bacteriophages in the evolution and transfer of bacterial virulence determinants. *Mol. Microbiol.* **1995**, *18*, 201–208. [CrossRef]
89. Gaillard, M.; Vallaeys, T.; Vorhölter, F.J.; Minoia, M.; Werlen, C.; Sentchilo, V.; Pühler, A.; van der Meer, J.R. The *clc* element of *Pseudomonas* sp. strain B13, a genomic island with various catabolic properties. *J. Bacteriol.* **2006**, *188*, 1999–2013. [CrossRef]
90. Zamarro, M.T.; Martín-Moldes, Z.; Díaz, E. The ICE_{XTD} of *Azoarcus* sp. CIB, an integrative and conjugative element with aerobic and anaerobic catabolic properties. *Environ. Microbiol.* **2016**, *18*, 5018–5031. [CrossRef]
91. Paliwal, V.; Raju, S.C.; Modak, A.; Phale, P.S.; Purohit, H.J. *Pseudomonas putida* CSV86: A candidate genome for genetic bioaugmentation. *PLoS ONE* **2014**, *9*, e84000. [CrossRef]
92. Obi, C.C.; Vayla, S.; De Gannes, V.; Berres, M.E.; Walker, J.; Pavelec, D.; Hyman, J.; Hickey, W.J. The integrative conjugative element *clc* (ICEclc) of *Pseudomonas aeruginosa* JB2. *Front. Microbiol.* **2018**, *9*, 1532. [CrossRef]
93. Ohtsubo, Y.; Ishibashi, Y.; Naganawa, H.; Hirokawa, S.; Atobe, S.; Nagata, Y.; Tsuda, M. Conjugal transfer of polychlorinated biphenyl/biphenyl degradation genes in *Acidovorax* sp. strain KKS102, which are located on an integrative and conjugative element. *J. Bacteriol.* **2012**, *194*, 4237–4248. [CrossRef]
94. Toussaint, A.; Merlin, C.; Monchy, S.; Benotmane, M.A.; Leplae, R.; Mergeay, M.; Springael, D. The biphenyl-and 4-chlorobiphenyl-catabolic transposon Tn4371, a member of a new family of genomic islands related to IncP and Ti plasmids. *Appl. Environ. Microbiol.* **2003**, *69*, 4837–4845. [CrossRef]
95. Hickey, W.J.; Chen, S.; Zhao, J. The *phn* island: A new genomic island encoding catabolism of polynuclear aromatic hydrocarbons. *Front. Microbiol.* **2012**, *3*, 125. [CrossRef]
96. Basu, A.; Phale, P.S. Conjugative transfer of preferential utilization of aromatic compounds from *Pseudomonas putida* CSV86. *Biodegradation* **2008**, *19*, 83–92. [CrossRef]
97. Annweiler, E.; Richnow, H.H.; Antranikian, G.; Hebenbrock, S.; Garms, C.; Franke, S.; Francke, W.; Michaelis, W. Naphthalene degradation and incorporation of naphthalene-derived carbon into biomass by the thermophile *Bacillus thermoleovorans*. *Appl. Environ. Microbiol.* **2000**, *66*, 518–523. [CrossRef]
98. Yen, K.M.; Gunsalus, I.C. Regulation of naphthalene catabolic genes of plasmid NAH7. *J. Bacteriol.* **1985**, *162*, 1008–1013.
99. Sota, M.; Yano, H.; Ono, A.; Miyazaki, R.; Ishii, H.; Genka, H.; Top, E.M.; Tsuda, M. Genomic and functional analysis of the IncP-9 naphthalene-catabolic plasmid NAH7 and its transposon Tn4655 suggests catabolic gene spread by a tyrosine recombinase. *J. Bacteriol.* **2006**, *188*, 4057–4067. [CrossRef]
100. Shamsuzzaman, K.M.; Barnsley, E.A. The regulation of naphthalene oxygenase in pseudomonads. *Microbiology* **1974**, *83*, 165–170. [CrossRef]
101. Park, W.; Padmanabhan, P.; Padmanabhan, S.; Zylstra, G.J.; Madsen, E.L. *nahR*, encoding a LysR-type transcriptional regulator, is highly conserved among naphthalene-degrading bacteria isolated from a coal tar waste-contaminated site and in extracted community DNA. *Microbiology* **2002**, *148*, 2319–2329. [CrossRef]
102. Dennis, J.J.; Zylstra, G.J. Complete sequence and genetic organization of pDTG1, the 83 kilobase naphthalene degradation plasmid from *Pseudomonas putida* strain NCIB 9816-4. *J. Mol. Biol.* **2004**, *341*, 753–768. [CrossRef]
103. Izmalkova, T.Y.; Sazonova, O.I.; Nagornih, M.O.; Sokolov, S.L.; Kosheleva, I.A.; Boronin, A.M. The organization of naphthalene degradation genes in *Pseudomonas putida* strain AK5. *Res. Microbiol.* **2013**, *164*, 244–253. [CrossRef]

104. Zuniga, M.C.; Durham, D.R.; Welch, R.A. Plasmid-and chromosome-mediated dissimilation of naphthalene and salicylate in *Pseudomonas putida* PMD-1. *J. Bacteriol.* **1981**, *147*, 836–843.
105. Bosch, R.; García-Valdés, E.; Moore, E.R. Genetic characterization and evolutionary implications of a chromosomally encoded naphthalene-degradation upper pathway from *Pseudomonas stutzeri* AN10. *Gene* **1999**, *236*, 149–157. [CrossRef]
106. Bosch, R.; García-Valdés, E.; Moore, E.R. Complete nucleotide sequence and evolutionary significance of a chromosomally encoded naphthalene-degradation lower pathway from *Pseudomonas stutzeri* AN10. *Gene* **2000**, *245*, 65–74. [CrossRef]
107. Connors, M.A.; Barnsley, E.A. Naphthalene plasmids in pseudomonads. *J. Bacteriol.* **1982**, *149*, 1096–1101.
108. Foght, J.M.; Westlake, D.W. Transposon and spontaneous deletion mutants of plasmid-borne genes encoding polycyclic aromatic hydrocarbon degradation by a strain of *Pseudomonas fluorescens*. *Biodegradation* **1996**, *7*, 353–366. [CrossRef]
109. Li, W.; Shi, J.; Wang, X.; Han, Y.; Tong, W.; Ma, L.; Liu, B.; Cai, B. Complete nucleotide sequence and organization of the naphthalene catabolic plasmid pND6-1 from *Pseudomonas* sp. strain ND6. *Gene* **2004**, *336*, 231–240. [CrossRef]
110. Heinaru, E.; Vedler, E.; Jutkina, J.; Aava, M.; Heinaru, A. Conjugal transfer and mobilization capacity of the completely sequenced naphthalene plasmid pNAH20 from multi plasmid strain *Pseudomonas fluorescens* PC20. *FEMS Microbiol. Ecol.* **2009**, *70*, 563–574. [CrossRef]
111. Kim, J.; Park, W. Genome analysis of naphthalene-degrading *Pseudomonas* sp. AS1 harboring the mega plasmid pAS1. *J. Microbiol. Biotechnol.* **2018**, *28*, 330–337. [CrossRef]
112. Sud, R.K.; Sud, A.K.; Gupta, K.G. Degradation of Sevin (1-naphthyl N-methyl carbamate by *Achromobacter* sp. *Arch. Mikrobiol.* **1972**, *87*, 353–358. [CrossRef]
113. Larkin, M.J.; Day, M.J. The metabolism of Carbaryl by three bacterial isolates, *Pseudomonas* spp. (NCIB 12042 & 12043) and *Rhodococcus* sp. (NCIB 12038) from garden soil. *J. Appl. Bacteriol.* **1986**, *60*, 233–242.
114. Hayatsu, M.; Nagata, T. Purification and characterization of Carbaryl hydrolase from *Blastobacter* sp. strain M501. *Appl. Environ. Microbiol.* **1993**, *59*, 2121–2125.
115. Hayatsu, M.; Hirano, M.; Nagata, T. Involvement of two plasmids in the degradation of Carbaryl by *Arthrobacter* sp. strain RC100. *Appl. Environ. Microbiol.* **1999**, *65*, 1015–1019.
116. Hashimoto, M.; Fukui, M.; Hayano, K.; Hayatsu, M. Nucleotide sequence and genetic structure of a novel Carbaryl hydrolase gene (*cehA*) from *Rhizobium* sp. strain AC100. *Appl. Environ. Microbiol.* **2002**, *68*, 1220–1227. [CrossRef]
117. Doddamani, H.P.; Ninnekar, H.Z. Biodegradation of Carbaryl by a *Micrococcus* species. *Curr. Microbiol.* **2001**, *43*, 69–73. [CrossRef]
118. Seo, J.S.; Keum, Y.S.; Li, Q.X. Metabolomic and proteomic insights into Carbaryl catabolism by *Burkholderia* sp. C3 and degradation of ten *N*-methyl carbamates. *Biodegradation* **2013**, *24*, 795–811. [CrossRef]
119. Trivedi, V.D.; Bharadwaj, A.; Varunjikar, M.S.; Singha, A.K.; Upadhyay, P.; Gautam, K.; Phale, P.S. Insights into metabolism and sodium chloride adaptability of Carbaryl degrading halotolerant *Pseudomonas* sp. strain C7. *Arch. Microbiol.* **2017**, *199*, 907–916. [CrossRef]
120. Zhu, S.; Wang, H.; Jiang, W.; Yang, Z.; Zhou, Y.; He, J.; Hong, Q. Genome analysis of Carbaryl-degrading strain *Pseudomonas putida* XWY-1. *Curr. Microbiol.* **2019**, *76*, 927–929. [CrossRef]
121. Cheesman, M.J.; Horne, I.; Weir, K.M.; Pandey, G.; Williams, M.R.; Scott, C.; Russell, R.J.; Oakeshott, J.G. Carbamate pesticides and their biological degradation: Prospects for enzymatic bioremediation. *Ration. Environ. Manag. Agrochem. Risk Assess. Monit. Remedial Act.* **2007**, *966*, 288–305.
122. Singh, R.; Trivedi, V.D.; Phale, P.S. Metabolic regulation and chromosomal localization of Carbaryl degradation pathway in *Pseudomonas* sp. strains C4, C5 and C6. *Arch. Microbiol.* **2013**, *195*, 521–535. [CrossRef]
123. Trivedi, V.D. Biochemical and Evolutionary Aspects of Carbaryl Metabolism in *Pseudomonas* sp. PhD Thesis, Indian Institute of Technology-Bombay, Mumbai, India, December 2016.
124. Trivedi, V.D.; Jangir, P.K.; Sharma, R.; Phale, P.S. Insights into functional and evolutionary analysis of Carbaryl metabolic pathway from *Pseudomonas* sp. strain C5pp. *Sci. Rep.* **2016**, *6*, 38430. [CrossRef]

125. Kamini; Sharma, R.; Punekar, N.S.; Phale, P.S. Carbaryl as a carbon and nitrogen source: An inducible methylamine metabolic pathway at the biochemical and molecular levels in *Pseudomonas* sp. strain C5pp. *Appl. Environ. Microbiol.* **2018**, *84*, e01866–e01918. [CrossRef]
126. Kamini; Shetty, D.; Trivedi, V.D.; Varunjikar, M.; Phale, P.S. Compartmentalization of the Carbaryl degradation pathway: Molecular characterization of inducible periplasmic Carbaryl hydrolase from *Pseudomonas* spp. *Appl. Environ. Microbiol.* **2018**, *84*, e02115–e02117.

© 2019 by the authors. Licensee MDPI, Basel, Switzerland. This article is an open access article distributed under the terms and conditions of the Creative Commons Attribution (CC BY) license (http://creativecommons.org/licenses/by/4.0/).

genes

MDPI

Review

Biodegradation of Tetralin: Genomics, Gene Function and Regulation

Belén Floriano [1], Eduardo Santero [2] and Francisca Reyes-Ramírez [2,*]

[1] Universidad Pablo de Olavide, Departamento de Biología Molecular e Ingeniería Bioquímica, 41013 Seville, Spain; bflopar@upo.es
[2] Centro Andaluz de Biología del Desarrollo, CSIC, Universidad Pablo de Olavide, Junta de Andalucía, Departamento de Biología Molecular e Ingeniería Bioquímica, 41013 Seville, Spain; esansan@upo.es
* Correspondence: freyram@upo.es; Tel.: +34-954349053

Received: 28 March 2019; Accepted: 30 April 2019; Published: 6 May 2019

Abstract: Tetralin (1,2,3,4-tetrahydonaphthalene) is a recalcitrant compound that consists of an aromatic and an alicyclic ring. It is found in crude oils, produced industrially from naphthalene or anthracene, and widely used as an organic solvent. Its toxicity is due to the alteration of biological membranes by its hydrophobic character and to the formation of toxic hydroperoxides. Two unrelated bacteria, *Sphingopyxis granuli* strain TFA and *Rhodococcus* sp. strain TFB were isolated from the same niche as able to grow on tetralin as the sole source of carbon and energy. In this review, we provide an overview of current knowledge on tetralin catabolism at biochemical, genetic and regulatory levels in both strains. Although they share the same biodegradation strategy and enzymatic activities, no evidences of horizontal gene transfer between both bacteria have been found. Moreover, the regulatory elements that control the expression of the gene clusters are completely different in each strain. A special consideration is given to the complex regulation discovered in TFA since three regulatory systems, one of them involving an unprecedented communication between the catabolic pathway and the regulatory elements, act together at transcriptional and posttranscriptional levels to optimize tetralin biodegradation gene expression to the environmental conditions.

Keywords: tetralin; *Sphingopyxis granuli* strain TFA; *Rhodococcus* sp. strain TFB; redox proteins; carbon catabolite repression

---

1. Introduction

The ability of microorganisms to degrade a variety of environmental pollutants and to use them as growth substrates is a powerful tool to clean up hazardous contaminants that are causing irreversible damage to the biosphere. Many hydrocarbons such as most alkanes can be efficiently degraded by several microorganisms [1] and many aromatic compounds can serve as growth substrates for a large variety of bacteria [2]. On the other hand, cycloalkanes seem to be more persistent to microbial attack [3]. Tetralin (1,2,3,4-tetrahydronaphthalene) is a bicyclic molecule composed of an aromatic and an alicyclic ring. It is naturally found in different crude oil reservoirs and also produced for industrial purposes from naphthalene by catalytic hydrogenation or from anthracene by cracking. It is widely used as a degreasing agent and solvent for fats, resins and waxes, as a substitute for turpentine in paints, lacquers, and shoe polishes, and in the petrochemical industry in connection with coal liquefaction [3]. Like other cyclic hydrocarbons, tetralin is toxic due to its lipophilic character, which facilitates its accumulation within the biological membranes causing changes in their structure and function, and to the formation of hydroperoxides inside the cells [4]. Its hydrophobic character, which reduces tetralin bioavailability, along with the fact that concentrations greater than 100 μM are toxic for microbial cultures might be the reasons why it has been difficult to isolate pure cultures of microorganisms able to grow on tetralin as the only carbon and energy source [3]. Tetralin utilization by microorganisms has long been reported by mixed cultures or by pure cultures through co-oxidation of tetralin in the presence of mixed substrates [5]. Reports in the literature describing poor growth on tetralin with pure cultures are from *Pseudomonas stutzeri* AS39 [6] and *Corynebacterium* sp. strain C125 [7]. By identifying accumulated intermediates, it was suggested that they might attack the molecule of tetralin in different ways. In one, the alicyclic ring is initially hydroxylated and then oxidized [6] whereas in the other the aromatic ring is dioxygenated and subsequently cleaved in the extradiol position [7]. The degradation pathway has not been further studied in any of these strains. The main source of knowledge about tetralin degradation has been provided *by Sphingopyxis granuli* strain TFA (formerly *Sphingomonas macrogolitabida* strain TFA) where the catabolic pathway has been completely characterized at biochemical, genetic and regulatory levels. A similar degradation pathway has been described in *Rhodococcus* sp. strain TFB based on the induction in tetralin-grown cells of homologous enzymes to the TFA ones. These two different bacteria were isolated from the river Rhine sediments after selection for growth on tetralin. TFA is a Gram-negative α-proteobacterium, whose genome is available and shows characteristics of oligotrophic bacteria, great plasticity and presence of other degradative genes that might confer new capabilities to this bacterium [8]. Moreover, TFA is the first of its genus, generally described as aerobic, in which the ability to grow in the absence of oxygen using nitrate as a terminal electron acceptor has been demonstrated [8]. This capacity might allow TFA to invade other niches. TFB is a Gram-positive Actinobacterium from the *Nocardiaceae* family refractory to genetic manipulation, although a proteomic approach has allowed the identification of all the tetralin gene clusters. In this review, we summarize the genetic organization, the biochemistry of the tetralin biodegradation pathway and its connection with the central metabolism via production of pyruvate and acetil-CoA in both strains. We also highlight the contributions that the study of the tetralin degradation genes regulation in each strain has made to understand the regulation of gene expression of other degradation pathways, and to the knowledge of bacterial mechanisms of gene regulation in general.

2. Identification and Organization of Tetralin Degradation Gene Clusters

As mentioned above, the tetralin biodegradation pathway has been elucidated only in two bacteria isolated from the same environmental niche, *Sphingophyxis granuli* strain TFA and *Rhodococcus* sp. strain TFB. In this section, we review the different methodology employed to identify *thn* genes and the organization of the *thn* gene clusters in both strains.

2.1. Organization of the Gene Clusters in TFA

A Tn5 or miniTn5*Km* transposon mutant isolation approach was employed to identify *thn* genes in TFA. A gene bank of TFA was constructed [9] and used to complement transposon insertion mutants. Cosmids containing genes that complemented the growth deficiency on tetralin of transposon mutants were obtained and sequenced ([10] and references therein). As shown in Figure 1, the regulatory and structural *thn* genes consist of 21 genes comprised in a 14.3-kb DNA fragment and organized in four closely linked operons (B, C, H and M). The divergent operons C and B have two internal promoters P_R and P_H.

Operon C contains five genes: *thnC*, encoding an extradiol dioxygenase, *thnA3* and *thnA4*, coding for the ferredoxin and the ferredoxin reductase components of the complex catalysing the initial dioxygenase reaction, and the regulatory genes *thnRY*. ThnR belongs to the LysR-type transcriptional regulators (LTTR) family. In most cases, the gene encoding the LTTR is divergently transcribed to the regulated operon [11]. Thus, *thn* genes have an unusual organisation in TFA being that the regulatory genes are part of a structural operon while also being transcribed from the internal and constitutive promoter P_R. In turn, *thnY* codes for a ferredoxin reductase-like protein, which has a transcriptional regulatory function. This is unusual for an electron transport protein, thus representing another particularity of the tetralin biodegradation genes of TFA (see below).

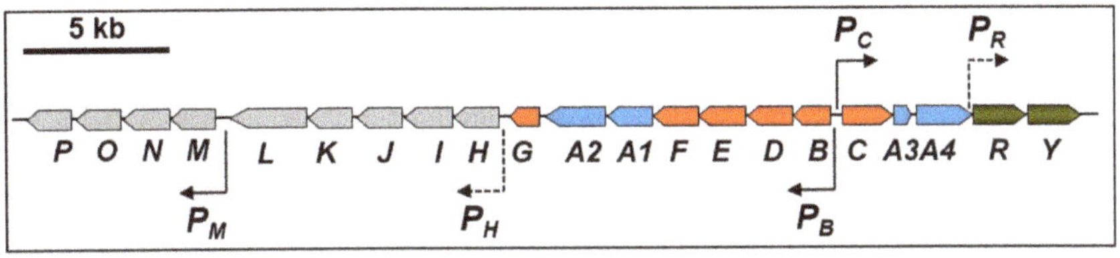

Figure 1. Organization of the chromosomal *thn* genes in TFA. The arrows show the directions of gene transcription. Bent arrows represent the identified promoters. Dashed lines represent internal promoters.

Operon B comprises 12 genes: *thnB*, which encodes a *cis*-dihydrodiol dehydrogenase, *thnD*, coding for the key hydroxylase that renders a linear compound, *thnE*, encoding a hydratase, *thnF*, encoding an aldolase, *thnA1* and *thnA2*, encoding the α- and β-subunits of the initial dioxygenase, respectively, and *thnG*, coding for a non-acylating aldehyde dehydrogenase. Five additional genes (*thnHIJKL*) are comprised within operon B although also transcribed from the internal promoter P_H. These genes encode proteins similar to those involved in β-oxidation pathways.

Finally, operon M comprises *thnM*, encoding a TonB-dependent protein of unknown function, *thnN*, encoding a putative glutaryl-CoA dehydrogenase and *thnO* and *thnP* which encode proteins similar to the α- and β-subunits of a flavoprotein, respectively [10].

A few years ago, the complete genome sequence of TFA was obtained and analysed [8]. The presence of large plasmids in strains of the sphingomonas group able to degrade different contaminants is very common. In fact, a number of "degradative megaplasmids" have been described in this group of bacteria that contribute to their catabolic flexibility. Some examples are pNL2, pCAR3, pSWIT02, pCHQ1, pISP0, and pISP1, which code for genes involved in the degradation of aromatic hydrocarbons, carbazole, dibenzo-p-dioxin and γ-hexachlorocyclohexane, respectively [12]. In contrast, no plasmids were detected in TFA by experimental approaches, which was later confirmed when the genome was sequenced. The *thn* genes were found within a chromosomal region identified as a putative genomic island and relatively close to a *tra/trb* gene cluster. Although more data are needed to be conclusive, those regions could be considered as integrative and conjugative elements (ICEs) [13] which might have played an important role in the final structure of the TFA genome. The analysis performed in 2016 to define the core genome of the *Sphingopyxis* genus showed that all tetralin degradation proteins, except ThnA4, ThnY and ThnM, are part of this core genome [8].

2.2. Organization of the Gene Clusters in TFB

TFB is a bacterium refractory to genetic manipulation and *thn* gene identification had to be carried out by reverse genetics using a proteomic approach [14]. This method has also been used to identify biodegradation genes in other bacteria of environmental interest [15]. Soluble proteins present in extracts of TFB cells grown on glucose or on tetralin were compared using 2D-DIGE (two-dimensional difference gel electrophoresis). Differentially expressed proteins present in tetralin cell extracts were identified by MALDI-MS(MS) (maldi-assisted laser desorption/ionization tandem mass spectrometry) or ESI-IT MS/MS (electrospray ionization-ion trap tandem mass spectrometry). Eight proteins with probable roles in the conversion of tetralin into a linear compound were identified, ThnD being one of them. Degenerated primers based on the ThnD-obtained peptide sequences were designed to amplify a DNA fragment of the putative *thnD* gene, which was used to screen a TFB cosmid library. A DNA fragment comprising 18 putative *thn* structural genes and two more coding for the putative two-component regulatory system *thnST* were sequenced. *thnS* encodes for a membrane sensor protein and ThnT is similar to members of the LuxR family of regulators. As shown in Figure 2, TFB *thn* structural genes are organized into two operons, *thnA1A2A3UA4B* and *thnA1'A2'CGQILWVDEF*, transcribed from P_{A1} and $P_{A1'}$ promoters, whereas the regulatory operon is divergently transcribed to the structural ones from the *Ps* promoter.

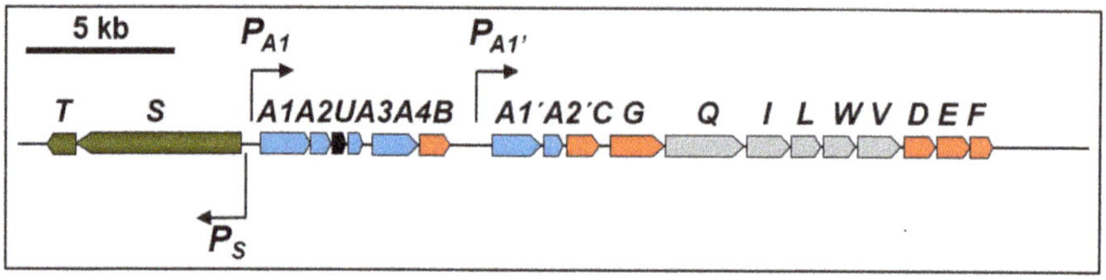

Figure 2. Organization of the chromosomal *thn* genes in TFB. The arrows show the directions of gene transcription. Bent arrows represent the identified promoters.

All genes encoding enzymes are similar to those found in TFA except *thnU*, coding for a putative sterol transfer protein, *thnQ*, encoding a pimeloyl-CoA synthase, *thnW* coding for a 3-hydroxyacyl-CoA dehydrogenase and *thnV* coding for an acyl-CoA dehydrogenase (see Table 1). TFB *thn* genes are highly similar to *bph* and *etb* genes in *R. jostii* sp RHA1, which are located on pRHL1 and pRHL2 plasmids. Similarly, using pulse field electrophoresis analysis, two large plasmids, pTFB1 and pTFB2, with estimated sizes of 1100 and 280 kb, respectively, have been identified in TFB, being the *thn* genes located on pTFB1 [14].

Table 1. Homology between TFA and TFB proteins.

Protein	Funtion	TFA (aa)	TFB (aa)	Identity (%)/ Coverage (%)
ThnA1	tetralin dioxygenase, α-subunit	460	456	62/94
ThnA2	tetralin dioxygenase, β-subunit	181	182	59/95
ThnA3	ferredoxin	104	121	46/92
ThnA4	ferredoxin reductase	339	406	50/3
ThnB	cis-dihydrodiol dehydrogenase	270	270	57/100
ThnC	extradiol dioxygenase	304	299	65/97
ThnD	hydrolase	284	285	54/98
ThnE	hydratase	267	285	60/97
ThnF	aldolase	255	253	50/95
ThnG	aldehyde dehydrogenase	475	476	36/94
ThnH	CoA transferase	405		
ThnQ	pimeloyl-CoA synthase		706	
ThnJ	pimeloyl-CoA dehydrogenase, large subunit	397		
ThnK	pimeloyl-CoA dehydrogenase, small subunit	379		
ThnV	acyl-CoA dehydrogenase		383	
ThnL	enoyl-CoA hydratase and 3-hydroxyacyl-CoA dehydrogenase	695	295*	42/28
ThnW	3-hydroxyacyl-CoA dehydrogenase		286*	32/38
ThnI	acetyl-CoA acetyl transferase	402	389	55/99
ThnM	TonB-dependent protein	723		
ThnN	glutaryl-CoA dehydrogenase	394		
ThnO	flavoprotein, α-subunit	249		
ThnP	flavoprotein, β-subunit	312		
ThnU	putative sterol transfer protein		131	

TFA proteins were used as query in BlastP analysis with default parameters. *Identity of $ThnL^{TFB}$ and $ThnW^{TFB}$ to $ThnL^{TFA}$ N and C-terminal part, respectively.

3. Tetralin Degradation Pathway and Enzymes Involved

In this section, the main features of the tetralin degradation pathway in TFA and TFB are reviewed. Although the degradation pathway is essentially the same in both bacteria (Scheme 1), differences have been found in some of the enzymes. The proteins involved in the degradation pathway to convert tetralin into pimelic acid were quite similar in both bacteria except for ThnA4 (Table 1). This similarity ranges from 65% of identity for ThnC proteins to 36% for ThnG enzymes. In all cases, coverage is higher than 90%. The main difference between TFA and TFB pathways was found in those proteins involved in pimelic acid conversion to central metabolites via a β-oxidation pathway [14,16]. It should be noted that the tetralin degradation pathway has been characterised in detail only in strain TFA. The existence of similar enzymes led to the conclusion that a similar pathway should operate in TFB.

Aerobic biodegradation of tetralin starts by a common mechanism quite well described for degradation of aromatic compounds, which is the insertion of oxygen into the aromatic ring of tetralin by a non-heme iron-containing dioxygenase [17]. These multicomponent ring-hydroxylating enzymes require Fe^{2+} and an electron supply in addition to molecular oxygen. These enzymes establish an electron-transport chain that allows the transfer of electrons from NAD(P)H, via flavin and [2Fe-2S] redox centers, to the terminal oxygenase containing the catalytic site. Bacterial oxygenases has been classified into classes I, II, and III according to the number of components and the nature of the redox centers of the electron transport chain [18]. Class I consists of two components, the reductase and the oxygenase. The reductase contains a chloroplast type [2Fe-2S] cluster and a flavin, FMN in class IA and FAD in class IB. Class II consists of three-component enzymes but in this case, the flavin and [2Fe-2S] redox centers of the reductase are on a separate flavoprotein and ferredoxin, respectively. Class III consists of three-component enzymes in which the reductase harbours both FAD and a chloroplast type [2Fe-2S] cluster and transfer electrons to an intermediary Rieske type ferredoxin rather than directly to the oxygenase component [18]. The tetralin dioxygenase system belongs to class III, composed of the dioxygenase complex itself, formed by ThnA1 and ThnA2, the α and β subunits of the initial tetralin dioxygenase, respectively, and two electron transfer components, the ferredoxin ThnA3 and the ferredoxin reductase ThnA4. Both $ThnA1^{TFA}$ and $ThnA1^{TFB}$ amino acid sequences contain two Cys and two His residues, ($CXHX_{15-17}CX_2H$), that co-ordinate the Rieske-type [2Fe-2S] cluster and the other catalytic residues that coordinate the mononuclear iron, identified in other crystallized dioxygenases, such as the one from *Pseudomonas* sp. NCIB 9816–4 [19]. ThnA3 is homologous to the Rieske-type ferredoxin components that are electron transfer intermediates between the NAD(P)H-dependent ferredoxin reductases and the dioxygenase components. $ThnA4^{TFA}$ TFA is homologous to the class III NAD(P)H-ferredoxin reductases of degradative pathways. The modular structure of these ferredoxin reductases is composed of a N-terminal domain that contains a conserved Cys-X_4-Cys-X_2-Cys-$X_{29/30}$-Cys motif that binds a plant-type [2Fe-2S] cluster, and a central and C-terminal domains with the conserved motifs for flavin and NAD(P)H binding, respectively [20]. However, $ThnA4^{TFB}$ is 81.5% identical to EtbAd from *Rhodococcus jostii* RHA1, a protein with Pyr_redox_2 domains which are present in pyridine nucleotide-disulphide oxdoreductases, a family that includes both class I and class II oxidoreductases.

Both $ThnA4^{TFA}$ and $ThnA3^{TFA}$ have been purified, their redox centres spectroscopically confirmed, the specific sequence in the electron transfer reaction from $ThnA4_{red}$ to $ThnA3_{ox}$ has been analysed using stopped-flow spectroscopy, and their midpoint reduction potentials have been determined. These results agree with the electron transfer from NADPH to FAD and then to [2Fe-2S]) in ThnA4, and subsequently to [2Fe-2S] cluster in ThnA3 [21]. Interestingly, this transfer of electrons is important not only for the initial oxidation of tetralin but also for the correct regulation of the *thn* genes expression (see Section 4), since a new electron transfer route from ThnA4 and ThnA3 to the regulatory protein ThnY may control the activity of the regulators of *thn* genes in response to particular substrates. The absence of ThnY in TFB and the different regulatory system for *thn* gene expression in this bacterium might explain the lack of ThnA4 conservation between TFA and TFB. Experimental evidence that $ThnA1A2A3A4^{TFA}$ are directly involved in the initial dioxygenation of tetralin is provided by accumulation of 2HT (2-hydroxy-5,6,7,8-tetrahydronaphthalene), which could be determined by HPLC, in the supernatant of a TFA mutant lacking all *thn* genes supplied with tetralin and overexpressing ThnA1A2A3 [17]. Although ThnA4 is not essential for 2HT production, maximal production of 2HT is achieved only in its presence. According with this, mutants in ThnA1, ThnA2, or ThnA3 are unable to grow on tetralin. ThnA4 is partially dispensable and its function is apparently substituted by other ferredoxin reductases present in TFA, as it has been described for other ring-hydroxylating dioxygenases [18].

The next reaction of the catabolic pathway is catalysed by the *thnB* gene product. ThnB is a *cis*-dihydrodiol dehydrogenase highly homologous to those involved in phenanthrene degradation. A ThnB^{TFA} mutant is unable to grow on tetralin. However, when this mutant is grown in minimal medium containing 8 mM β-hydroxybutyrate (βHB) and tetralin as carbon sources, a condition that allows growth of Thn-mutants while expressing the *thn* genes, a mixture of 1HT (1-hydroxy-5,6,7,8-217 tetrahydronaphthalene) and 2HT (2-hydroxy-5,6,7,8-tetrahydronaphthalene) were detected by high-performance liquid chromatography (HPLC) in the supernatant of the growth culture after acidification. Formation of 1HT and 2HT indicated that the metabolite accumulated in ThnB mutant corresponds to *cis*-dihydrodiol 1,2-dihydroxy-1,2,5,6,7,8-hexahydronaphthalene (1,2-DHT) since dihydrodiols are unstable under acidic conditions and this results in the formation of rearomatized monohydroxylated derivatives [17].

ThnC is the extradiol dioxygenase required for tetralin utilization. Extradiol dioxygenases are key enzymes in the aerobic degradation of many aromatics compounds. They function by cleaving the meta-position of the C–C bond of catechol derivatives, leading to the aromatic ring opening with the concomitant incorporation of two oxygen atoms [22]. ThnC^{TFA} has been purified, and biochemically characterized. The estimated molecular mass of ThnC indicated that it is a decamer [23], which is different from the octameric structures reported for other extradiol dioxygenases closely related to ThnC, such as BphC [24], NahC [25], and EtbC [26]. TFA *thnC* mutants are unable to grow on tetralin and accumulate 1,2-DHT in the supernatant when growing on 8 mM βHB and tetralin [23]. The product of the reaction catalysed by ThnC is 4-(2-oxocyclohexyl)-2-hydroxy-buta-2,4-dienoic acid (OCHBDA), a yellow ring product resulting from the cleavage proximal to the alicyclic ring of 1,2-DHT that exhibits the typical behaviour of a hydroxymuconic semialdehyde. In addition to 1,2-DHT, the range of substrates of ThnC is relatively wide. It showed activity towards 1,2-dihydroxynaphthalene (1,2-DHN), the intermediate in the naphthalene biodegradation pathway, and to a lesser extent, towards 2,3-dihydroxybiphenyl (2,3-DHB), the intermediate in the biphenyl biodegradation pathways, and the methylated catechols, 3-methylcatechol or 4-methylcatechol [23,27]. This broad substrate specificity of ThnC is also shown by other 1,2-dihydroxynaphthalene dioxygenases [25]. ThnC and these enzymes cluster together in group 2 of the I.3.E subfamily of two-domain extradiol dioxygenases, which preferentially cleave bicyclic substrates [23,28]. As shown by a metagenomic survey, dioxygenases of the I.3.E subfamily are the ones with the broadest substrate specificity [29]. Mutational analysis of ThnC^{TFA} showed that it is possible to obtain altered ThnC proteins with modified positions surrounding the substrate-binding site (Q198H, G206M, N213H, A282R and A282G) that improved activity and/or specificity for 1,2-DHT, thus indicating that ThnC activity is not maximized for this substrate.

Hydrolases of *meta*-cleavage pathways are a class of β-ketolases which catalyse the hydrolysis of linear C-C bonds of vinylogous 1,5-diketones formed by the dioxygenative *meta*-ring cleavage of arenes. Most of them belong to the *α/β*-hydrolase-fold superfamily of enzymes. These enzymes are structurally, functionally and mechanistically related and generally utilize a nucleophilic residue (a serine in most cases)-acidic residue-histidine catalytic triad to hydrolyse the C-C bond adjacent to a carbonyl [30]. The serine hydrolase ThnD from TFA has been purified and biochemically characterized [31]. ThnD^{TFA} is an octameric enzyme with an optimum reaction temperature at 65 °C. Moreover, it shows high stability at room temperature and heat treatment at 70 °C for 2 h resulted in 78% of the initial activity. ThnD^{TFB} has also been heterologously expressed in *Escherichia coli* and purified, and its optimal temperature is 70 °C [14]. This stability at high temperature is a biotechnological interesting property given the use of this class of enzymes in organic synthesis [32]. ThnD^{TFA} and ThnD^{TFB} catalyse the hydrolysis of

OCHBDA rendering the dienol 2-hydroxydec-2,4-diene-1,10-dioicacid as the reaction product. In fact, a TFA insertion mutant defective in ThnD accumulates OCHBDA [31]. It has been suggested that ThnD could be involved in biphenyl meta-cleavage pathways since it is highly similar to BphD enzymes and, in addition to OCHBDA, ThnD also efficiently hydrolyses 2-hydroxy-6-oxo-6-phenylhexa-2,4-dienoic acid (6-phenyl-HODA), the ring fission product of the biphenyl meta-cleavage pathway. In fact, the K_m for OCHBDA and for 6-phenyl-HODA are 26.3 mM and 34.8 mM, respectively [31]. ThnD^{TFB} hydrolase is the most similar to BphD hydrolases and even more active towards 6-phenyl-HODA than towards OCHBDA, which showed just 40% of the activity obtained with 6-phenyl-HODA [14]. ThnD^{TFA} does not hydrolyse 6-methyl-HODA, aldehydes HODA or 5-methyl-HODA, ring fission products of 3-methylcatechol, catechol and 4-methylcatechol, equivalent intermediates of *meta*-cleavage pathways of mono-aromatic compounds. These compounds have in common the small substituents in C-6. Interestingly, since the bond of the vinylogous β-diketone connecting C-5 and C-6 in OCHBDA is within the alicyclic ring of tetralin derivative, hydrolysis of this bond by ThnD renders a single linear dicarboxylate product instead of the two products usually generated by these hydrolases (Scheme 1). Concomitantly, this hydrolysis breaks open the recalcitrant alicyclic ring, thus rendering a more easily catabolised product. This confers unique properties to the tetralin degradation pathway, since ThnD makes unnecessary the recruitment of enzymatic activities involved in degradation of alicyclic rings for tetralin complete mineralization.

Both ThnE and ThnF showed similarities with the corresponding hydratases and aldolases involved in the catabolic degradation of homo-protocatechuate, and their substrates and products of the reactions they catalyse have been identified [33]. GC-MS analysis of the reaction products using OCHBDA as a substrate and crude extracts of an *E. coli* strain overproducing ThnD and ThnE, identified a C10 dicarboxylic compound with two hydroxyl groups, the enol tautomer 2,4-dihydroxydec-2-ene-1,10-dioic acid (DHDDA) as the major product resulting from the hydration reaction [33]. Although it has been suggested that both the dienol and the keto tautomers of the ThnD product could be substrates of ThnE, in TFA the enol form is the most probable since when purified hydrolase ThnD^{TFA} was used to transform OCHBDA, only the dienol tautomer reaction product was detected [31].

ThnF is a class II aldolase encoded by the *thnF* gene, whose start codon in TFA is just six nucleotides downstream of the stop codon of *thnE* gene [33]. A similar situation is found in TFB where *thnF* start codon is within the *thnE* coding sequence [14]. This indicates that translational coupling is highly probable for these genes. Aldolases are an evolutionarily diverse class of enzymes that are typically divided into two groups based on their catalytic mechanism. The Class I aldolases utilize a Schiff base mechanism to stabilize a carbanion intermediate, whereas the Class II enzymes utilize divalent metal ions to stabilize an enolate intermediate. Within each mechanistic class of aldolases, further diversity exists with respect to the sequence, structure, and substrate specificities. Pyruvate-specific enzymes are within the Class II containing an octahedral coordinated divalent metal cofactor, which is usually Mg^{2+}, Mn^{2+} or Co^{2+} [34]. Transformation of OCHBDA by *E. coli* extracts of enriched in ThnD, ThnE, and ThnF enzymes from TFA resulted in accumulation of two compounds as a result of the cleavage reaction, pimelic semialdehyde, identified by GC-MS, and pyruvate, which was spectrophotometrically estimated [33].

For complete oxidation, pimelic semialdehyde needs to be further metabolized into compounds that can enter the Krebs cycle. A first oxidation of pimelic semialdehyde leading to the synthesis of pimelic acid is enzymatically carried out by ThnG, which is similar to the non-acylating NAD-dependent aldehyde dehydrogenases involved in the biodegradation pathways of aromatic compounds. Aldehyde dehydrogenases (ALDHs) catalyse the oxidation of aldehydes to their corresponding carboxylic acids and all of them require either NAD or NADP as a cofactor [35]. The catalytic activity of ThnG has been assayed in crude extract of TFA and in a ThnG^{TFA} *E. coli* overproducing strain [10]. ThnG shows ALDH activity depending on pimelic semialdehyde and NAD as an oxidant but not with NADP, and its activity does not show coenzyme A or dithiothreitol dependence, supporting the non-acylating ALDH activity of ThnG. Indeed, the primary structure of ThnG^{TFA} contains the four invariant residues

(Gly228, Gly281, Glu381, and Phe383) found in this type of dehydrogenases, which are critical for binding of the nicotinamide ring of NAD and catalytic activity. These residues are also present in $ThnG^{TFB}$ in similar locations. The participation of ThnG in the catabolism of tetralin is clear since ThnG activity and *thnG* gene expression as part of the B operon is induced in TFA cells growing on tetralin. However, as it happens in *thnA4* mutants, TFA *thnG* mutants are able to grow on tetralin. Therefore, it seems that this role can be assumed by other aldehyde dehydrogenases present in TFA that could also use pimelic semialdehyde as a substrate [10]. Since ThnG is not an acylating enzyme, the formation of pimeloyl-CoA to enter a β-oxidation pathway needs an intermediate step consisting in the activation of the pimelic acid molecule.

Based on the homology of the *thnHIJKL* gene product with enzymes that participate in β-oxidation pathways, a complete β-oxidation pathway for pimelic acid has been described in TFA. The steps involved are: initially, ThnH, a protein similar to enzymes that belong to the "CaiB/BaiE" family of CoA transferases that catalyse the reversible transfer of CoA groups from CoA thioesters to free acids, would activate pimelic acid yielding pimeloyl-CoA. In TFB, the same step seems to be catalysed by ThnQ, a protein similar to pimeloyl-CoA synthases that catalyses the binding of a CoA-SH molecule to pimelic acid. In TFA, the resulting pimeloyl-CoA molecule would be the substrate of ThnJK, which encodes proteins similar to the large and small subunits of a pimeloyl-CoA dehydrogenase, respectively. This enzyme might oxidise pimeloyl-CoA to pimeloyl-2,3-dehydroacyl-CoA acid. In TFB, ThnV could be the acyl-CoA dehydrogenase responsible of catalysing this step. In TFA, a bifunctional enzyme (ThnL) carrying the enoyl-CoA hydratase and 3-hydroxyacyl-CoA dehydrogenase domains located at its N and C termini, respectively, could convert the pimeloyl-2,3-dehydroacyl-CoA acid into pimeloyl-3-ketoacyl-CoA in two steps. In TFB, ThnL would render pimeloyl-2,3-hydroxyacyl-CoA, which would be oxidised to the ketoacyl-CoA by ThnW. Finally, in both bacteria, ThnI, encoding an acetyl-CoA acetyl transferase, would cleave pimeloyl-3-ketoacyl-CoA into the central metabolites acetyl-CoA and glutaryl-CoA. In TFA, ThnN, encoding a putative glutaryl-CoA dehydrogenase, could catalyse the oxidative decarboxylation of glutaryl-CoA to crotonyl-CoA, which could be further converted into acetyl-CoA. Moreover, two extra proteins, ThnO and ThnP encode proteins similar respectively to the α- and β-subunits of flavoproteins, which serve as specific electron acceptors for primary dehydrogenases of the β-oxidation pathway and transfer electrons to the respiratory chain. Proteins similar to ThnN, ThnO and ThnP have not been found associated to *thn* genes in TFB. Mutations in *thnH, thnI, thnJ* or *thnK*, did not affect the growth of TFA on tetralin, or on sebacic or pimelic acids [10]. Thus, redundant β-oxidation pathways should be present in TFA cells. Actually, there are 39 proteins annotated in TFA genome within the GO terms GO:0006635 (fatty acid beta-oxidation) and GO:009062 (fatty acid catabolic process) [8].

Taking all the steps together, degradation of one molecule of tetralin by this pathway renders one molecule of pyruvate, one of acetyl-CoA and one of glutaryl-CoA (Scheme 1). These molecules can be used either as energy or carbon source. The use of the same degradation strategy by two non-phylogenetically related bacteria and different gene arrangements could mean that it is the most efficient pathway to obtain carbon and energy source from tetralin.

Interestingly, *thn* genes are also induced by naphthalene in both TFA and TFB [36,37]. However, and despite the ThnC activity towards 1,2 DHN, TFA is unable to grow on naphthalene while TFB can use it as the only carbon and energy source [37]. In fact, Thn proteins are the most prominent in extracts of naphthalene-grown TFB cells. Because of that, it has been proposed that ThnA1A2A3A4, ThnB and ThnC proteins also participate in the naphthalene degradation pathway in TFB and that the TFB naphthalene degradation ability actually results from the combination of different catabolic pathways [37].

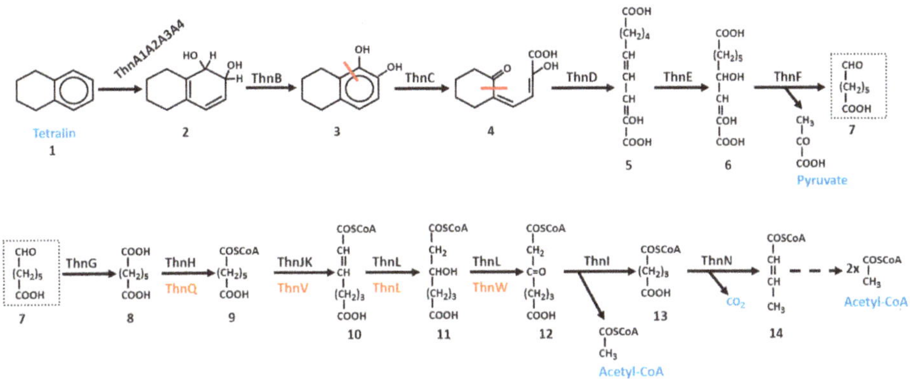

Scheme 1. Tetralin catabolic pathway. Compound designations: **1**, 1,2,3,4-tetrahydronaphthalene (Tetralin); **2**, 1,2-dihydroxy-1,2,5,6,7,8-hexa-hydronaphthalene; **3**, 1,2-dihydroxytetralin (DHT); **4**, 4-(2-oxocyclohexyl)-2-hydroxybuta-2,4-dienoic acid (OCHBDA); **5**, 2-hydroxydeca-2,4-dienedioic acid; **6**, 2,4-dihydroxydec-2-ene-1,10-dioic acid (DHDDA); **7**, 7-Oxoheptanoic acid (pimelic semialdehyde); **8**, heptanedioic acid (pimelic acid); **9**, pimeloyl-CoA; **10**, 2,3-dehydropimeloyl-CoA; **11**, 3-hydroxypimeloyl-CoA; **12**, 3-oxopimeloyl-CoA; **13**, glutaryl-CoA; **14**, crotonyl-CoA. Proteins in brown correspond to those identified in TFB. Red lines show the cleavage of the rings.

4. Regulation of the *thn* Gene Expression

In this section we review the regulatory elements and their function controlling the expression of the tetralin degradation genes in TFA and TFB. In both strains, tetralin degradation genes are subjected to a specific induction by tetralin but also to an over-imposed global control, which dictates the use of other preferential carbon sources when present. The regulatory mechanisms behind these responses are summarized.

4.1. Specific Regulation: Elements Involved in Tetralin Induction

4.1.1. Tetralin Induction in TFA

When TFA cells are growing on tetralin as the sole source of carbon and energy, the expression of the *thn* genes from P_C and P_B promoters is induced more than 100-fold compared to when growing on the preferential carbon source βHB [38]. In many degradative routes, induction of the pathway requires an intermediate in the catabolism of the substrate [39–41]. However, induction of *thn* genes is exerted directly in response to tetralin, the substrate of the catabolic pathway, since some mutants unable to transform it are still able to activate *thn* gene transcription [36]. Three regulatory proteins achieve this specific response to tetralin. ThnR, a LysR-type regulator involved in the recognition of potential effector molecules, ThnY, the ferredoxin like reductase that has evolved to function as a ThnR co-activator, both essential for *thn* gene expression, and ThnA3, the ferredoxin that delivers electrons to the tetralin dioxygenase and that prevents promiscuous induction by molecules that are not substrates of the catabolic pathway [38]. Details of each regulatory element are described below.

ThnR has exclusive properties. Firstly, unlike other LysR-type regulators, it does not repress its own synthesis, but activates its own expression in a positive circuit of regulation, by being part of the same transcriptional unit as the catabolic genes *thnCA3A4* located upstream (Figure 1) [38]. It has been proposed that these positive auto-regulation circuits may allow a faster switch on/off of the regulatory systems in response to the inducer molecules, for example, in environments where the amount of substrate is limited [42]. This might be the case for tetralin considering its tendency to accumulate into biological membranes [4]. Secondly, ThnR needs the regulatory protein ThnY for transcription activation [17]. In this way, ThnR is one of the few examples of LysR-type activator that requires an additional protein to activate transcription [43]. Additionally, ThnY allows the system to integrate other signals, such us the presence of potential inducers that, being similar to tetralin, are not substrates of the pathway and thus preventing a wasteful production of useless catabolic enzymes (see ThnA3 modulation below).

The molecular mechanism responsible for *thn* gene induction in the presence of tetralin has been characterized. According to it, basal levels of ThnR and ThnY are provided by the weak constitutive promoter P_R, which in turn are responsible for the activation of the four promoters, P_B, P_C, P_H and P_M, when tetralin is present [10]. Footprinting analyses have shown that ThnR binds to contiguous high affinity and low affinity sites in each promoter region [16]. A meticulous analysis of ThnR binding to the *thnB–thnC* intergenic region, (P_B–P_C promoters), led to a detailed molecular characterization of the activation mechanism mediated by ThnR and the importance that both primary and secondary binding sites have on transcription activation from each promoter. These results showed that mutations in the primary binding sites of each promoter can affect not only the transcriptional activation of its respective promoter but transcription from the other one, and vice versa. This transcriptional coordination is mediated by a complex structure of ThnR molecules that interact among them when bound to their respective sites in both promoters via a DNA loop, which is detected as a higher order structure in electrophoretic mobility shift assays (EMSA) [10]. For this DNA loop to be formed, primary binding sites of each promoter have to be aligned in the same face of the helix, since insertions of 4 and 6 bp between them prevented the formation of the complex structure and cooperative transcription activation. Furthermore, the DNA loop functions by compensating non-productive ThnR-DNA interactions at an activating site [16,44].

Similarly to ThnR, P_R also provides basal levels of ThnY, which are increased from the inducible P_C promoter in a ThnR-dependent manner when tetralin is present [45]. As mentioned above, ThnY is essential for expression of each of the four *thn* promoters controlled by ThnR. ThnY is homologous to the Class III of ferredoxin reductases of degradation of aromatic compounds [18], such as benzoate 1,2-dioxygenase reductase [46], naphthalene dioxygenase reductase [47] and carbazole dioxygenase reductase [48]. These ferredoxin reductases function by accepting electrons from NAD(P)H and transferring them through the FAD and [2Fe-2S] cluster, to an intermediary Rieske-type ferredoxin, which finally transfer the electrons to the dioxygenase. Purified ThnY contains a non-covalently attached FAD and one plant-type [2Fe-2S] cluster characteristic of this class of ferredoxin reductases [45]. Regarding these features, ThnY is similar to the ferredoxin reductase ThnA4 of the tetralin dioxygenase [21]. However, despite having these redox centers, ThnY does not behave like a NAD(P)H-dependent ferredoxin reductase, since neither can be reduced directly by NAD(P)H nor does it have the enzymatic activities of these enzymes, such as diaphorase and/or cytochrome *c* reductase activity. Actually, ThnY lacks a proper NAD(P)H binding in its C-terminal domain [45], such as the conserved glycine signature (GGXGXXP) proposed to be involved in the binding of the adenosine-5-phosphate groups of NADH or 2′-phospho-AMP of NADPH [20]. Knowing why ThnY is necessary for ThnR-mediated transcription activation has been one of the most interesting findings on *thn* genes regulation. Footprint and DNA binding assays revealed that when oxidised, ThnY does not alter the footprint pattern generated by ThnR but it does increase the affinity of ThnR for the P_C promoter region and stabilize the structure of the protein-DNA complex, as a new defined complex is detected in its presence in EMSA assays. It has been proposed that ThnY co-activates transcription by interacting with ThnR [45]. This regulatory function of ThnY is of remarkable interest since it is the first example of a ferredoxin like reductase of the dioxygenases systems that works as an additional sensor regulating gene expression [45,49].

The role of ThnY as a new redox sensor linking cellular metabolism to gene expression has also come from the experiments carried out with the ferredoxin ThnA3 [16,21,50]. As mentioned above in Section 3, ThnA3 is a Rieske-type ferredoxin that functions by transferring electrons to the tetralin dioxygenase, but in addition to this catalytic function, ThnA3 has a regulatory role as a discriminator. There are several lines of evidences that show that ThnA3 reports to ThnY whether a molecule is a good or a bad substrate of the tetralin dioxygenase through its redox state and that this communication is established by modifying the redox state of ThnY. Firstly, ThnA3 has a discriminatory role since in *thnA3* mutants, when only ThnR and ThnY are present, the *thn* promoters are activated promiscuously, producing high levels of wasteful expression with effector molecules that are not substrates of the first step of the tetralin catabolic pathway, such as *cis*-decalin, cyclohexane, *trans*-decalin, or benzene [16]. This phenomenon of gratuitous induction has also been described in other degradative pathways [51]. Secondly, reduced ThnA3 has a negative effect on *thn* expression since mutants strains lacking either the α (ThnA1) or β (ThnA2) subunits of the dioxygenase, a condition resulting in an accumulation of the reduced form of ThnA3, failed to induce properly the expression of *thn* genes even in the presence of the substrate tetralin [16]. Under this physiological condition, reduced ThnA3 is fully available to interact with ThnY. Thirdly, ThnA3 is the electron partner of ThnY. ThnY variants, ThnY-C40S and ThnY-N201G,S206P, affected in the iron sulfur cluster and in the degenerated NAD(P)H binding domain, respectively, render ThnA3 useless for signal transduction as the discrimination capacity is completely lost and the expression of *thn* genes occurs at high levels even in the presence of molecules other than tetralin. Thus, the discriminatory function of ThnA3 is exerted through the redox protein ThnY [50]. By using stopped-flow spectrophotometry and determination of midpoint reduction potentials of each redox cofactor within ThnA4, ThnA3 and ThnY, it has been possible to show that ThnY can actually be reduced by ThnA3, according to the following ordered steps during the electron transfer reactions [21]:

$$\text{NADH} \rightarrow \text{ThnA4FAD} \rightarrow \text{ThnA4[2Fe-2S]} \rightarrow \text{ThnA3[2Fe-2S]} \rightarrow \text{ThnYFAD} \rightarrow \text{ThnY[2Fe-2S]}.$$

Purified ThnA4 behaves as a typical ferredoxin reductase of dioxygenase systems, able to be reduced preferentially by NADH, able to reduce the ferredoxin ThnA3, and with midpoint potentials for $E_{ThnA4FADox/hq}$ and $E_{ThnA4SFeox/red}$ of −200 mV to −150 mV ranges respectively [21], similar to other reported NAD(P)H ferredoxin reductases of the dioxygenase systems [52]. Similarly, ThnA3 behaves like the one electron carrier Rieske-type [2Fe-2S] ferredoxin although with an $E_{ThnA3SFeox/red}$ (−112 ± 5 mV) [21] less negative than reported ferredoxin [53]. $E_{ThnYFADox/hq}$ and $E_{ThnYSFeox/redwith}$ with values of −131 ± 8 mV and −136 ± 8 mV, respectively, are less negative too than those of other ferredoxin reductases counterparts and slightly more negative than that of ThnA3. However, ThnY can only be reduced by ThnA3$_{red}$ but not by NAD(P)H. The redox potentials also highlight that in order to reduce ThnY, the reduced form of ThnA3 has to be predominant, thus allowing the equilibrium of the redox reaction to change toward ThnY [21]. Figure 3 shows the regulatory model for *thn* genes expression which implies two electron transport routes: (i) NAD(P)H-ThnA4–ThnA3–ThnA1/ThnA2 that occurs in the presence of tetralin, in which tetralin dioxygenase acts as an electron sink through the dioxygenation reaction. Reduction of ThnY by ThnA3 is minimal thus allowing ThnR and ThnY$_{ox}$ to activate the *thn* promoters and (ii) the regulatory NAD(P)H–ThnA4–ThnA3–ThnY that occurs in the presence of non-efficient substrates of the dioxygenase, thus leaving an abundant ThnA3$_{red}$ form fully available to interact with ThnY and to impair *thn* expression. This represents an unprecedented regulatory system in which components of an electron transfer system of a catabolic pathway prevent its gratuitous induction by molecules that cannot be effectively metabolized.

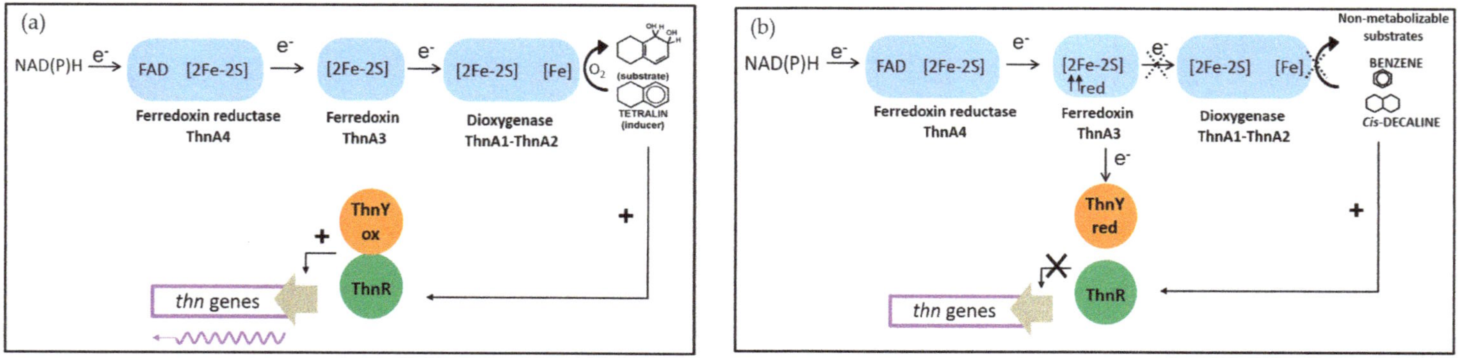

Figure 3. Model for specific regulation and modulation of *thn* genes transcription in TFA. (**a**) In the presence of the tetralin, electrons are preferentially delivered to the dioxygenase and $ThnY_{ox}$ and ThnR can activate transcription from the inducible *thn* promoters. (**b**) In the presence of compounds that are non-substrates of the dioxygenase, ThnA3 is accumulated in the reduced state and electrons are redirected towards ThnY instead of the dioxygenase impairing *thn* genes transcription.

4.1.2. Tetralin Induction in TFB

Semiquantitative RT-qPCR analysis of *thn* genes expression in TFB showed the induction of structural and regulatory genes in tetralin-grown cells in three operons (Figure 2). In glucose- or glucose plus tetralin-grown cells, regulatory genes were expressed at low level while transcription of structural genes was undetected in glucose-grown cells and partially repressed in glucose plus tetralin-grown cells [14]. Primer extension experiments allowed to establish the transcription start points for *thnS* and *thnA1* (which is the same as *thnA1′*) and the putative sequences of the divergent promoters [54]. An 18 bp sequence, similar to those present in biphenyl-inducible promoters, was found in *thnA1* and *thnA1′* promoters. Point mutations in this sequence negatively affected tetralin induction of a *thnA1::gfp* translational fusion indicating that those nucleotides could be recognised by a transcriptional activator. Moreover, introduction of extra copies of *thnST* into TFB resulted in higher expression of *thnA1::gfp* in all conditions. Taken together, those results suggested that the membrane sensor ThnS recognises a signal in the presence of tetralin, and that, in response, the regulator ThnT binds the promoters to induce *thn* gene expression in TFB [54].

4.2. Carbon Catabolite Repression

One of the major drawbacks of microbial degradation of pollutant compounds is the repression of the catabolic genes in the environment due to the presence of preferential carbon and energy sources. However, from an energetic point of view, this phenomenon can be regarded as a control system, which prevents expression of biodegradation capabilities when they are dispensable, thus improving the bacterial adaptation to its nutritional and energetic needs. This regulatory phenomenon is known as carbon catabolite repression (CCR) and can be exerted in different ways. Regulation of gene expression at both transcriptional and translational levels has been described in different biodegradation pathways [55]. CCR of *thn* genes has been demonstrated in TFA and TFB [14,38]. However, it is exerted by different carbon sources, which are metabolized via distinct central pathways, and through different mechanisms.

4.2.1. Carbon Catabolite Repression in TFA

In TFA, the presence of βHB or fatty acids, such as sebacic acid, in the medium containing tetralin prevents the induction of *thn* genes [38]. Regarding the regulatory mechanisms, despite all the efforts invested in its hunt, like transposon mutagenesis or affinity binding assays, no transcriptional regulators responsible of CCR have been found in TFA. However, a clear connection between CCR and the energetic status of the cell has been demonstrated in this bacterium (Figure 4). Transposon mutants affecting in *phaC* gene, coding for a poly(3-hydroxybutyrate) synthase, are unable to synthetize the carbon storage PHB (polyhydroxybutyrate) granule and show a partial release of *thn* genes CCR [56]. Conversely to what happens in the wild-type, *phaC* mutants prematurely induce *thn* genes in the presence of βHB, although maximal induction is never reached. Homo- and heterologous complementation restores the wild type phenotype indicating that the amount of PHB granule is an internal signal of the energetic status of the cell that prevents *thn* genes expression in CCR conditions. However, CCR in TFA does not only rely on PHB synthesis.

Recently, a new element involved in CCR of TFA *thn* genes has been identified. A trans-encoded small non-coding RNA (sRNA), 61-nt long, named SuhB, and highly conserved in *Sphingopyxis* genus, was annotated in TFA genome and experimentally identified by dRNA-seq and Northern blot experiments [8,57]. SuhB deletion mutants exhibit *thn* gene de-repression in CCR conditions [57]. sRNAs are 50–250 bp long transcripts acting as bacterial riboregulators of mRNAs encoded elsewhere in the genome. The presence of hundreds of these molecules has been demonstrated in a range of bacteria post-transcriptionally controlling the expression of genes involved in a wide range of important processes, including cell division, responses to abiotic stimuli, quorum sensing and virulence [58,59]. sRNAs affect translation and/or stability of their mRNA targets and most of them are Hfq-dependent.

Hfq is a RNA-binding protein that facilitates the binding of the sRNA to its target and/or protects it from degradation [60,61]. In TFA growing under high growth rate conditions, SuhB binds the 5′ UTR of *thnR* mRNA, thus negatively affecting its translation in an Hfq-dependent manner. As result, *thn* genes are silenced even if tetralin is present in the medium. The lack of SuhB or Hfq provokes a premature induction of *thn* genes in CCR conditions. However, only partial induction, even lower than that reached in *phaC* mutants, is achieved [56,57]. It has been demonstrated that *suhB* expression is regulated by the growth rate, being elevated at high growth rates, which would imply a high-energy status of the cells. In a *phaC* mutant, the absence of PHB granule could be detected as a low-energy status, *suhB* will not be transcribed and *thn* genes would be induced in the presence of tetralin.

The high conservation of SuhB in other *Sphingopyxis* strains that do not bear *thn* genes suggests that this sRNA must be controlling other targets in these bacteria and that its involvement in the regulation of the tetralin degradation genes should have appeared more recently. This is another interesting example of how existing bacterial posttranscriptional regulatory mechanisms can be incorporated into the regulatory system of probably horizontally acquired metabolic pathways.

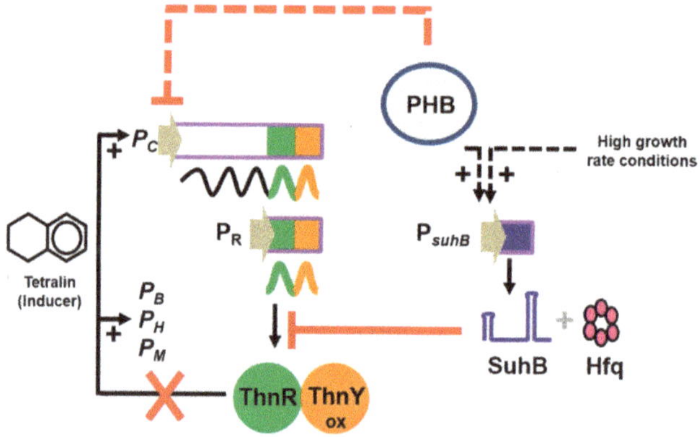

Figure 4. Elements involved in carbon catabolite repression in TFA. High growth rate conditions like growth on a preferential carbon source like βHB and the presence of PHB granules have a positive effect on *suhB* transcription. This sRNA, together with Hfq, affects negatively ThnR translation in those conditions impeding induction of *thn* promoters (P_C, P_B, P_H and P_M) in the presence of tetralin. The presence of PHB granules themselves have a negative effect on *thn* induction. Doted and solid lines represent unknown and known mechanism, respectively.

4.2.2. Carbon Catabolite Repression in TFB

CCR in TFB seems to be exerted in a more canonical way (Figure 5). In the presence of glucose, a FNR/CRP-like protein binds to the *thnS-thnA1* intergenic region [54]. A palindromic nucleotide sequence (CTGTGT-N6-TCACAG) similar to the *E. coli* CRP binding site was found overlapping the -10 and +1 sequences of *thnA1*. Thus, the binding of the FNR/CRP-like protein would prevent effective transcription of *thn* genes. In fact, point mutants in this sequence prevents binding of the FNR/CRP-like protein and provokes a partial relief of CCR of a *thnA1::gfp* translational fusion in TFB [54]. Partial de-repression could be due to the presence of a similar sequence at the *thnS* transcription start point to which the FNR/CRP could also bind, thus preventing *thnST* transcription. In fact, *thnS* expression, measured by RT-qPCR, is 5-fold higher in tetralin-grown cells than in glucose or glucose plus tetralin grown cells. The CCR regulatory protein of strain TFB would then act as a transcriptional repressor of *thn* genes in CCR conditions.

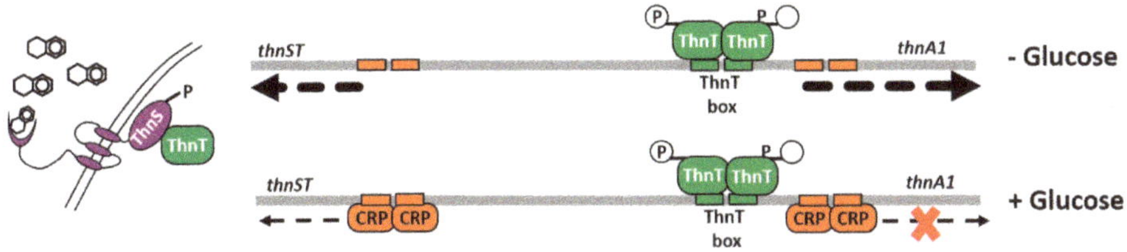

Figure 5. Elements involved in carbon catabolite repression in TFB. In the absence of glucose and presence of tetralin, phosphorylated ThnS transfer the P group to ThnT which binds the ThnT box and induces *thnA1* transcription. In the presence of glucose, a CRP-like protein binds CRP-like binding sites present at *thnS* and *thnA1* promoters provoking a reduction in *thnS* transcription and impeding *thnA1* transcriptional activation by phosphorylated ThnT.

5. Concluding Remarks

The organic solvent tetralin is composed of an aromatic and an alicyclic ring, none of which is easily biodegraded in nature. The biodegradation pathways described to degrade alicyclic and aromatic rings are completely different and use different enzymatic activities. However, a similar strategy for tetralin biodegradation has evolved in unrelated bacteria by which a single enzymatic team, typically involved in biodegradation of one aromatic ring, is able to break open both rings of the tetralin in two subsequent steps, thus rendering products easily catabolised by many bacteria. Apparently, the *Rhodococcus* strain also uses part of this enzymatic team to degrade naphthalene, thus representing an interesting way of reutilising bacterial enzymatic activities for different purposes. Although the biodegradation strategy and the enzymatic activities are the same, elements regulating the expression of the *thn* genes are completely different in both strains, thus supporting the view that catabolic pathways and their regulatory systems evolve independently. Particularly worth highlighting is the exquisite and sophisticated *thn* gene regulation observed in strain TFA, in which three different regulatory circuits act together at transcriptional and posttranscriptional levels to optimise the gene expression levels to the environmental conditions. One of the circuits involves transcription activation of different promoters in a non-canonical way for a LysR-type regulator, which implies cooperative transcription from divergent promoters. The other circuit affecting transcription constitutes an unprecedented regulatory system aimed to prevent gratuitous induction by related molecules that are not substrate of the catabolic pathway. Finally, the third system uses a less abundant type of regulators, an sRNA, to control *thn* genes' expression. Sophistication of these regulatory systems suggests that this biodegradation pathway has not been assembled recently as a response to human activity but bacteria might have been exposed to tetralin a long time ago, possibly as a component in the oil reservoirs.

The work presented here shows the usefulness of using a combination of different methodological approaches such as mutagenesis, mutant isolation, gene cloning, functional complementation, biochemical characterization and purification of enzymes, high throughput proteome analysis and chemical identification of metabolites to discover and characterize new mechanisms of microbial degradation. The knowledge gained substantially contributes to improving our understanding of important processes such as new metabolic capabilities and their regulatory mechanisms that control integrated physiological responses in bacteria other than the main models.

Author Contributions: Conceptualization, F.R.-R.; Original draft preparation, B.F. and F.R.-R.; Writing—review B.F., E.S. and F.R.-R.

Funding: This research was funded by Ministerio de Ciencia, Innovación y Universidades, grant numbers BIO2011-24003 and BIO2014-57545-R, and The APC (Article Processing Charge) was waived by MDPI.

Acknowledgments: We wish to thank Guadalupe Martín and Nuria Pérez for their technical support and all members of the laboratory for their insights and suggestions.

Conflicts of Interest: The authors declare no conflict of interest.

References

1. Rojo, F. Degradation of alkanes by bacteria. *Environ. Microbiol.* **2009**, *11*, 2477–2490. [CrossRef]
2. Fuchs, G.; Boll, M.; Heider, J. Microbial degradation of aromatic compounds—From one strategy to four. *Nat. Rev. Microbiol.* **2011**, *9*, 803–816. [CrossRef]
3. Sikkema, J.; de Bont, J.A.; Poolman, B. Mechanisms of membrane toxicity of hydrocarbons. *Microbiol. Rev.* **1995**, *59*, 201–222.
4. Sikkema, J.; Poolman, B.; Konings, W.N.; de Bont, J.A. Effects of the membrane action of tetralin on the functional and structural properties of artificial and bacterial membranes. *J. Bacteriol.* **1992**, *174*, 2986–2992. [CrossRef]
5. Sikkema, J.; de Bont, J.A. Isolation and initial characterization of bacteria growing on tetralin. *Biodegradation* **1991**, *2*, 15–23. [CrossRef]
6. Schreiber, A.F.; Winkler, U.K. Transformation of tetralin by whole cells of Pseudomonas stutzeri AS 39. *Eur. J. Appl. Microbiol. Biotechnol.* **1983**, *18*, 6–10. [CrossRef]
7. Sikkema, J.; de Bont, J.A. Metabolism of tetralin (1,2,3,4-tetrahydronaphthalene) in Corynebacterium sp. strain C125. *Appl. Environ. Microbiol.* **1993**, *59*, 567–572. [PubMed]
8. Garcia-Romero, I.; Perez-Pulido, A.J.; Gonzalez-Flores, Y.E.; Reyes-Ramirez, F.; Santero, E.; Floriano, B. Genomic analysis of the nitrate-respiring Sphingopyxis granuli (formerly Sphingomonas macrogoltabida) strain TFA. *BMC Genomics* **2016**, *17*, 93. [CrossRef]
9. Hernaez, M.J.; Reineke, W.; Santero, E. Genetic analysis of biodegradation of tetralin by a Sphingomonas strain. *Appl. Environ. Microbiol.* **1999**, *65*, 1806–1810.
10. Lopez-Sanchez, A.; Floriano, B.; Andujar, E.; Hernaez, M.J.; Santero, E. Tetralin-induced and ThnR-regulated aldehyde dehydrogenase and beta-oxidation genes in Sphingomonas macrogolitabida strain TFA. *Appl. Environ. Microbiol.* **2010**, *76*, 110–118. [CrossRef] [PubMed]
11. Schell, M.A. Molecular biology of the LysR family of transcriptional regulators. *Annu. Rev. Microbiol.* **1993**, *47*, 597–626. [CrossRef]
12. Stolz, A. Degradative plasmids from sphingomonads. *FEMS Microbiol. Lett.* **2014**, *350*, 9–19. [CrossRef]
13. Johnson, C.M.; Grossman, A.D. Integrative and conjugative elements (ICEs): What they do and how they work. *Annu. Rev. Genet.* **2015**, *49*, 577–601. [CrossRef] [PubMed]
14. Tomas-Gallardo, L.; Santero, E.; Camafeita, E.; Calvo, E.; Schlomann, M.; Floriano, B. Molecular and biochemical characterization of the tetralin degradation pathway in Rhodococcus sp. strain TFB. *Microb. Biotechnol.* **2009**, *2*, 262–273. [CrossRef] [PubMed]
15. Ghosal, D.; Ghosh, S.; Dutta, T.K.; Ahn, Y. Current state of knowledge in microbial degradation of polycyclic aromatic hydrocarbons (PAHs): A review. *Front. Microbiol.* **2016**, *7*, 1369. [CrossRef]
16. Lopez-Sanchez, A.; Rivas-Marin, E.; Martinez-Perez, O.; Floriano, B.; Santero, E. Co-ordinated regulation of two divergent promoters through higher-order complex formation by the LysR-type regulator ThnR. *Mol. Microbiol.* **2009**, *73*, 1086–1100. [CrossRef] [PubMed]
17. Moreno-Ruiz, E.; Hernaez, M.J.; Martinez-Perez, O.; Santero, E. Identification and functional characterization of Sphingomonas macrogolitabida strain TFA genes involved in the first two steps of the tetralin catabolic pathway. *J. Bacteriol.* **2003**, *185*, 2026–2030. [CrossRef]
18. Mason, J.R.; Cammack, R. The electron-transport proteins of hydroxylating bacterial dioxygenases. *Annu. Rev. Microbiol.* **1992**, *46*, 277–305. [CrossRef] [PubMed]
19. Kauppi, B.; Lee, K.; Carredano, E.; Parales, R.E.; Gibson, D.T.; Eklund, H.; Ramaswamy, S. Structure of an aromatic-ring-hydroxylating dioxygenase-naphthalene 1,2-dioxygenase. *Structure* **1998**, *6*, 571–586. [CrossRef]
20. Karplus, P.A.; Daniels, M.J.; Herriott, J.R. Atomic structure of ferredoxin-NADP+ reductase: Prototype for a structurally novel flavoenzyme family. *Science* **1991**, *251*, 60–66. [CrossRef]
21. Ledesma-Garcia, L.; Sanchez-Azqueta, A.; Medina, M.; Reyes-Ramirez, F.; Santero, E. Redox proteins of hydroxylating bacterial dioxygenases establish a regulatory cascade that prevents gratuitous induction of tetralin biodegradation genes. *Sci. Rep.* **2016**, *6*, 23848. [CrossRef]
22. Lipscomb, J.D. Mechanism of extradiol aromatic ring-cleaving dioxygenases. *Curr. Opin. Struct. Biol.* **2008**, *18*, 644–649. [CrossRef] [PubMed]

23. Andujar, E.; Hernaez, M.J.; Kaschabek, S.R.; Reineke, W.; Santero, E. Identification of an extradiol dioxygenase involved in tetralin biodegradation: gene sequence analysis and purification and characterization of the gene product. *J. Bacteriol.* **2000**, *182*, 789–795. [CrossRef]
24. Taira, K.; Hayase, N.; Arimura, N.; Yamashita, S.; Miyazaki, T.; Furukawa, K. Cloning and nucleotide sequence of the 2,3-dihydroxybiphenyl dioxygenase gene from the PCB-degrading strain of Pseudomonas paucimobilis Q1. *Biochemistry* **1988**, *27*, 3990–3996. [CrossRef]
25. Kuhm, A.E.; Stolz, A.; Ngai, K.L.; Knackmuss, H.J. Purification and characterization of a 1,2-dihydroxynaphthalene dioxygenase from a bacterium that degrades naphthalenesulfonic acids. *J. Bacteriol.* **1991**, *173*, 3795–3802. [CrossRef] [PubMed]
26. Hauschild, J.E.; Masai, E.; Sugiyama, K.; Hatta, T.; Kimbara, K.; Fukuda, M.; Yano, K. Identification of an alternative 2,3-dihydroxybiphenyl 1,2-dioxygenase in Rhodococcus sp. strain RHA1 and cloning of the gene. *Appl. Environ. Microbiol.* **1996**, *62*, 2940–2946.
27. Andujar, E.; Santero, E. Site-directed mutagenesis of an extradiol dioxygenase involved in tetralin biodegradation identifies residues important for activity or substrate specificity. *Microbiology* **2003**, *149*, 1559–1567. [CrossRef] [PubMed]
28. Eltis, L.D.; Bolin, J.T. Evolutionary relationships among extradiol dioxygenases. *J. Bacteriol.* **1996**, *178*, 5930–5937. [CrossRef]
29. Terron-Gonzalez, L.; Martin-Cabello, G.; Ferrer, M.; Santero, E. Functional metagenomics of a biostimulated petroleum-contaminated soil reveals an extraordinary diversity of extradiol dioxygenases. *Appl. Environ. Microbiol.* **2016**, *82*, 2467–2478. [CrossRef]
30. Fischer, F.; Kunne, S.; Fetzner, S. Bacterial 2,4-dioxygenases: New members of the alpha/beta hydrolase-fold superfamily of enzymes functionally related to serine hydrolases. *J. Bacteriol.* **1999**, *181*, 5725–5733.
31. Hernaez, M.J.; Andujar, E.; Rios, J.L.; Kaschabek, S.R.; Reineke, W.; Santero, E. Identification of a serine hydrolase which cleaves the alicyclic ring of tetralin. *J. Bacteriol.* **2000**, *182*, 5448–5453. [CrossRef] [PubMed]
32. Pokorny, D.; Steiner, W.; Ribbons, D.W. β-Ketolases-forgotten hydrolytic enzymes? *Trends Biotechnol.* **1997**, *15*, 291–296. [CrossRef]
33. Hernaez, M.J.; Floriano, B.; Rios, J.J.; Santero, E. Identification of a hydratase and a class II aldolase involved in biodegradation of the organic solvent tetralin. *Appl. Environ. Microbiol.* **2002**, *68*, 4841–4846. [CrossRef] [PubMed]
34. Samland, A.K.; Sprenger, G.A. Microbial aldolases as C-C bonding enzymes—Unknown treasures and new developments. *Appl. Microbiol. Biotechnol.* **2006**, *71*, 253–264. [CrossRef]
35. Lindahl, R. Aldehyde dehydrogenases and their role in carcinogenesis. *Crit. Rev. Biochem. Mol. Biol.* **1992**, *27*, 283–335. [CrossRef]
36. Martinez-Perez, O.; Lopez-Sanchez, A.; Reyes-Ramirez, F.; Floriano, B.; Santero, E. Integrated response to inducers by communication between a catabolic pathway and its regulatory system. *J. Bacteriol.* **2007**, *189*, 3768–3775. [CrossRef] [PubMed]
37. Tomas-Gallardo, L.; Gomez-Alvarez, H.; Santero, E.; Floriano, B. Combination of degradation pathways for naphthalene utilization in Rhodococcus sp. strain TFB. *Microb. Biotechnol.* **2014**, *7*, 100–113. [CrossRef]
38. Martinez-Perez, O.; Moreno-Ruiz, E.; Floriano, B.; Santero, E. Regulation of tetralin biodegradation and identification of genes essential for expression of thn operons. *J. Bacteriol.* **2004**, *186*, 6101–6109. [CrossRef]
39. Barnsley, E.A. The induction of the enzymes of naphthalene metabolism in pseudomonads by salicylate and 2-aminobenzoate. *J. Gen. Microbiol.* **1975**, *88*, 193–196. [CrossRef] [PubMed]
40. Parsek, M.R.; Shinabarger, D.L.; Rothmel, R.K.; Chakrabarty, A.M. Roles of CatR and cis,cis-muconate in activation of the catBC operon, which is involved in benzoate degradation in Pseudomonas putida. *J. Bacteriol.* **1992**, *174*, 7798–7806. [CrossRef]
41. McFall, S.M.; Chugani, S.A.; Chakrabarty, A.M. Transcriptional activation of the catechol and chlorocatechol operons: variations on a theme. *Gene* **1998**, *223*, 257–267. [CrossRef]
42. Wall, M.E.; Hlavacek, W.S.; Savageau, M.A. Design of gene circuits: lessons from bacteria. *Nat. Rev. Genet.* **2004**, *5*, 34–42. [CrossRef]
43. Tropel, D.; van der Meer, J.R. Bacterial transcriptional regulators for degradation pathways of aromatic compounds. *Microbiol. Mol. Biol. Rev.* **2004**, *68*, 474–500. [CrossRef]
44. Rivas-Marin, E.; Floriano, B.; Santero, E. Genetic dissection of independent and cooperative transcriptional activation by the LysR-type activator ThnR at close divergent promoters. *Sci. Rep.* **2016**, *6*, 24538. [CrossRef]

45. Garcia, L.L.; Rivas-Marin, E.; Floriano, B.; Bernhardt, R.; Ewen, K.M.; Reyes-Ramirez, F.; Santero, E. ThnY is a ferredoxin reductase-like iron-sulfur flavoprotein that has evolved to function as a regulator of tetralin biodegradation gene expression. *J. Biol. Chem.* **2011**, *286*, 1709–1718. [CrossRef]
46. Yamaguchi, M.; Fujisawa, H. Purification and characterization of an oxygenase component in benzoate 1,2-dioxygenase system from Pseudomonas arvilla C-1. *J. Biol. Chem.* **1980**, *255*, 5058–5063. [PubMed]
47. Haigler, B.E.; Gibson, D.T. Purification and properties of ferredoxinNAP, a component of naphthalene dioxygenase from Pseudomonas sp. strain NCIB 9816. *J. Bacteriol.* **1990**, *172*, 465–468. [CrossRef]
48. Nam, J.W.; Nojiri, H.; Noguchi, H.; Uchimura, H.; Yoshida, T.; Habe, H.; Yamane, H.; Omori, T. Purification and characterization of carbazole 1,9a-dioxygenase, a three-component dioxygenase system of Pseudomonas resinovorans strain CA10. *Appl. Environ. Microbiol.* **2002**, *68*, 5882–5890. [CrossRef] [PubMed]
49. Mettert, E.L.; Kiley, P.J. Fe-S proteins that regulate gene expression. *Biochim. Biophys. Acta* **2015**, *1853*, 1284–1293. [CrossRef]
50. Ledesma-Garcia, L.; Reyes-Ramirez, F.; Santero, E. The ferredoxin ThnA3 negatively regulates tetralin biodegradation gene expression via ThnY, a ferredoxin reductase that functions as a regulator of the catabolic pathway. *PloS ONE* **2013**, *8*, e73910. [CrossRef]
51. Shingler, V. Integrated regulation in response to aromatic compounds: From signal sensing to attractive behaviour. *Environ. Microbiol.* **2003**, *5*, 1226–1241. [CrossRef] [PubMed]
52. Gassner, G.T.; Ludwig, M.L.; Gatti, D.L.; Correll, C.C.; Ballou, D.P. Structure and mechanism of the iron-sulfur flavoprotein phthalate dioxygenase reductase. *FASEB J.* **1995**, *9*, 1411–1418. [CrossRef]
53. Colbert, C.L.; Couture, M.M.; Eltis, L.D.; Bolin, J.T. A cluster exposed: structure of the Rieske ferredoxin from biphenyl dioxygenase and the redox properties of Rieske Fe-S proteins. *Structure* **2000**, *8*, 1267–1278. [CrossRef]
54. Tomas-Gallardo, L.; Santero, E.; Floriano, B. Involvement of a putative cyclic amp receptor protein (CRP)-like binding sequence and a CRP-like protein in glucose-mediated catabolite repression of thn genes in Rhodococcus sp. strain TFB. *Appl. Environ. Microbiol.* **2012**, *78*, 5460–5462. [CrossRef]
55. Rojo, F. Carbon catabolite repression in Pseudomonas: Optimizing metabolic versatility and interactions with the environment. *FEMS Microbiol. Rev.* **2010**, *34*, 658–684. [CrossRef]
56. Martin-Cabello, G.; Moreno-Ruiz, E.; Morales, V.; Floriano, B.; Santero, E. Involvement of poly(3-hydroxybutyrate) synthesis in catabolite repression of tetralin biodegradation genes in Sphingomonas macrogolitabida strain TFA. *Environ. Microbiol. Rep.* **2011**, *3*, 627–631. [CrossRef]
57. Garcia-Romero, I.; Forstner, K.U.; Santero, E.; Floriano, B. SuhB, a small non-coding RNA involved in catabolite repression of tetralin degradation genes in Sphingopyxis granuli strain TFA. *Environ. Microbiol.* **2018**, *20*, 3671–3683. [CrossRef] [PubMed]
58. Gottesman, S.; Storz, G. Bacterial small RNA regulators: versatile roles and rapidly evolving variations. *Cold Spring Harb. Perspect. Biol.* **2011**, *3*. [CrossRef] [PubMed]
59. Wagner, E.G.H.; Romby, P. Small RNAs in bacteria and archaea: Who they are, what they do, and how they do it. *Adv. Genet.* **2015**, *90*, 133–208. [CrossRef] [PubMed]
60. Saramago, M.; Barria, C.; Dos Santos, R.F.; Silva, I.J.; Pobre, V.; Domingues, S.; Andrade, J.M.; Viegas, S.C.; Arraiano, C.M. The role of RNases in the regulation of small RNAs. *Curr. Opin. Microbiol.* **2014**, *18*, 105–115. [CrossRef]
61. Updegrove, T.B.; Zhang, A.; Storz, G. Hfq: The flexible RNA matchmaker. *Curr. Opin. Microbiol.* **2016**, *30*, 133–138. [CrossRef] [PubMed]

© 2019 by the authors. Licensee MDPI, Basel, Switzerland. This article is an open access article distributed under the terms and conditions of the Creative Commons Attribution (CC BY) license (http://creativecommons.org/licenses/by/4.0/).

genes

MDPI

Review

Steroids as Environmental Compounds Recalcitrant to Degradation: Genetic Mechanisms of Bacterial Biodegradation Pathways

Elías R. Olivera * and José M. Luengo

Departamento Biología Molecular (Área Bioquímica y Biología Molecular), Universidad de León, 24007 León, Spain

* Correspondence: erodo@unileon.es; Tel.: +34-987-29-1229

Received: 2 June 2019; Accepted: 3 July 2019; Published: 6 July 2019

Abstract: Steroids are perhydro-1,2-cyclopentanophenanthrene derivatives that are almost exclusively synthesised by eukaryotic organisms. Since the start of the Anthropocene, the presence of these molecules, as well as related synthetic compounds (ethinylestradiol, dexamethasone, and others), has increased in different habitats due to farm and municipal effluents and discharge from the pharmaceutical industry. In addition, the highly hydrophobic nature of these molecules, as well as the absence of functional groups, makes them highly resistant to biodegradation. However, some environmental bacteria are able to modify or mineralise these compounds. Although steroid-metabolising bacteria have been isolated since the beginning of the 20th century, the genetics and catabolic pathways used have only been characterised in model organisms in the last few decades. Here, the metabolic alternatives used by different bacteria to metabolise steroids (e.g., cholesterol, bile acids, testosterone, and other steroid hormones), as well as the organisation and conservation of the genes involved, are reviewed.

Keywords: sterols; bile acids; steroid hormones; biodegradation; 9,10-*seco* pathway; 4,5-*seco* pathway; 2,3-*seco* pathway

---

1. Introduction

Steroids are tetracyclic triterpenoid lipids containing a perhydro-1,2-cyclopentanophenanthrene structure, and include sterols, bile acids, steroid hormones, cardenolides, sapogenins, saponins, and vitamin D derivatives. Classically, they have been studied based on their physiological role. In mammals, the steroid hormones, bile acids, and other essential steroids are produced from cholesterol (Figure 1), a molecule that is also an important component of membranes, maintaining their fluidity, and participating in cell differentiation and proliferation. The major constituents of plant sterols are sitosterol, stigmasterol, campesterol, and brassicasterol, while in yeast and filamentous fungi, ergosterol is an important component of cell walls [1–3].

It is generally accepted that the biosynthesis of steroids is one of the hallmarks of the evolutionary progression of eukaryotes [4]. However, reports on the biosynthesis of sterols and modified sterols in methanotrophic bacteria (*Methylococcus capsulatus, Methylosphaera hansonii*, and *Methylobacterium organophilum*), *Gemmata obscuriglobus, Eudoraea adriatica*, and in a variety of myxobacteria have been published, though there is limited information about the metabolic machinery involved [5–15].

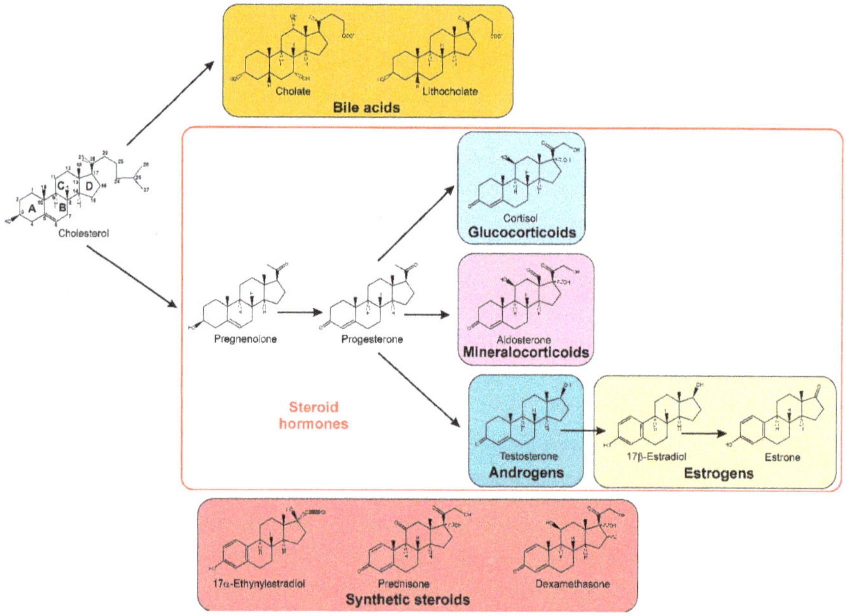

Figure 1. Chemical structure of cholesterol and some of its mammalian derivatives, and selected synthetic steroids.

From an ecological point of view, the release into the environment of vertebrate steroids, such as androgens, estrogens, progestogens, cholesterol, and bile acids, during excreta or decomposition of biomass is constantly occurring [16–22]. In addition, not only natural steroids are released. Along with antibiotics, synthetic steroids represent a significant sector of the global pharmaceutical market, and they have found their way into environmental niches, as in the case of synthetic hormones (ethynylestradiol and anti-inflammatory drugs like dexamethasone) (Figure 1) [19,20,23–27]. With their widespread presence, they can affect endocrine activity, even at low concentrations, with potential adverse effects for both the environment and human health. As such, there is an increasing need for new approaches in the bioremediation of steroids from the environment.

Although steroid synthesis is mostly restricted to eukaryotic organisms [4], the biodegradation of these compounds seems to be carried out exclusively by bacteria. Even so, steroids are highly recalcitrant to microbial degradation because of the low number of functional groups present in their structure and their extremely low solubility in water. Nevertheless, the persistence of these molecules in the environment has triggered a response by some bacteria, either efficiently detoxifying these hydrophobic molecules or metabolising their carbon and energy-rich scaffold. Thus, the isolation of microorganisms able to degrade or modify steroids has been reported from soil [28–30], and from freshwater [31] and marine [32–34] environments. This indicates the importance of bacteria in reducing the adverse impact of environmentally released steroids, as well as emphasising their importance in introducing these molecules into the carbon cycle. Moreover, for some pathogenic bacteria, such as *Mycobacterium tuberculosis* and *Rhodococcus equii*, catabolism of cholesterol has been identified as a trait involved in their pathogenicity and persistence in the host [35,36].

In 1913, Söhngen reported that a *Mycobacterium* strain could use phytosterols as a sole carbon and energy source [37]. Thereafter, many bacterial strains have been isolated that are able to degrade steroids, including species initially classified as belonging to genera *Mycobacterium, Gordonia, Tsukamurella, Rhodococcus, Azotobacter, Nocardia, Flavobacterium, Arthrobacter, Bacillus, Brevibacterium, Corynebacterium, Streptomyces, Microbacterium, Serratia, Achromobacter, Protaminobacter*, and *Pseudomonas* [29,38–54].

However, it has not been until the last two decades of the 20th century that studies have been initiated to elucidate the physiological and metabolical determinants, and the mechanisms involved in the aerobic catabolism of different steroid molecules. These studies have included sterols, chiefly cholesterol, testosterone and 17β-estradiol, and bile acids, mainly cholic acid, using selected model organisms from Actinobacteria, comprising different strains of *Mycobacterium, Rhodococcus*, and *Gordonia* [36,55–59] (Table 1), and some selected Gram negative strains, the α-proteobacteria *Comamonas testosteroni* and a few *Pseudomonas* strains [28,29,60–62]. In addition, the catabolism of steroids under anoxic conditions has been studied in *Sterolibacterium* (β-proteobacteria) and in *Steroidobacter* (Y-proteobacteria) [63–65] (Table 1).

Table 1. Model strains widely used in the studies that have allowed the characterization of the metabolic mechanisms used for steroid degradation. Steroids used in these studies are also indicated.

		Strain	Steroids Used	Reference Genome
9,10-*seco* pathway	*Mycobacterium*	*tuberculosis* H37Rv *smegmatis* mc^2 155	Cholesterol Cholesterol	NC_000962 NC_008596
	Rhodococcus	*neoaurum* ATCC 25795	Cholesterol β-sitosterol Stigmasterol Campesterol	NZ_JMDW00000000
		equi 103S	Cholesterol	NC_014659
		rhodochrous DSM43269	Cholesterol β-sitosterol Campesterol	unpublished
		jostii RHA1	Cholesterol Cholic acid	NC_008268
		erythropolis SQ1	Cholesterol	unpublished
	Gordonia	*neofelifaecis* NRRL B-59395	Cholesterol	NZ_AEUD00000000
		cholesterolivorans Chol-3	Cholesterol Ergosterol Stigmasterol	unpublished
	Comamonas	*testosteroni* TA441	Testosterone Cholic acid	NZ_CP006704
		thiooxidans CNB-1	Testosterone	NC_013446
	Pseudomonas	*stutzeri* Chol-1	Cholic acid	NZ_AMSL00000000
		putida DOC21	Bile acids Testosterone	unpublished
4,5-*seco* pathway	*Sphingomonas*	sp. KC8	17β-estradiol	NZ_AFMP01000000
2,3-*seco* pathway	*Sterolibacterium* *Steroidobacter*	*denitrificans* Chol1S *denitrificans* DSMZ18526	Cholesterol Testosterone	LT837803 NZ_CP011971

2. Uptake of Steroids

The catabolism of steroids implies their selective uptake and internalisation inside microbial cells. However, there is limited information about these processes. Regarding steroid uptake, given the physiological and structural surface differences between bacterial groups, its mechanisms in most bacteria have remained elusive. This is especially evident in Gram-negative bacteria, which possess an outer membrane where the lipopolysaccharide leaflet on the outer surface impedes access of steroids to the cytoplasm by passive diffusion [66].

In mycolic acid-containing Actinobacteria, the transport of steroids is carried out by different transport systems, depending on the particular steroid to be assimilated. The sterol uptake into cells of *M. tuberculosis, M. smegmatis, Rhodococcus jostii, R. equi*, and *Gordonia cholesterolivorans* is performed using a set of proteins encoded by the *mce4* locus [35,67–71]. The ten genes contained in the *mce4* locus, referred to in short-hand by the names *yrbE4ABmce4ABCDEFmas4AB* in *Mycobacterium* spp. [71–73], and *supABmce4ABCDEFHI* in *Rhodococcus* and *Gordonia* spp. [67–69] (Figure 2), codify multicomponent ATP-dependent sterol uptake systems. Thus, these genes are upregulated in *M. tuberculosis, M. smegmatis*, and *R. jostii* during growth on cholesterol [55,73]. The deletion of some or all of the genes of the cluster prevents mutants of *M. smegmatis* [72] and *R. jostii* [67] growing on cholesterol as the sole carbon and energy source. Surprisingly, analogous mutants of *M. tuberculosis* are still able to grow using cholesterol, suggesting such strains have an alternative cholesterol transport system [67].

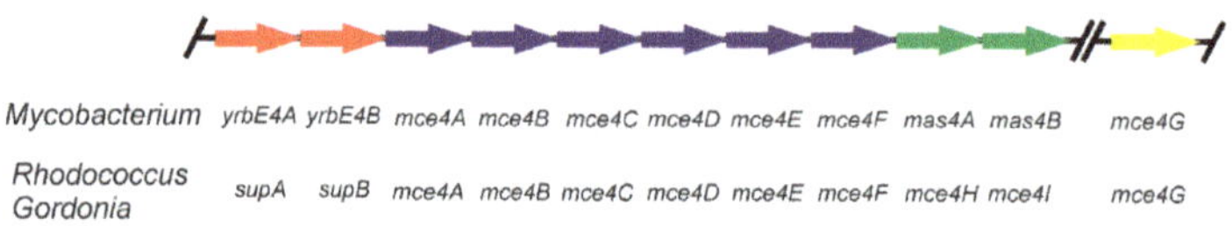

Figure 2. Genetic organisation of the *mce4* cluster involved in sterol uptake in Actinobacteria.

Further analysis revealed that *yrbE4AB* (*supAB*) encode transmembrane proteins similar to permease components of ABC (ATP-binding cassette) transporters [74]. These putative permease subunits should be anchored into the cytoplasmic membrane to facilitate cholesterol translocation across this cellular barrier. They are associated with Mkl ATPase, encoded by *mceG*, which is not linked with the cluster [35,75] (Figure 2). In *M. tuberculosis*, this protein acts as the ATPase component, not only with Mce4 proteins in the uptake of sterols, but also with Mce1, another transport system that is linked to fatty acid transport rather than to steroid catabolism [35,76]. Even so, both transport systems, Mce1 and Mce4, participate in mycobacterial infection [35,75–78]. Thus, MceG ATPase hydrolyses ATP, providing energy for the substrate import for both Mce4 in cholesterol uptake and for Mce1 in fatty acid internalisation [35,67]. Mce4ABCDEF are abundant in cell envelope protein fractions, hence their characterisation as cell wall-associated proteins. They probably facilitate cholesterol transport across the mycolic acid layer and/or pseudoperiplasmic space. Signal sequences for protein export via the general secretory pathway can be found in the amino acidic sequences of Mce4ABCDEF. Moreover, Mce4F has a putative transmembrane domain in the N-terminus of the protein, which probably allows for its embedding in the cytoplasmic membrane or cell wall [74,79,80]. In addition, Mas4AB (Mce4HI), both required for cholesterol import [74], have been suggested to be accessory subunits for their respective systems, with a putative function in stabilising or assembling Mce4ABCDEF complexes [76,81].

Other proteins, codified by genes not included in the *mce4* operon, have been connected to cholesterol uptake in *Mycobacterium*. OmamA has been proposed to be a stabiliser of Mce1 and Mce4 complexes, playing a role in cholesterol and fatty acid uptake in this bacterium [81]. In the same way, LucA also participates in the uptake of both cholesterol and fatty acids, having a cell membrane or cell envelope location. LucA interacts with Mas4A, Mas4B, OmamA, and probably MceG, potentially protecting these proteins, and hence the uptake complexes, from proteolytic attack [76].

As previously indicated, the Mce4 system is required for the uptake of sterols in Actinobacteria. However, this system is not used by *R. jostii* when growing on cholic acid as the sole carbon and energy source [82]. The mutation of the gene coding the porin RjpA impairs growth on metabolising cholic acid, however, it had no effect when sterols were the molecules used to support growth, indicating the importance of this porin for cholic acid uptake [82]. Although the system for the entry of this bile acid through plasma membrane has not been identified, a proteomic study of cholic acid growing cells suggested there might be two systems, CamM, a major facilitator superfamily transporter, and CamABCD, an ATP-binding cassette (ABC) transporter [83]. However, a parallel study demonstrated that both systems act in the reassimilation of transiently accumulated cholic acid intermediates, previously secreted to the culture broth by bacteria during bile acid catabolism [84]. The accumulation of metabolic intermediates in culture broths by steroid-degrading bacteria has been documented [85]; *Pseudomonas stutzeri* strain Chol1 metabolising bile acids extracellularly accumulates intermediates, probably to regulate their intracellular levels or due to overflowing of the downstream metabolic pathway [85]. Later, during growth, these metabolites disappear from the culture broth, which implies: (i) the existence in these bacteria of an exclusion system, and (ii) a catabolic route involved in the reassimilation of these compounds [85].

In the anaerobic cholesterol-degrading *Sterolibacterium denitrificans* DSMZ 13999, it has been suggested that assimilation of cholesterol could proceed via direct adhesion, followed by outer membrane transport mediated by a FadL-like transport system. Once in the periplasm, initial reactions in cholesterol catabolism occur. Notably, the Fad-like system exhibited high substrate specificity for C27 sterols, but not for C19 androgens [86].

3. The Aerobic 9,10-*seco* Pathway

The most complete biochemical and genetic information about steroid catabolic pathways is based on the investigation of a few Actinobacteria and Proteobacteria species, which can mineralise steroids under aerobic conditions. Thus, testosterone and bile acid degradation have been studied in *C. testosteroni* strain TA441 and *C. thiooxidans* CNB-1 [60], *P. stutzeri* strain Chol1 [85,87], and *Pseudomonas putida* strain DOC21 [29,62]. Cholesterol degradation has been studied in *R. jostii* strain RHA1 [55], *R. equi* strain 103S [36], *Rhodococcus rhodocrous* strain DSM 43269 [88], *Rhodococcus ruber* strain Chol-4 [54], *M. tuberculosis* strain H37Rv [89,90], *M. smegmatis* strain mc^2 155 [56,91], *G. cholesterolivorans* strain Chol-3 [53,69], and *Gordonia neofelifaecis* strain NRRL B-59395 [59,92]. Bile acid degradation has also been studied in *R. jostii* RHA1 [57,84]. In all these bacteria, the degradation of the chemical nucleus of the different steroids follows similar steps, using the 9,10-*seco* pathway. Results from the laboratory of Mohn [34,93] suggest that only Actinobacteria degrade sterols with the 9,10-*seco* pathway, while Proteobacteria degrade bile acids and other less structurally complex steroids using this pathway.

3.1. *Oxidation of 3-hydroxyl-substituent from Sterols and Bile Acids*

The degradation of steroid compounds by the 9,10-*seco* pathway is initiated by oxidative attack of 3-oxo steroid. Therefore, steroids, such as sterols or bile acids, carrying a 3-hydroxyl-substituent, need to be oxidised to the corresponding ketone as a preliminary step. However, in *R. jostii* strain RHA1 it has been reported that the catabolism of sterols starts through the ω-oxidation of the C_{17} side chain to the corresponding carboxylic acid, which is later catabolized through the classic pathway (see below). However, it is not clear if the formation of the carboxylic acid occurs before or simultaneously with the initial oxidation of the sterane rings [94].

For sterols, the initial reaction involves the oxidation of the 3β-hydroxy group, concomitantly with isomerisation of the C5–C6 double bond present in these molecules, to yield 3-keto-4-ene structures. Thus, considering cholesterol as a sterol model, the first reaction in its catabolism results in its transformation into cholest-4-en-3-one. This double step, oxidation of the hydroxyl group and isomerisation of the double bond, is catalysed either by a cholesterol oxidase requiring molecular oxygen or by NAD(P)-dependent 3-β-hydroxy-$\Delta^{(5)}$-steroid dehydrogenase (Figure 3) [95].

Most of the organisms that possess a cholesterol oxidase produce it as an extracellular form, either released into the culture broth or linked to the cell surface. Cholesterol oxidases are monomeric enzymes containing FAD, and belong to two different classes. Class I includes enzymes, belonging to the glucose/methanol/choline oxidoreductase family, which fix FAD into their active site cavity through non-covalent bonds [96]. However, Class II enzymes belong to the vanillyl-alcohol oxidase family, covalently binding the FAD cofactor [97].

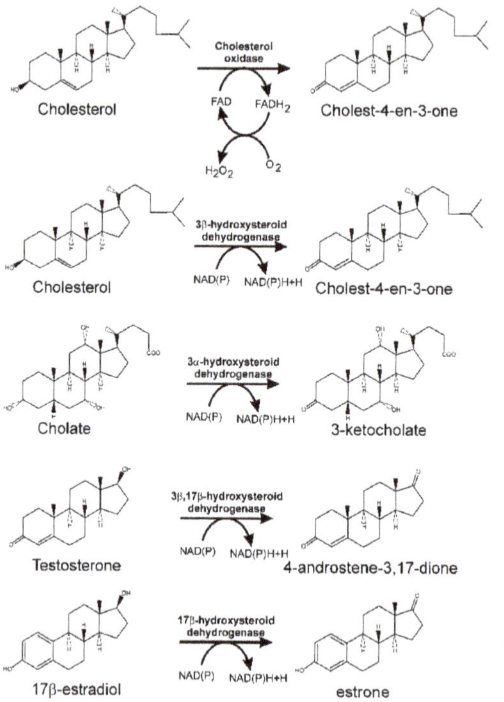

Figure 3. Initial reactions catalysed by cholesterol oxidase and different hydroxysteroid dehydrogenases in steroid degradation through the 9,10-*seco* pathway.

A NAD(P)-dependent cholesterol dehydrogenase has been isolated from cells of *Nocardia* sp. Ch2-1 [98]. In *M. tuberculosis*, the formation of cholest-4-en-3-one is carried out by a 3-β-hydroxysteroid dehydrogenase (HsdD, Rv1106c) [99]. Although a gene for cholesterol oxidase (ChoD, Rv3409c) has also been found in this bacterium, its activity has not been confirmed in vitro, and its homology to many described cholesterol oxidases is rather limited. Moreover, its disruption does not result in a marked change in the ability of the mutant strain to grow using cholesterol [100].

In *M. smegmatis*, a constitutive cholesterol oxidase (Msmeg1604) similar to ChoD from *M. tuberculosis*, and two cholesterol inducible enzymes, Msmeg5228 (a cholesterol oxidase) and Msmeg5233 (a cholesterol dehydrogenase/isomerase), have been identified. Msmeg1604 does not seem to play a critical role in the mineralisation of cholesterol, since a specific mutation in this protein does not affected the production of cholest-4-en-3-one. However, an Msmeg5228 defective mutant shows a drastic reduction in the formation of this intermediate [56]. In addition, double mutants affecting HsdD and ChoD in *M. tuberculosis*, and Msmeg5228 and Msmeg5233 in *M. smegmatis*, are still able to grow on cholesterol, suggesting there are other dehydrogenase/isomerases that could replace them in the first reaction of the cholesterol degradation pathway [56,101].

In *R. ruber*, Chol-4, a constitutive extracellular cholesterol oxidase has been characterised [58]. However, although the deletion of this gene delayed bacterial growth when cholesterol was the sole carbon source, it did not completely prevent it, suggesting the existence of genes with overlapping activities in this bacterium [58].

Bile acids also have a hydroxyl group in the C_3 position that should be oxidised to a 3-oxo group prior to oxidation of the steroid nucleus. However, this hydroxyl group is in an α-configuration, and its oxidation in those strains able to use bile acids as a carbon and energy source is performed by 3α-hydroxysteroid dehydrogenase, which is a NAD(P)-dependent enzyme belonging to the short chain dehydrogenase/reductase superfamily [102]. The first description of this kind of enzymes corresponds to a 3α-hydroxysteroid dehydrogenase/carbonyl reductase in *C. testosteroni* [103]. It was found to be functional as an oxidoreductase towards a variety of 3α-steroid substrates [104,105]. The enzyme also catalyses the reduction of non-steroidal aldehydes and ketones, and consequently, has been named 3α-hydroxysteroid dehydrogenase/carbonyl reductase [105]. Its gene has been cloned from *C. testosteroni* ATCC 11996 [106], and it is also involved in steroid catabolism in *C. testosteroni* strain TA441 [107].

NAD-dependent 3α-hydroxysteroid dehydrogenase activity has been detected in *P. stutzeri* Chol1 cell-free extracts when this bacterium was cultured in a medium containing cholic acid as the carbon source [85]. Moreover, the encoding gene has been annotated in the draft genome of this bacterium within the 79-Kb gene cluster containing the ORFs required for steroid assimilation (Figure 4) [108].

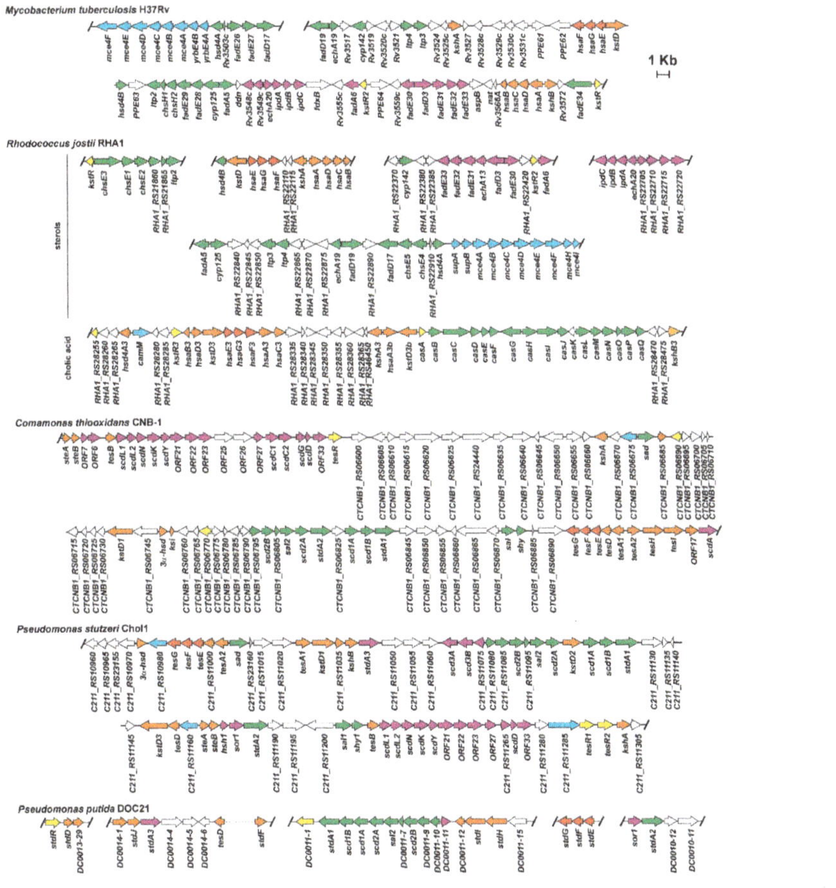

Figure 4. Genetic organisation of the genes encoding the 9,10-*seco* pathway involved in cholesterol (*Mycobacterium tuberculosis* H37Rv, *Rhodococcus jostii* RHA1), or in cholic acid and testosterone (*R. jostii* RHA1, *Comamonas thiooxidans* CNB-1 (formerly, *Comamonas testosteroni* CNB-2), *Pseudomonas stutzeri* Chol1 (formerly *Pseudomonas* sp. Chol1), and *Pseudomonas putida* DOC21 catabolism. Genes coding enzymes involved in cholesterol or bile acids side chain degradation are shown in green; genes participating in ring A/B degradation are in orange; in purple are shown genes coding for ring C/D degradation; blue color indicates genes coding transport systems.

In *P. putida* DOC21, a Tn5 insertion in a gene encoding a 3α-hydroxysteroid dehydrogenase revealed that this mutant was unable to further metabolise any of the tested bile acids, cholic, lithocholic, chenodeoxycholic, ursodeoxycholic, and deoxycholic acid [62].

3.2. Side Chain Degradation of Sterols and Bile Acids

After this initial step, the catabolism of steroids proceeds through two sub-pathways, which involve C_{17} side chain cleavage and/or steroid nucleus oxidation. Degradation of the alkane side chain of cholesterol has been proposed to proceed through a β-oxidation-like process analogous to that of the catabolism of fatty acyl-CoA in human mitochondria and peroxisomes (Figure 5) [109–111]. Thus, cholest-4-en-3-one is used as substrate by Cyp125 P450 cytochrome, catalysing the oxidation of the C_{26} or C_{27} terminal methyl group [94,112–114], followed by further oxidation to provide the initial carboxylate. Although in *M. tuberculosis* strain CDC1551 this protein is absolutely required for bacterial growth on cholesterol as the sole carbon source [115], in *M. tuberculosis* strain H37Rv, loss of this activity is compensated for by another P450 cytochrome, Cyp142 [116]. However, substrate specificities of the two P450 cytochromes are different, with Cyp125 being specific for cholesterol, and although able to oxidise cholesteryl sulfate at a low rate, it is unable to oxidise cholesteryl propionate. By contrast, Cyp142 can efficiently metabolise cholesteryl-sulfate as well as cholesteryl-propionate. This suggests that Cyp142 enzymes may play an important role during *M. tuberculosis* infection, by providing access to additional reservoirs of esterified intracellular cholesterol that would not otherwise be available to the pathogen [117].

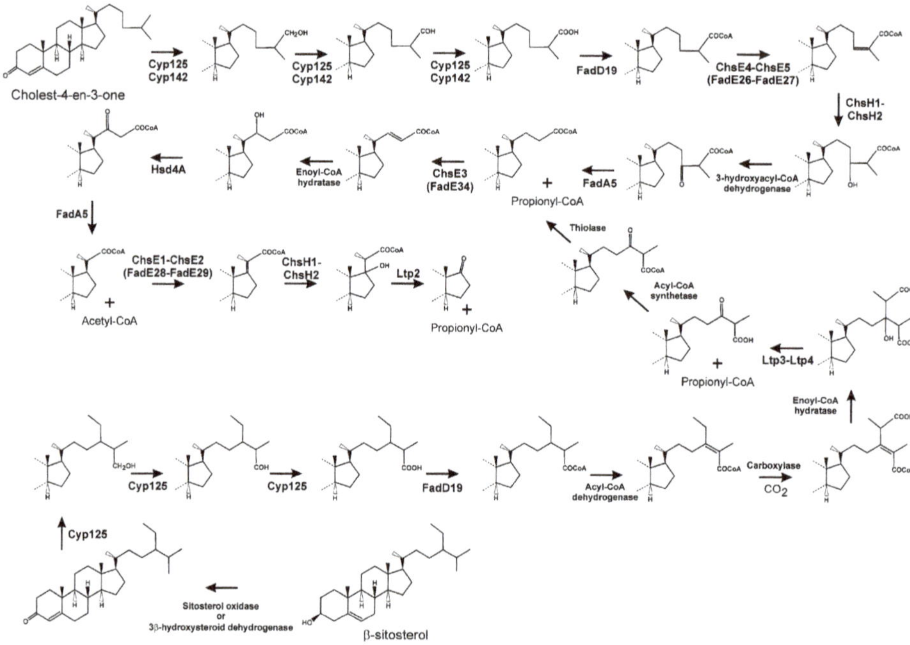

Figure 5. Catabolism of the cholesterol side chain and C_{24}-branched chain of β-sitosterol in Actinobacteria.

The carboxyl-functionalised side chain is then converted to its CoA thioester by FadD19, producing 3-oxocholest-4-en-26-oyl-CoA (Figure 5) [118,119]. Later, an $\alpha_2\beta_2$ heterotetrameric acyl-CoA dehydrogenase, ChsE4-ChsE5, also named FadE26-FadE27, catalyses the α,β-unsaturation of this acyl-CoA thioester, to form 3-oxocholest-4,24-dien-26-oyl-CoA (Figure 5) [90,120,121]. Notably, those human and bacterial acyl-CoA dehydrogenases involved in β-oxidation form α4 homotetramers or α2 homodimers [122], in contrast to the unusual quaternary structure of acyl-CoA dehydrogenases acting in cholesterol side chain catabolism. 3-Oxocholest-4,24-dien-26-oyl-CoA is the substrate of a MaoC-like enoyl-CoA hydratase, ChsH1-ChsH2, being converted into 24-hydroxy-3-oxocholest-4-en-26-oyl-CoA (Figure 5) [123]. Transformation of this compound to 3,24-dioxocholest-4-en-26-oyl-CoA should be carried out by a β-hydroxyacyl-CoA dehydrogenase. Although there is no experimental evidence, it has been proposed that Hsd4A in *M. tuberculosis* could be the enzyme involved in this process [124,125]. Next, FadA5, a thiolase, catalyses the cleavage of this last CoA-esterified oxo-derivative, forming 3-oxochol-4-en-24-oyl-CoA and propionyl-CoA (Figure 5) [126,127]. FadA5 has been categorised as a member of the trifunctional enzyme-like thiolases, type-1 class, which contains a predicted binding site for a bulky fatty acid tail [128,129]. At this point, a second β-oxidation cycle starts, with the introduction of a *trans* double-bond in 3-oxochol-4-en-24-oyl-CoA by ChsE3 (FadE34), followed by the hydration of the double bond by an as yet unknown enoyl-CoA hydratase, and then a dehydrogenation performed by HsdA4, generating 3,22-dioxochol-4-en-24-oyl-CoA [130]. Finally, FadA5 acts again, releasing acetyl-CoA and 3-oxo-pregne-20-carboxyl-CoA (Figure 5) [124,126,127].

FadA5 has been proposed as the thiolase involved in finishing the first two β-oxidation-like rounds. However, although the gene encoding this activity has been shown to be upregulated in the presence of cholesterol in *M. tuberculosis* strain H37Rv, the *fadA5* homologue in *R. jostii* strain RHA1 was not similarly upregulated [55]. Conversely, when *fadA5* was inactivated in a *R. rhodochrous* DSM 43269 derivative, the resulting mutant was not impaired in cholesterol or β-sitosterol side chain degradation, indicating that FadA5 is not essential for the degradation of the sterol side chain in this strain [126,131].

The elimination of the side chain ends by a mechanism resembling a new β-oxidation cycle (Figure 5), starting with dehydrogenation of 3-oxo-4-pregnene-20-carboxyl-CoA to 3-oxo-4,17-pregna diene-20-carboxyl-CoA in a process catalysed by another heterotetrameric $\alpha_2\beta_2$ acyl-CoA dehydro genase complex, ChsE1-ChsE2 (FadE28-FadE29; Figure 5) [90,120,121]. This last compound is then the substrate of an enoyl-CoA hydratase (ChsH1-ChsH2) [90], and the hydrated molecule undergoes an aldol-lyase cleavage reaction catalysed by Lpt2, producing androst-4-en-3,17-dione (AD) and releasing another propionyl-CoA molecule (Figure 5) [90,109,110,132]. Ltp2, at the protein sequence level, is more related to thiolases and acetoacetyl-CoA synthases than to aldolases from other metabolic pathways. It has been shown that Lpt2 interacts with ChsH2, forming a complex ChsH1-ChsH2-Ltp2, catalysing the last two steps of side chain removal from cholesterol [132]. In summary, the complete metabolism of the cholesterol side chain results in a 17-keto steroid intermediate, AD, as well as one acetyl-CoA and two propionyl-CoA molecules.

Removal of the C_{17} side chains of β-sitosterol and campesterol are processes less characterised than cholesterol side chain degradation. Both β-sitosterol and campesterol contain branched side chains that impede the first β-oxidation round of the side chain; therefore, these need to be eliminated to allow full degradation of the side chain. Initially, the C_{24}-branched side chain is oxidised at position C_{26} by Cyp125, followed by CoA activation by FadD19 (Figure 5) [118]. Next, the bond between C_{24}/C_{25} is desaturated. Removal of the C_{24}-branches then starts by carboxylation of the C_{28} carbon, followed by a hydration reaction of the double bond, and finally the release of acetyl-CoA, from campesterol, or propionyl-CoA, from β-sitosterol, via cleavage of the C_{24}–C_{25} bond by a heteromeric aldol-lyase (Ltp3-Ltp4) (Figure 5) [131]. It is evident that after aldolytic cleavage of the branch, thioesterification of the resulting carboxylate should be mandatory for degradation of the side chain (Figure 5). At the end of side chain degradation from β-sitosterol and campesterol, AD is formed.

Catabolism of the side chain of bile acids is a process in which not all the enzymatic steps are fully characterised. After oxidation of their hydroxy-group at C_3 to an oxo-group in these carboxylated molecules, the degradation of the lateral chain begins. Firstly, the carboxy group from the side chain in C_{17} is activated to a coenzyme A thioester through a reaction catalysed by a bile acid-CoA ligase (StdA1 in *P. putida* DOC21 and CasG in *R. jostii*; Figure 6) [56,62,118]. Once activated, a β-oxidation-like process occurs, although it is different to the two initial rounds in the cholesterol side chain degradation. In the first place, a dehydrogenation of C_2/C_3 of the acyl-CoA side chain occurs. Based on bioinformatics analysis of clusters coding for bile acid catabolism in *P. stutzeri* Chol1, the participation of a heteromeric acyl-CoA dehydrogenase, Scd1AB, has been proposed [108]. In a second step, the hydration of the α,β-double bond takes place. The introduction of a hydroxy group at the third carbon of the lateral chain in *P. stutzeri* Chol1 is catalysed by the enoyl-CoA hydratase Shy1 (Figure 6) [87]. Until this step, the process is identical to a canonical β-oxidation. However, the next enzymatic step is not the expected oxidation of the hydroxy function to a keto group, followed by a thiolytic cleavage mediated by the introduction of a CoA molecule; instead, it is an aldolic cleavage of the C–C bond, catalysed by Sal1, generating an aldehyde and releasing an acetyl-CoA molecule (Figure 6) [87,133]. Thus, cleavage of these two carbons in unbranched side chains of bile acids proceeds through a retroaldol reaction, as a retro-Claisen reaction, instead of the typical reaction catalysed by thiolases. The resulting aldehyde is then oxidised to the corresponding carboxylic acid by a specific aldehyde dehydrogenase (Sad in *P. stutzeri* Chol1; Figure 6) [87], and later, a second acyl-CoA synthetase (StdA2 in *P. putida* DOC21 and CasI in *R. jostii*) activates it to a CoA derivative [56,62,119]. The subsequent degradation of the remaining side chain is believed to occur through a similar mechanism, where an acyl-CoA dehydrogenase, putatively Scd2AB, introduces an α,β-desaturation in the CoA-activated C_3 side chain (Figure 6) [85,108]. Hydration of the double bond takes place, resulting in the formation of a hydroxyl group at C_{17}, and then an aldolytic cleavage of the molecule occurs, yielding a molecule of propionyl-CoA and a steroid derivative with a keto function at C_{17} (Figure 6). The enzymes catalysing these two steps have not been fully characterised yet [108].

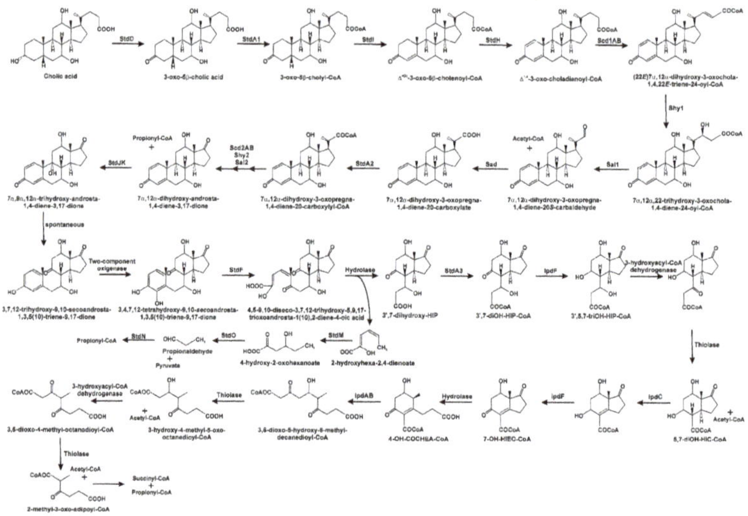

Figure 6. Cholic acid metabolism in *Pseudomonas putida* DOC21 and *P. stutzeri* Chol1 through the 9,10-*seco* pathway. Preliminary evidence suggests that oxidation of the A-ring occurs simultaneously with C_{17} side chain degradation. Hydroxyl groups at C_7 and C_{12} are maintained during degradation of the molecule, the affecting 3aα-H-4α(3′-propanoate)7a-β-methylhexahydro-1,5- indanedione-hydroxylated derivatives metabolism. 3′,7-diOH-HIP, 3aα-*H*-4α(3′(*R*)-hydroxy-3′propanoate)-7-hydroxy-7aβ-methylhexahydro-1,5-indanedione; 3′,7-diOH-HIP-CoA, 3aα-*H*-4α(3′(*R*)-hydroxy-3′propanoyl-CoA)-7-hydroxy-7aβ-methylhexahydro- 1,5-indanedione; 3′,5,7-triOH-HIP-CoA, 3aα-*H*-(3′(*R*)-hydroxy-3′propanoyl-CoA)-5,7-dihydroxy-7aβ-methylhexahydro-1-indanone; 4OH-COCHEA-CoA, 2-(2-carboxyethyl)-4-hydroxy-3-methyl-6-oxocyclohex-1-ene-1-carboxyl-CoA; 7-OH-HIEC-CoA, (7a*S*)-7a-methyl-7-hydroxy-1,5-dioxo-2,3,5,6,7,7a-hexahydro-1*H*-indene-4-carboxyl-CoA; 5,7-diOH-HIC-CoA, 3aα-*H*-4α(3′-carboxyl-CoA)-5,7-dihydroxy-7aβ-methylhexahydro-1-indenone.

3.3. Testosterone Catabolism Convergence

For the catabolic convergence of testosterone into this pathway, the 17β-hydroxy substituent needs to be oxidised to a 17-oxo-derivative. The oxidation of this substituent to a keto group produces androst-4-en-3,17-dione [134]. Currently, the best characterised bacterial enzyme performing this reaction is 3β,17β-hydroxysteroid dehydrogenase (3,17β-HSD) from *C. testosteroni*, encoded by the gene *βhsd* (Figure 3). This NAD(H)-dependent enzyme belongs to the short-chain dehydrogenases/reductases superfamily. It was first purified from strain ATCC11996, and was initially defined as 3β-HSD; however, the cloned enzyme was revealed to act on both the 3β-hydroxyl and 17β-hydroxyl groups of androgens, estrogens, and iso-bile acids [134–136]. Horinouchi et al. [107] suggested that the main role of 3,17β-HSD in *C. testosteroni* TA441 cells is 3β-dehydrogenation, and there is at least one more dehydrogenase acting on the 17β-hydroxyl group, since a knock out mutation in the coding gene did not prevent its growth on testosterone. Thus, *C. testosteroni* may have more than one enzyme with 17β-dehydrogenating/hydrogenating activities, to deal with intermediate compounds having considerable structural differences [107]. However, the degradation of testosterone by the 3,17β-HSD mutant is highly affected. Moreover, 3,17β-HSD gene expression is highly induced by testosterone, but not by estradiol and cholesterol, suggesting that this enzyme is a key component in the degradation of testosterone [137].

In *P. putida* DOC21, a microorganism isolated from soil based on its capacity to degrade bile acids and testosterone, a gene encoding 17β-hydroxysteroid dehydrogenase has been identified (unpublished results). This enzyme only has the ability to oxidise the 17β-hydroxy-group of testosterone, because, although this bacterium efficiently metabolises testosterone and androstanolone, it is unable to metabolise closely related compounds with a 3β-hydroxy group (i.e., *trans*-androsterone) [29]. However, current studies are being performed to more completely characterise this enzyme. Other bacterial 17β-hydroxysteroid dehydrogenases have been identified; however, they have been linked to estrogen degradation (Figure 3) through the 4,5-*seco* pathway (see below).

3.4. Oxidation of A/B Rings

Because of the initial processing of the steroid nucleus and elimination of the C_{17} side-chain, a common intermediate, AD, arises from sterols and testosterone. Starting from this compound, a common pathway is used, under aerobic conditions, for the complete assimilation of carbon atoms with the concomitant production of metabolic energy (Figure 7). This starts with the transformation of AD into androst-1,4-dien-3,17-dione (ADD), by the desaturation of the bond between C_1 and C_2, with trans axial removal of the hydrogen atoms C_1 (α) and C_2 (β), in a reaction catalysed by a 3-ketosteroid-$\Delta^{1(2)}$-dehydrogenase [138,139]: TesH in *C. testosteroni* [140], StdH in *P. putida* DOC21 [141], KstD in *M. tuberculosis* and *Rhodococcus erythropolis* [142–145], and KsdD in *M. smegmatis* [146].

For the degradation of bile acids in *P. putida* DOC21, concomitant with the elimination of the C_{17} side chain, oxidation of the ring A starts with desaturation at C_4 (catalysed by a 3-ketosteroid-$\Delta^{4(5\alpha)}$-dehydrogenase, StdI) and at C_1 (catalysed by StdH; Figure 6) [62,141]. After the elimination of the acyl-side chain and the introduction of two double bonds in the A ring of the cholic acid molecule, 7α,12α-dihydroxy-androsta-1,4-diene-3,17-dione is obtained (Figure 6). Notably, these hydroxyl substituents are maintained during the catabolic process (see below). However, the degradation of lithocholic acid (lacking the hydroxy groups at C_7 and C_{12}) results in ADD, which is a catabolite of convergence when testosterone and AD degradation occurs. Thus, ADD, as well as the hydroxylated derivatives generated from other bile acids, seems to represent a convergence point for the degradative pathway of steroids in *Pseudomonas* species.

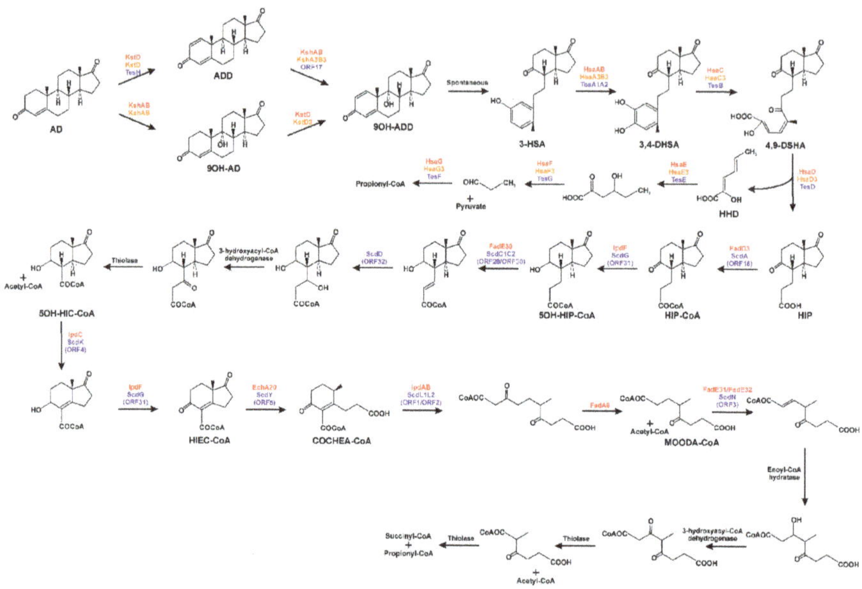

Figure 7. Proposed 9,10-*seco* pathway from androst-4-en-3,17-dione to central metabolites, pyruvate, acetyl-CoA, propionyl-CoA, and succinyl-CoA. AD, androst-4-en-3,17-dione; ADD, androst-1,4-dien-3, 17-dione; 9OH-AD, 9-hydroxy-androst-4-en-3,17-dione; 9OH-ADD, 9-hydroxy-androst-1,4-dien-3,17-di one; 3-HAS, 3-hydroxy-9,10-secoandrosta-1,3,5(10)-triene-9,17-dione; 3,4-DHSA, 3,4-dihydroxy-9,10-secoandrosta-1,3,5(10)-triene-9,17-dione; 4,9-DSHA, 4,5,9,10-diseco-3-hydroxy-5-9-17- trioxoandrosta-1(10),2-diene-4-oic acid; HHD, 2-hydroxy-2,4-hexadienoic acid; HIP, 3aα-H-4α(3′-propanoate)7a-β-me thylhexahydro-1,5-indanedione; 5OH-HIP-CoA, 3aα-H-4α(3′-propanoyl-CoA)-5-hydroxy-7a-β-meth ylhexahydro-1-indanone; 5OH-HIC-CoA, 3aα-H-4α(3′-carboxyl-CoA)-5-hydroxy-7a-β-methylhexahy dro-1-indanone; HIEC-CoA, (7a*S*)-7a-methyl-1,5-dioxo-2,3,5,6,7,7a-hexahydro-1*H*-indene-4-carboxyl-CoA; COCHEA-CoA, 2-(2-carboxyethyl)-3-methyl-6-oxocyclohex-1-ene- 1-carboxyl-CoA; MOODA-C oA, 4-methyl-5-oxo- octanedioyl-CoA. Actinobacterial enzymes from cholesterol metabolism are indicated in red, actinobacterial enzymes involved in cholic acid catabolism are indicated in orange, and those from catabolism of testosterone and cholic acid from *Comamonas* spp. are written in blue.

Moreover, a 3-ketosteroid-$\Delta^{4(5a)}$-dehydrogenase, TesI, has been identified in *C. testosteroni* strains ATCC17410 and TA441 (Figure 4) [107,147]. This enzyme is needed by this bacterium to efficiently catabolise androsterone, androstanolone, androstenedione, and bile acids. Along with its gene sequence, this enzyme has also been characterised in *R. jostii* RHA1 [148]. It is interesting to note that *stdI* and *stdH* from *P. putida* DOC21, as well as *tesH* and *tesI* from *C. testosteroni* TA441, are adjacent to each other in their respective genomes, being part of the steroid-degrading gene clusters (Figure 4) [60,141].

Once ADD, or the hydroxylated derivatives originating from catabolism of different bile acids catabolism, has been synthesised, a monooxygenase/reductase complex (3-ketosteroid 9α-hydroxylase) introduces an α-hydroxy group at C_9 with KshAB (Figure 7) in *R. erythropolis* [149,150] and in *M. tuberculosis* [151,152]. The resulting 9-hydroxy-androst-1,4-dien-3,17-dione is an unstable molecule that undergoes abiotic cleavage of the B-ring and aromatisation of the A-ring to form 3-hydroxy-9,10-secoandrosta-1,3,5(10)-triene-9,17-dione (Figure 7).

While the formation of 9α-hydroxy-4-androstene-3,17-dione, and later desaturation in the A ring has been described, it is worth mentioning that the order of desaturation and hydroxylation reactions in Actinobacteria is unclear [149–151]. Moreover, the introduction of desaturation between C_1 and C_2 has been proposed to occur at different stages of side chain degradation in some Actinobacteria, i.e., in *R. ruber* Chol4, *R. erythropolis* SQ1, and different strains of *Nocardia, Arthrobacter,* and *Mycobacterium* [153,154]. In fact, different 3-ketosteroid-$\Delta^{1(2)}$-dehydrogenase paralogs showing differences in their substrate specificities have been characterised in some of these strains. For example, in *R. ruber* Chol-4, three different KstD isoenzymes (KstD1, KstD2, and KstD3) differing in their respective substrate profiles have been described, with KstD2 being the isoenzyme mainly involved in AD degradation in this strain [155]. In *Mycobacterium neoaurum* ATCC 25795, three different paralogs of this dehydrogenase have also been identified. In this strain, KstD1 showed a higher affinity for 9α-hydroxy-4-androstene-3,17-dione, while KstD3 preferred AD [156].

In Actinobacteria, it is not surprising to find different paralog genes coding KshA subunits, ranging from one to six. In *M. tuberculosis* H37Rv, only one *kshA* homolog has been found, in *Mycobacterium* spp. VKM Ac-1815D and 1816D, two different versions have been reported, and in *Mycobacterium* VKM Ac-1817, five different paralogs have been described [157,158]. In *R. rhodochrous* DSM43269 and *R. erythropolis* SQ1 chromosomes, five different paralogs of *kshA* have been identified [157,159], and the coded proteins showed different substrate specificities. Thus, KshA5 from *R. rhodochrous* appears to have the broadest substrate range, but without a clear substrate preference. By contrast, KshA1 seems to be specifically involved in cholic acid catabolism [159]. The reductase component of 3-ketosteroid 9α-hydroxylase in Actinobacteria, KshB, is generally present as a single copy gene, with some exceptions for those bacteria able to degrade sterols and bile acids, suggesting that each copy of this gene could be involved in the specific degradation of a particular compound.

Conversely, steroid-degrading Proteobacteria seem to only have single copies of 3-ketosteroid-$\Delta^{1(2)}$-dehydrogenase and both subunits of 3-ketosteroid 9α-hydroxylase, suggesting a possible reason for their rare ability to catabolise a wide range of steroids. Although it has been proposed that the multiplicity of genes encoding 3-ketosteroid-$\Delta^{1(2)}$-dehydrogenase and the oxygenase subunits of 3-ketosteroid 9α-hydroxylase in some Actinobacteria could represent an environmental advantage, their role could also be in the maintaining metabolic flux throughout the pathway, avoiding the accumulation of intermediates, and so facilitating a dynamic and fine-tuned steroid catabolism.

Catabolism continues through the hydroxylation of the aromatised A-ring in 3-hydroxy-9,10-secoandrosta-1,3,5(10)-triene-9,17-dione by a two-component oxygenase, TesA1A2 in *C. testosteroni* [160], and HsaAB in *M. tuberculosis* and *R. jostii* RHA1 [161], leading to the formation of a catecholic derivative, 3,4-dihydroxy-9,10-secoandrosta-1,3,5(10)-triene-9,17-dione (Figure 7). In this type of monooxygenase, the reductase utilises NADH to reduce a flavin, which is then transferred to the oxygenase. The dihydroxylated ring is then opened by *meta*-cleavage by an extradiol dioxygenase (TesB in *C. testosteroni* [162], and HsaC in *M. tuberculosis* [163] and *R. jostii* RHA1 [55]) that introduces two oxygen atoms. In *P. putida* DOC21 the coding gene for this meta-cleavage dioxygenase, *stdF*, was identified by Tn5 transposon mutagenesis and was sequenced [29]. The product of the cleavage, 4,5,9,10-diseco-3-hydroxy-5-9-17-trioxoandrosta- 1(10),2-diene-4-oic acid, is hydrolysed by TesD in *C. testosteroni* [164], and by HsaD in *M. tuberculosis* [165,166] and *R. jostii* [55], yielding 3aα-H-4α(3′-propanoate)7a-β-methylhexahydro-1,5-indanedione (HIP) and 2-hydroxy-2,4-hexadienoic acid (Figure 7).

In *C. testosteroni*, the catabolism of 2-hydroxy-2,4-hexadienoic acid is carried out by the products of the genes *tesEFG* [167]. Thus, 2-hydroxy-2,4-hexadienoic acid is the substrate of the hydratase TesE leading to the formation of 4-hydroxy-2-oxohexanoate, which then undergoes an aldol cleavage catalysed by TesG aldolase, yielding pyruvate and propionaldehyde. Finally, propionaldehyde is the substrate of an acylating aldehyde dehydrogenase, TesF, producing propionyl-CoA, which enters into the central metabolism (Figure 7). Orthologous genes have been identified in the *M. tuberculosis* H37Rv genome, *hsaEFG*, and it has been proposed that the aldolase and dehydrogenase, HsaF and HsaG, interact as a complex to efficiently perform the catalysis [168].

3.5. Degradation of B/C Rings

The degradation of HIP starts with its CoA thioesterification by a specific acyl-CoA synthetase (Figure 7), StdA3 in *P. putida* DOC21 [62], ScdA (ORF18) in *C. testosteroni* [169] and FadD3 in Actinobacteria [119,170]. It has been proposed that in *M. tuberculosis*, HIP-CoA is the substrate of IpdF (ScdG/ORF31 in *C. testosteroni*), which reduces the 5′-keto substituent to a hydroxyl group, leading to the formation of 5-hydroxy-HIP (Figure 7). This compound is then subjected to a β-oxidative process for the elimination of the propionyl-CoA side chain, yielding 3aα-*H*-4α(3′-carboxyl-CoA)-5-hydroxy-7aβ-methylhexahydro-1-indanone (Figure 7). The enzymes involved in this β-oxidation in *M. tuberculosis* are unknown, with the exception of FadE30, which has been proposed to be the acyl-CoA dehydrogenase involved in this process [36,171]. However, in *C. testosteroni*, it has been proposed that this enzymatic reaction is carried out by a heteromeric acyl-CoA dehydrogenase, ScdC1C2 (ORF28,30), and the product of this reaction is the substrate of hydration of the double bond, in a reaction catalysed by an enoyl-CoA hydratase, ScdD (ORF32; Figure 7) [172–174]. (7aS)-7a-methyl-1,5-dioxo-2,3,5,6,7,7a-hexahydro-1*H*-indene-carboxyl-CoA, in which the rings C and D remain intact, is then produced by two reactions catalysed by IpdC in *M. tuberculosis* (ScdK/ORF4 in *C. testosteroni*), which introduces a double bond in the C ring, and IpdF in *M. tuberculosis* (ScdG/ORF31 in *C. testosteroni*), which oxidises the 5-OH group (Figure 7) [171–174]. However, the order of these two reactions has not been determined. Ring D is now opened, previously to ring C, through a hydrolytic reaction mediated by a crotonase, EchA20 (ScdY/ORF5 in *C. testosteroni*) [171,174], producing (*R*)-2-(2-carboxyethyl)-3-methyl-6-oxocyclohex-1-ene- 1-carboxyl-CoA. This molecule is the substrate for the hydrolytic cleavage of ring C, catalysed by IpdAB (ScdL1L2/ORF1,2 in *C. testosteroni*) by a retro-Claisen hydrolysis (Figure 7) [175,176]. The product of the opening of both the C and D rings is the substrate of a thiolase (putatively Fad6), resulting in a molecule of acetyl-CoA and 4-methyl-5-oxo-octanedioyl-CoA (Figure 7) [171,177]. It has been proposed that this last intermediate undergoes a β-oxidation process, starting with a desaturation catalysed by an acyl-CoA dehydrogenase, FadE32, or by the heteromeric Fad31-FadE32 in *Mycobacterium* [171]. In *C. testosteroni*, the enoyl-CoA hydratase involved in the next reaction of this β-oxidation process, ScdN (ORF3), has been identified [177]. As final products of the β-oxidation, a molecule of acetyl-CoA would be released together with 2-methyl-β-ketoadipyl-CoA, which could be cleaved to propionyl-CoA and succinyl-CoA (Figure 7) by a mechanism analogous to the final step in the catabolism of aromatic compounds through the β-ketoadipate pathway [178].

It is important to mention that in the degradation of bile acids in *P. putida* DOC21, the hydroxyl substituents present at C_7 and/or C_{12} in the different bile acids are maintained during the degradative process. Thus, a *P. putida* DOC21 mutant lacking StdA3, the ATP-depending acyl-CoA synthetase activating HIP, when cultured in the presence of chenodeoxycholic acid accumulates 3′(*R*)-hydroxy-HIP in the culture broth. When deoxycholic acid is used, 7β-hydroxy-HIP accumulation is observed, and in the presence of cholic acid, 3′(*R*),7β-dihydroxy-HIP is accumulated [62]. For *R. jostii* RHA1, it has been proposed that *echA13*, a gene included in the HIP catabolic gene cluster, could codify a function for the removal of the hydroxyl group at position 7β [57]. On the other hand, maintaining the hydroxyl group in the 3′ position in the aliphatic chain of HIP could potentially generate, after CoA thiosterification of the molecule, a C_3-hydroxy-intermediate from the first β-oxidation round, avoiding the need for an acyl-CoA dehydrogenase, FadE30, and enoyl-CoA hydratase, in the formation of 3aα-*H*-4α(3′-carboxyl-CoA)-5-hydroxy-7aβ-methylhexahydro-1-indanone.

A bioinformatics approach, using selected characteristic genes from this pathway described from model organisms against genomes deposited in public databases, made it possible to identify 265 putative steroid degraders within only the Actinobacteria and Proteobacteria, from different habitats, including eukaryote hosts, soil, and aquatic environments [93]. This study allowed for a comparison of the organisation of the genetic clusters encoding this pathway in different organisms, and with different steroid use profiles (sterols, testosterone and bile acids). This study showed that most of the genes coding the whole 9,10-*seco* pathway are randomly located in different low number large clusters in the bacterial chromosome. However, there are some exceptions, such as *M. tuberculosis* H37Rv, where the predicted catabolic genes are mainly located in a single cluster (80 genes in a region of about 50 kb; Figure 4) [55,93]. *Rhodococcus* spp. have their cholic acid catabolism genes grouped in a separate cluster not close to the cholesterol catabolism genes, and lacking C/D-rings catabolic genes (i.e., *R. jostii* RHA1, Figure 4). Even so, the *R. equii* genome has a single cluster containing the putative cholic acid and cholesterol catabolic genes [93]. In general, the location of these clusters is chromosomal, however, Bergstrand et al. [93] identified some exceptions, where the genes are located on plasmids, such as the putative cholic acid A/B-ring degradation genes located on pRHL1 from *R. jostii*, and a gene cluster encoding putative enzymes involved in testosterone and/or cholic acid catabolism in two strains belonging to the genus *Novosphingobium*. The association of genes coding this pathway opens the possibility of the horizontal transfer of catabolism of steroids between bacteria.

Genes involved in the catabolism of cholesterol through the 9,10-*seco* pathway are part of the core genome characterising most of the genera of the Corynebacterinae. However, the cholic acid pathway, although conserved in *Rhodococcus* spp., is seldom found in closely related genera. This restricted distribution of the actinobacterial bile acid pathway suggests that its origin could be through the duplication of a pre-existing cholesterol pathway in an ancestor of the genus *Rhodococcus*. This could be the reason why genes encoding A/B-ring degradation are found in gene clusters for both cholesterol and bile acid pathways. Surprisingly, genes for C/D-ring degradation are found only in cholesterol pathway clusters, indicating that they have not been duplicated, or have been lost through the evolution of the strains.

Proteobacteria are unable to degrade sterols with alkyl side chains [29,60]. This could be due to the absence of Cyp125 and/or Cyp142 orthologs allowing side chain functionalisation and degradation, as well as the lack of a *mce*-like transport system for sterol uptake. The wide distribution of 9,10-*seco* pathway genes in Actinobacteria contrasts with the scarcity of androgen/bile acids pathway genes in Proteobacteria genomes. However, an exception to this could be the genus *Comamonas*, in which this pathway seems to be part of the genome core [93].

Although the bile acid and the androgen catabolic pathways of Proteobacteria are highly similar to the proposed pathways for actinobacterial cholesterol and cholic acid catabolism, the encoding genes show low sequence similarity. This could be explained by: (i) convergent evolution, where these genes evolved independently in both taxonomic groups [179]; or (ii) divergent evolution, with the genes having an actinobacterial origin, and being disseminated through horizontal gene transfer to the other taxon [93]. Although some interesting articles assess the phylogeny of some critical genes from these pathways [93,141,158], complementary studies will be necessary to clarify this issue.

3.6. Regulation of 9,10-seco Pathway

In *M. tuberculosis*, the regulation of cholesterol degradation is carried out by two TetR-like transcriptional repressors, KstR and KstR2 [73,180]. These repressors regulate the expression of particular genes after binding specific intermediates from cholesterol degradation. Thus, KstR unlocks the expression of the genes encoding the membrane transport system of cholesterol, as well as enzymes involved in side chain catabolism and the opening and removal of steroidal A and B rings through binding to 3-hydroxy-cholest-5-en-26-oyl-CoA [181]. This regulatory protein is not only involved in cholesterol catabolism, as the KstR regulon comprises 74 genes in *M. tuberculosis*, some of them involved in growth on palmitate, suggesting that KstR may control metabolism of lipids in this bacterium [73]. KstR may act as a de-repressor by binding to molecules other than 3-hydroxy-cholest-5-en-26-oyl-CoA. Moreover, in *M. smegmatis*, it has been demonstrated that cholesterol and its first catabolic intermediate, cholest-4-en-3-one, were unable to induce the release of KstR from proposed promoter regions in the regulon. Furthermore, in *M. smegmatis*, this regulator remains unaffected by other cholesterol catabolic intermediates, such as AD and ADD [182]. By contrast, KstR2 also downregulates the pool of genes needed for C and D ring degradation, and de-represses their expression by binding HIP-CoA as a ligand [183,184].

The regulation of the 9,10-*seco* pathway mediated steroid catabolism in Proteobacteria has mainly been studied in *C. testosteroni*. A gene called *tesR* in strain TA441, and *teiR* in strain ATCC11996, in the steroid degradation gene clusters, encode a transcriptional regulator needed for induction of

3,17β-HSD dehydrogenase, 3α-hydroxysteroid dehydrogenase, 3-ketosteroid-$\Delta^{1(2)}$-dehydrogenase, 3-ketosteroid-$\Delta^{4(5\alpha)}$-dehydrogenase and most of the identified steroid degradation genes [185–188]. It has been reported that TesR/TeiR is a membrane protein with a polar location in cells involved in chemotaxis, and mediates steroid sensing and metabolism via kinase activity, which likely triggers a cascade of phosphorylation events that induce the expression of steroid catabolising enzymes [187].

In addition, two negative regulator genes, *repA* and *repB*, have been identified close to *hsdA*, the gene coding for 3α-hydroxysteroid dehydrogenase. The products of these genes repress *hsdA* expression: (i) at the transcriptional level, by binding to *hsdA* promoter sequences in response to the presence of steroids (RepA) [189]; and (ii) by interfering with its translation, by binding to the mRNA of 3α-hydroxysteroid dehydrogenase (RepB) [190]. Moreover, HsdR, a LysR-type transcriptional repressor, activates transcription of the *hsdA* gene in *C. testosterone*, dependent on decreased repression by RepA [191,192].

By contrast, in *C. testosteroni* ATCC11996, a complex network regulating the expression of the 3,17β-HSD gene, *βhsd*, has been proposed, with several transcriptional repressors, PhaR [193], LuxR [194], TetR [195], and BRP [196] having been described. In knock-out mutants for *phaR*, *luxR*, *tetR*, and *brp*, the basal expression levels of *βhsd* did not increase with reference to the wild type strain; however, on the addition of testosterone there was a several-fold increase in expression levels compared to the wild type, suggesting that these regulators function as repressors of *βhsd* expression [193–196]. This implies the existence of a complex controlling *βhsd* expression, which is regulated by testosterone. In addition, the expression of 3α-HSD was also increased in the *brp* knock-out mutant, indicating that BRP represses the expression of both *hsdA* and *βhsd* in the presence of testosterone [196].

4. The 4,5-*seco* Pathway

Compared with the 9,10-*seco* pathway for sterols, bile acids, and androgen aerobic degradation, current knowledge on estrogen degradation pathways is very limited. The partial characterisation of the aerobic degradation of 17β-estradiol by *Sphingomonas* sp. strain KC8, an obligate aerobic alpha-proteobacterium isolated from wastewater [197], has allowed a new pathway for aerobic estrogen degradation, the 4,5-*seco* pathway, to be proposed [198].

Sphingomonas KC8 codes for a gene for 3β,17β-hydroxysteroid dehydrogenase (*oecA*) that is responsible for the transformation of 17-hydroxy from 17β-estradiol and testosterone to their respective oxo-derivatives. This gene was similarly expressed when this strain was cultured on 17β-estradiol and testosterone, although testosterone is degraded by this strain through the 9,10-*seco* pathway [199]. OecA converts 17β-estradiol into estrone (Figure 8). However, this gene does not cluster with other steroid-degrading genes in the genome of *Sphingomonas* KC8. An ortholog of this gene has been characterised in *P. putida* SJTE-1, a strain isolated from sludge, which can use 17β-estradiol as a sole carbon source [200,201].

Additionally, other genetic clusters involved in estrogen catabolism have been identified in the genome of *Sphingomonas* KC8 (Figure 8). In gene cluster I, an *oecB* gene encodes a flavin-dependent estrone 4-hydroxylase, which introduces a hydroxyl group at C_4 in the aromatic ring, yielding 4-hydroxyestrone. In gene cluster II, *oecC* codes for a 4-hydroxyestrone 4,5-dioxygenase that catalyses the *meta*-cleavage of the aromatic ring, creating 4-norestrogen-5(10)-en-3-oxo-carboxylic acid. This meta-cleavage product is unstable and undergoes an abiotic recyclisation, producing pyridinestrone acid, a dead-end byproduct that accumulates in the culture broth of *Sphingomonas* KC8 when grown on 17β-estradiol (Figure 8) [198]. However, the metabolic pathway continues by using a 2-oxoacid oxidoreductase that catalyses the decarboxylation of this meta-cleavage product, as well as the thioesterification of the resulting molecule with CoA, to produce 4-norestrogen-5(10)-en-3-oyl-CoA (Figure 8). The gene encoding this 2-oxoacid oxidoreductase is also found in gene cluster II. This cluster also includes genes encoding β-oxidation-like enzymes. Metabolic intermediates in estrone degradation have allowed the identification of products of these genes involved in the transformation of the cleavage product from 4-hydroxyestrone to HIP (Figure 8). In addition, the existence of a gene encoding for a 3-hydroxy-3-methylglutaryl-CoA synthase-like protein in this cluster suggests its involvement in estrogen catabolism [202].

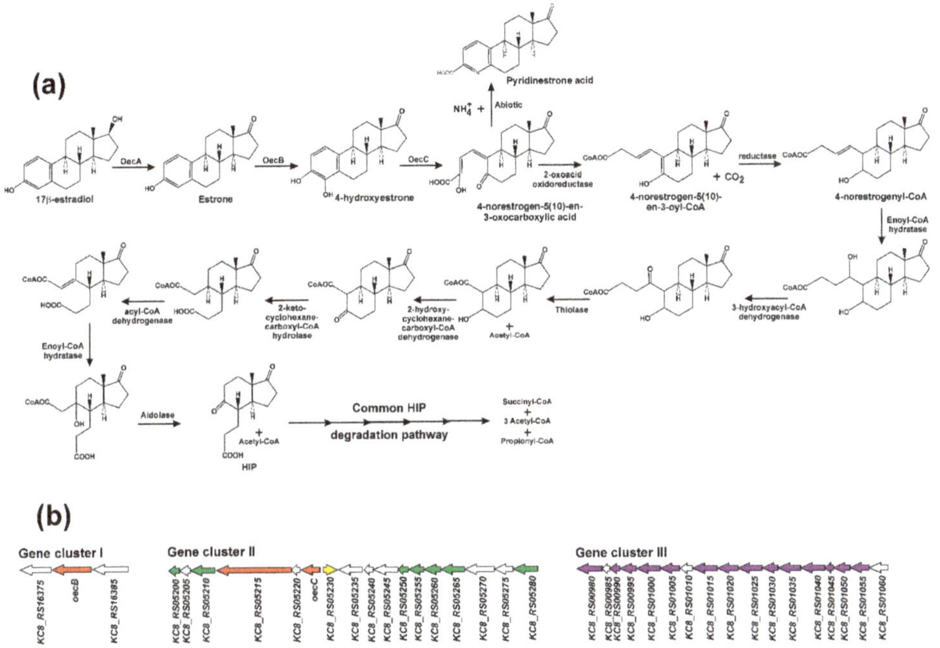

Figure 8. Metabolism of 17β-estradiol in *Sphingomonas* sp. KC8. (**a**) Reactions of the 4,5-*seco* pathway. HIP, 3aα-H-4α(3′-propanoate)7a-β-methylhexahydro-1,5-indanedione. (**b**) Genetic organisation of the three clusters codifying the enzymes required for 17β-estradiol assimilation (Genome Accession NZ_CP016306). Annotation from the genome: KC8_RS16375, putative dioxygenase; KC8_RS16385, Rieske (2Fe 2S) protein; KC8_RS05200, MaoC dehydratase; KC8_RS05205, hypothetical protein; KC8_RS05210, acyl-CoA dehydrogenase; KC8_RS05215, ferredoxin oxidoreductase; KC8_RS05220, VOC family protein; KC8_RS05230, TetR transcriptional regulator; KC8_RS05235, cytochrome P450; KC8_RS05240, hypothetical protein; KC8_RS05245, lipid-transfer protein; KC8_RS05250, MaoC dehydratase; KC8_RS05255 and KC8_RS05260, enoyl-CoA hydratases; KC8_RS05265, acetyl-CoA acetyltransferase; KC8_RS05270, 3-hydroxy-3-mthylglutaryl-CoA synthase; KC8_RS05275, Short-chain oxidoreductase; KC8_RS05280, acyl-CoA dehydrogenase; KC8_RS00980, CoA acyltransferase; KC8_RS00985, steroid Δ-isomerase; KC8_RS00990, MaoC dehydratase; KC8_RS00995, short-chain dehydrogenase/reductase; KC8_RS01000, acyl-CoA dehydrogenase; KC8_RS01005, short-chain oxidoreductase; KC8_RS01010, phenylacetic acid degradation protein PaaY; KC8_RS01015, acetyl-CoA acetyltransferase; KC8_RS01020, acyl-CoA dehydrogenase; KC8_RS01025, acyl-CoA dehydrogenase; KC8_RS01030, enoyl-CoA hydratase; KC8_RS01035, monooxygenase; KC8_RS01040, lipid-transfer protein (Ltp); KC8_RS01045, thiolase; KC8_RS01050, CoA transferase, β-subunit; KC8_RS01055, CoA transferase, α-subunit; KC8_RS01060, *meta*-dioxygenase. Genes encoding key enzymes for A ring degradation are depicted in red; genes coding β-oxidation related enzymes putatively catalyzing degradation of A/B ringare indicated in green; genes for C/D-ring degradation are depicted in purple.

In summary, *oecA* and genes from clusters I and II are likely to be involved in estrogen A/B-ring degradation to HIP in *Sphingomonas* KC8. In the genome of this strain, a third cluster (cluster III) related to steroid degradation has also been identified (Figure 8). It contains genes similar to those proposed in *C. testosteroni* for C/D-ring degradation. Thus, orthologs of *echA20*, *ipdB*, and *ipdA* from *M. tuberculosis* have been found (Figure 8), and they are expressed during the aerobic growth of

Sphingomonas KC8 on testosterone and 17β-estradiol, being involved in the transformation of HIP into general catabolites [198,202]. Similar clusters have been identified in other estrogen-degrading aerobes, such as *Altererythrobacter estronivorus* MH-B5 [203] and *Novosphingobium tardaugens* NBRC 16725 [204].

It has been proposed that the 4,5-*seco* pathway is highly prevalent, since: (i) the dead-end product pyridinestrone is found in wastewater treatment plants exposed to estrogens, and (ii) the gene *oecC* has been identified in bacteria isolated from these environments [198,205]. Alternative estrogen degradation pathways and metabolites have been proposed (Figure 9), although there is no biochemical or genetic evidence identifying their significance. Thus, in *Sphingomonas* sp. strain ED8, a pathway has been proposed, involving hydroxylation at different positions of the saturated ring of 17β-estradiol. This is supported by the detection of hydroxyl-17β-estradiol, keto-17β-estradiol, keto-estrone and 3-(4-hydroxyphenyl)-2-hydroxy-prop-2-enoate in the culture broth of this bacterium when 7β-estradiol was used as the sole carbon source (Figure 9). The appearance of this last compound suggested that catabolism of 17β-estradiol through this pathway takes place by opening the rings B, C, or D [206]. Another metabolic alternative for 17β-estradiol mineralisation has been proposed in *Nitrosomonas europaea*, based on the appearance of estra-1,3,5(10),16-tetraen-3-ol (estratetraenol) after dehydratation of ring D at C17 position (Figure 9). It was suggested that this strain could further metabolise this intermediate to non-estrogenic compounds [207]. The finding, in activated sludge, of a new intermediate from 17β-estradiol derived from estrone and containing a lactone in ring D, opens the possibility of a new estrogen catabolic pathway used for the mineralisation of this kind of steroid (Figure 9) [208].

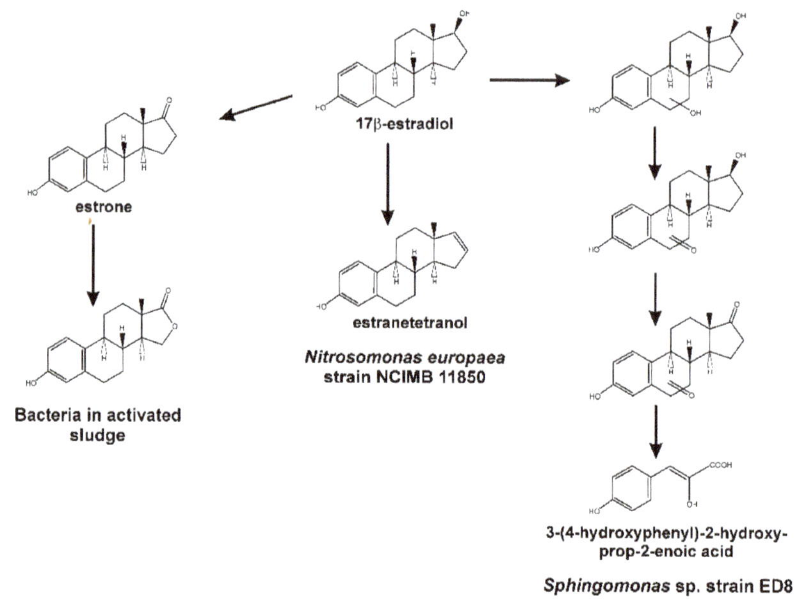

Figure 9. Metabolic alternatives to the 4,5-*seco* pathway proposed for estrogen mineralisation/biotransformation in different bacterial species.

5. The Anaerobic 2,3-*seco* Pathway

Several denitrifying Proteobacteria that degrade steroids, cholesterol, and testosterone under anaerobic conditions have been characterised [63,65,209]. Among them, *Sterolibacterium denitrificans* (*Stl. denitrificans*) and *Steroidobacter denitrificans* (*Std. denitrificans*) have been used as model organisms for studying anaerobic steroid metabolism [63,65]. The pathway involved in anoxic steroid catabolism, although having some similarities with the 9,10-*seco* pathway, uses dioxygen-independent reactions to degrade the steroidal core. It is referred to as the 2,3-*seco* pathway [210–212]. Similar to the aerobic metabolism performed through the 9,10-*seco* pathway, the anaerobic degradation of cholesterol by *Stl. denitrificans* is initiated by the oxidation of ring A using AcmA, a dehydrogenase/isomerase belonging to the short-chain dehydrogenase/reductase superfamily, yielding 4-cholesten-3-one (Figure 10) [64,213].

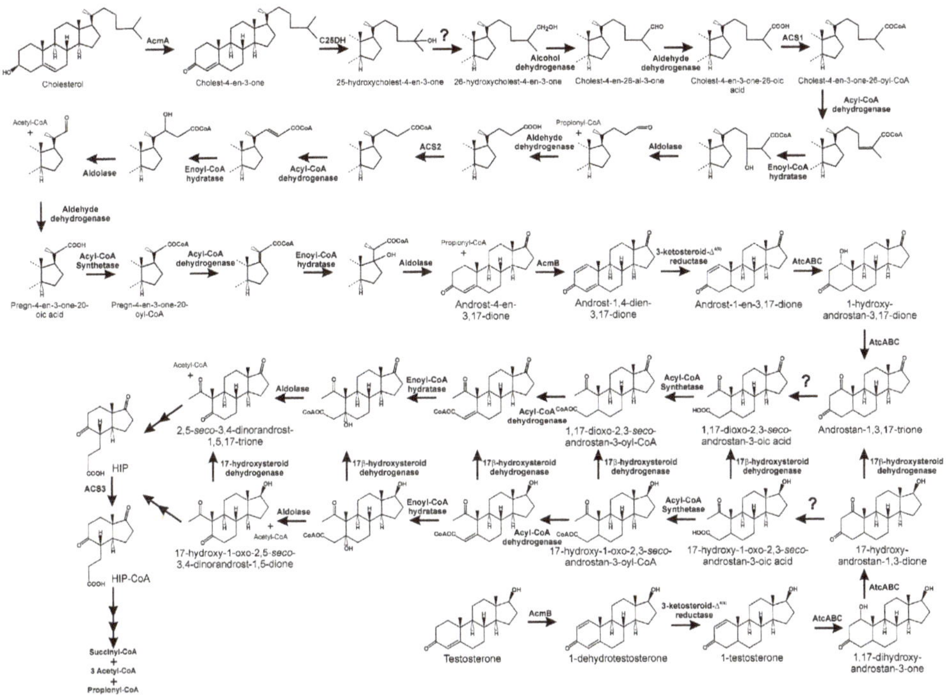

Figure 10. Anaerobic catabolism of cholesterol, in *Sterolibacterium denitrificans*, and testosterone, in *Steroidobacter denitrificans*, by the 2,3-*seco* pathway. Putative points of convergence between both metabolic mechanisms are also suggested.

In a second step, the anaerobic degradation of cholesterol C_{17} side chain starts with an oxygen-independent water-dependent hydroxylation at the tertiary C_{25} atom of the side chain of C_{27} esteroid substrates, resulting in the formation of a tertiary alcohol, 25-hydroxy-4-cholesten-3-one (Figure 10) [213]. This hydroxylation is carried out by a molybdopterin-containing enzyme, C25DH, belonging to the dimethyl sulfoxide dehydrogenase molybdoenzyme family. This enzyme consists of three different subunits, an α-subunit containing the molybdo-*bis*(pyranopterin guanine dinucleotide) cofactor in the active site and an iron–sulfur cluster [4Fe-4S], associated with electron-transfer machinery composed of β-subunits that contain three [4Fe-4S] and one [3Fe-4S] iron–sulfur clusters, and Y-subunits associated with one heme b. Notably, the genome of *Stl. denitrificans* Chol1-S contains seven paralogs that code for putative α-subunits of S25DH-like dehydrogenases, although there are few genes putatively encoding the β- and Y-subunits, indicating that molybdoenzymes share common βY-components, but have different α-subunits [214,215]. *Stl. denitrificans* is able to not only degrade cholesterol, but also β-sitosterol, stigmasterol, and ergosterol. Surprisingly, S25DH used in cholesterol catabolism is unable to hydroxylate any of the 4-en-3-one analogues of these sterols [214,216]. However, several S25DHs carrying different α-subunits have been functionally characterised, and are able to activate those sterols [217].

The next step in the anoxic degradation of cholesterol involves an unprecedented isomerisation of the hydroxyl group from the tertiary C_{25} to the primary C_{26}, catalysed by a yet unknown enzyme (Figure 10) [86,218]. Further degradation proceeds via oxidation of C_{26} primary alcohol to a carboxylate by the action of a putative cholesterol-induced alcohol dehydrogenase and aldehyde dehydrogenase (Figure 10) [216]. This carboxylic derivative is activated to a C_{26}-oyl-CoA thioesterifed molecule by a specific ATP-depending acyl-CoA synthetase, followed by a modified β-oxidation-like reaction sequence, with some steps similar to those of bile acid catabolism through the 9,10-*seco* pathway, yielding AD, two propionyl-CoAs, and one acetyl-CoA (Figure 10) [216,219]. Thus, in the genome of *Stl. denitrificans*, two gene clusters, *acd1* and *acd2*, contain the genes that, after induction by cholesterol, code for functions needed for cholesterol side chain elimination (Figure 11). Amazingly, although more of the intermediates of side chain degradation have been identified, no 3-oxoacyl-CoA intermediates have been found, which correlates with the lacking of genes specifically involved in the thiolytic β-oxidative processes (3-hydroxyacyl-CoA dehydrogenases and thiolases). Alternatively, the two aldolases and an aldehyde dehydrogenase in the *acd1* and *acd2* clusters indicate that the release of propionyl-CoA and acetyl-CoA molecules from degradation of the side chain occurs via aldolytic cleavage [216,219].

In vitro assays using cell-free extracts indicate the existence of an FAD-containing 4-cholesten-3-one-$\Delta^{1(2)}$-dehydrogenase (AcmB) in *Stl. denitrificans*, catalysing $\Delta^{1(2)}$-desaturation of 4-cholesten-3-one to 1,4-cholestadien-3-one (Figure 10) [220]. However, the topological location of AcmB on the cytoplasmic side of the inner membrane and 4-cholesten-3-one in the cellular periplasm puts this proposed reaction in doubt. Thus, AcmB may be a 3-ketosteroid-$\Delta^{1(2)}$-dehydrogenase, with AD, obtained after complete removal of the lateral side chain from 4-cholesten-3-one, being its physiological substrate, leading to the formation of ADD. This later compound is then specifically reduced to 1-androsten-3,17-dione, which acts as a substrate for AtcABC, a bifunctional molybdopterin-containing hydratase/dehydrogenase, which introduces a water molecule to the double bond between C_1 and C_2, leading to the formation of a hydroxyl group at C_1. Later, this hydroxyl group is oxidised to an oxo-group, releasing androstan-1,3,17-trione (Figure 10) [212]. The A ring of this compound is then cleaved, resulting in 1,17-dioxo-2,3-*seco*-androstan-3-oic acid, which would be activated by a yet unknown acyl-CoA synthetase (Figure 10) [86,212,218]. It has been suggested that acetyl-CoA could be released from 1,17-dioxo-2,3-*seco*-androstan-3-oyl-CoA through an aldolytic cleavage, to produce 2,5-*seco*-3,4-dinorandrost-1,5,17-trione [213]. Although the mechanism for cleavage of the B-ring remains unknown, HIP, the predicted product generated from 2,5-*seco*-3,4-dinorandrost-1,5,17-trione, has been identified (Figure 10) [216].

In *Stl. denitrificans*, genes involved in ring A cleavage have been found in the cluster *acd3* (Figure 11), although no candidates encoding a B-ring cleaving hydrolase or its degradation to HIP have been identified. Clusters *acd4* and *acd5* contain all the genes required for the integration of HIP into the central metabolism (Figure 11) [216].

The anaerobic degradation of the androgen testosterone has mainly been studied in model microorganism *Std. denitrificans* DSMZ 18526 [210,211,221–223]. The pathway used by this denitrifying strain generates analogous intermediates to those of the 2,3-*seco* pathway involved in cholesterol catabolism in *Stl. denitrificans*, and common steps in these two pathways are catalysed by orthologous enzymes (Figure 10) [212].

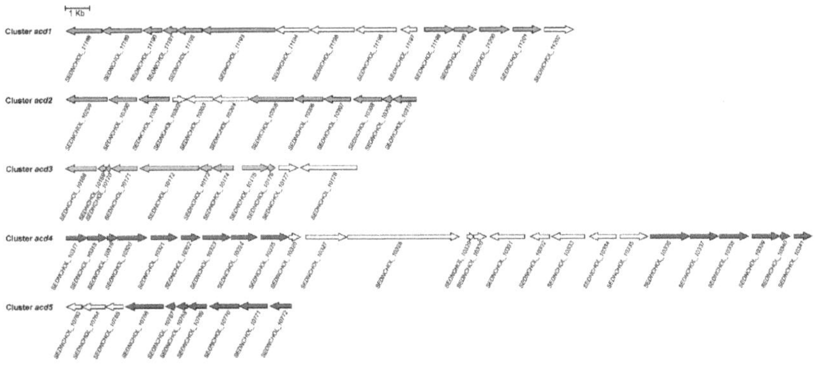

Figure 11. Genetic organisation of the genes encoding 2,3-*seco* pathway functions for cholesterol degradation in *Sterolibacterium denitrificans*. Annotation in the genome (accession LT837803): SEDNCHOL_11188, aldehyde dehydrogenase; SEDNCHOL_11189, acyl-CoA synthetase 1 (ACS1); SEDNCHOL_11190, short-chain dehydrogenase; SEDNCHOL_11191, SEDNCHOL_11192, and SEDNCHOL_11193 steroid C25 dehydrogenase, γ-, β- and α- subunit, respectively; SEDNCHOL_11194 and SEDNCHOL_11195, proteins with unknown functions; SEDNCHOL_11196, AcmB; SEDNCHOL_11197, putative transcriptional regulatory protein; SEDNCHOL_11198, putative aldolase; SEDNCHOL_11199, enoyl-CoA hydratase; SEDNCHOL_11200 and SEDNCHOL_11201, acyl-CoA dehydrogenases; SEDNCHOL_11202, oxidoreductase; SEDNCHOL_10299, acyl-CoA synthetase 2 (ACS2); SEDNCHOL_10300 and SEDNCHOL_10301, acyl-CoA dehydrogenases; SEDNCHOL_10302, protein of unknown function; SEDNCHOL_10303, putative metallo-β-lactamase; SEDNCHOL_10304, phytoene dehydrogenase-like protein; SEDNCHOL_10305, putative C_{22} acyl-CoA synthetase; SEDNCHOL_10306 and SEDNCHOL_10307, acyl-CoA dehydrogenases; SEDNCHOL_10308, aldolase; SEDNCHOL_10309 and SEDNCHOL_10310, enoyl-CoA hydratases; SEDNCHOL_10168, CoA transferase; SEDNCHOL_10169, plasmid stabilization system; SEDNCHOL_10170 and SEDNCHOL_10171, proteins of unknown function; SEDNCHOL_10172, SEDNCHOL_10173, and SEDNCHOL_10174, A, B, and C subunits of AtcABC, respectively; SEDNCHOL_10175 and SEDNCHOL_10176, proteins of unknown function; SEDNCHOL_10177; putative electron-transfer flavoprotein, β-subunit; SEDNCHOL_10178, protein of unknown function; SEDNCHOL_10317, SEDNCHOL_10318, and SEDNCHOL_10321, IpdABC-like proteins; SEDNCHOL_10319, probably subunit of benzoylsuccinyl-CoA thiolase; SEDNCHOL_10320, Propanoyl-CoA C-acyltransferase; SEDNCHOL_10322, enoyl-CoA hydratase; SEDNCHOL_10323 and SEDNCHOL_10324, acyl-CoA dehydrogenases; SEDNCHOL_10325, thiolase; SEDNCHOL_10326, MarR family transcriptional regulator; SEDNCHOL_10327, outer membrane protein; SEDNCHOL_10328, putative Filamentous hemagglutinin family N-terminal domain containing protein; SEDNCHOL_10329, protein of unknown function; SEDNCHOL_10330, N-acetyltransferase; SEDNCHOL_10331, RseB; SEDNCHOL_10332, enoyl-CoA hydratase; SEDNCHOL_10333 and SEDNCHOL_10334, proteins of unknown function; SEDNCHOL_10335, CoA transferase; SEDNCHOL_10336, putative IpdC; SEDNCHOL_10337 and SEDNCHOL_10338, CoA transferases; SEDNCHOL_10339, acetyl-CoA acetyltransferase; SEDNCHOL_10340, protein of unknown function; SEDNCHOL_10341, enoyl-CoA hydratase; SEDNCHOL_10763, putative 2-phospho-L-lactate guanylyltransferase; SEDNCHOL_10764, 2-phospho-L-lactate transferase; SEDNCHOL_10765, Coenzyme F420:L-glutamate ligase; SEDNCHOL_10766, acyl-CoA synthetase 3 (ACS3); SEDNCHOL_10767, steroid Δ–isomerase; SEDNCHOL_10768, enoyl-CoA hydratase; SEDNCHOL_10769, short chain dehydrogenase; SEDNCHOL_10770 and SEDNCHOL_10771, acyl-CoA dehydrogenases; SEDNCHOL_10772, short chain alcohol dehydrogenase. Genes encoding enzymes involved in cholesterol side chain degradation are shown in green; genes participating in A ring degradation are shown in orange; and in purple is shown genes coding for ring C/D degradation.

Thus, testosterone is initially transformed into 1-dehydrotestosterone by a 3-ketosteroid $\Delta^{1(2)}$-dehydrogenase/reductase (Figure 10). This compound is then transformed to 1-testosterone in a process catalysed by a 3-ketosteroid Δ^4-dehydrogenase/reductase (Figure 10). AtcABC, the analogous bifunctional molybdoenzyme, in this case a 1-testosterone hydratase/dehydrogenase, catalyses the C1-C2 hydration reaction and the subsequent oxidation, leading to the formation of 17-hydroxy-andro stan-1,3-one (Figure 10) [211,212]. It has been proposed that the process continues through the hydrolytic cleavage of this last compound, giving rise to 17-hydroxy-1-oxo-2,3-*seco*-androstan-3-oic acid (Figure 10). However, the hydrolase that would catalyse this process remains unidentified. Taking into account that 17-hydroxy-2,5-*seco*-3,4-dinorandrost-1,5-dione has been identified in cultures of *Std. denitrificans* growing in media containing testosterone as a carbon source, the metabolic steps for the transformation of 17-hydroxy-1-oxo-2,3-*seco*-androstan-3-oic acid into this compound have been proposed. Thus, the acid is first activated to a coenzyme A thioester, and after the introduction of a double bond and its hydration, a retroaldolic reaction, a molecule of acetyl-CoA is released, producing 17-hydroxy-2,5-*seco*-3,4-dinorandrost-1,5-dione (Figure 10). Notably, the use of acyl-CoA dehydrogenase inhibitors impairs the biotransformation of 17-hydroxy-1-oxo-2,3-*seco*-androstan-3-oic acid to 17-hydroxy-2,5-*seco*-3,4-dinorandrost-1,5-dione, reinforcing the participation of a β-oxidation-like mechanism in the processing of the A-ring cleavage product [211]. It has been proposed that, as occurs in cholesterol degradation, this compound will be converted to HIP.

It has to be noted that proposed intermediates in testosterone anaerobic catabolism conserve a hydroxyl group at C_{17} until their convergence at the HIP level. However, this only reflects the fact that these compounds are more prominent in the culture broth than their 17-keto structures, which appear at lower levels. This has resulted in a proposal that a 17β-hydroxysteroid dehydrogenase, or a group of enzymes with this activity, could interconvert the different 17β-hydroxyl compounds and their 17-keto intermediates [211,223]. Undoubtedly, if the catabolism of steroids through this pathway occurs by the formation of HIP at some stage, the keto present in the D-ring should be evident.

Taking into account that the aerobic and anaerobic degradation of steroids converge at HIP, it might be expected that the same metabolic mechanisms are used for its subsequent degradation. This hypothesis is reinforced by the conservation of genes in the genomes of microbial steroid degraders [216,219].

6. Biotechnological Interest in Steroid-Degrading Microorganisms

In addition to the relevance that steroid-degrading microorganisms have for the environment and maintaining the carbon cycle, their potential use in the pharmaceutical industry has to be highlighted. Classical pharmaceutical production processes for steroids have been carried out by extraction from plant or animal sources, by full organic synthesis, or by a combination of chemical and enzymatic synthesis. With exceptions, extraction from animals or plants is an inefficient large-scale production method. Full chemical and chemical-enzymatic syntheses are often multistage and very expensive in time, labor, and energy, as well as posing a potential risk for the environment.

As an alternative, the bioconversion of steroids from low-cost precursors (i.e., phytosterols) with genetically tailored microorganisms is becoming the method of choice for industrial production. The development of these biotechnological approaches, which are more cost-effective and environmentally friendly, have been stimulated due to the growing demand for steroid pharmaceuticals. This approach is classically developed with heterologous gene expressing transgenic bacteria, yeast, and fungi, or with specific strains empirically showing a specific reaction. Steroid degrading strains, or specific mutants tailored from them, has given rise to a new source of syntons (mainly AD, ADD, or 9-hydroxy-derivatives) or genes/enzymes useful for steroid transformation.

As far as we know, today there have been no studies focused in the development of biodegradative strategies for environmental steroid amendment. Despite the great interest that the elimination of steroids can have due to the risk that the release of these compounds to the environment has for flora and wild animal reproduction, and the potential effect on human beings, the biotechnological approaches to these applications have not been developed yet.

Although our knowledge about the metabolic mechanisms for catabolism of sterols, bile acids, testosterone, or 17β-estradiol has increased in recent years, some black holes need to be clarified. The transport of steroids across cell walls and membranes in some of the bacterial groups able to metabolize steroids, rate-limiting steps in the pathways, or global regulation and integration of these pathways at intermediate metabolism should be more deeply elucidated. Moreover, there are many steroid compounds released to the environment for which it is not yet known how they could be integrated into carbon cycles.

Study of metagenomics communities will allow the identification of novel steroid biodegraders. The isolation and characterization of individual microorganisms with the ability for steroid degradation will increase the pool of genes, enzymes, and metabolic strategies for environmental steroid amendment. It is to be expected that this set of knowledge, together with technologies developed in other disciplines, will allow the design of new bacteria, or communities of them, suitable for use in the biodegradation of these molecules. Thus, for instance, system biology will allow the design of interacting communities of microorganisms for the biodegradation of different steroids or the integration of new metabolic circuits inside a bacteria; metabolomics will permit the improvement of metabolic fluxes, ensuring a high efficiency in the catabolism of steroids; and synthetic biology would allow the design and expression of improved or new activities based on mammalian, fungi, or gut microbiota steroid metabolism. In sum, the open future for a biodegradative approach to remove environmentally released steroids is expected to be a promising reality.

Author Contributions: Both authors contributed similarly to the organization and writing of the manuscript.

Funding: Our research on the biodegradation of steroid compounds by bacteria has been supported by the Ministerio de Economía y Competitividad (Madrid, España), grants BIO2012-39695-C02-02, RTC2014-2249-1 and BIO2015-66960-C3-3-R, and by grants from the Junta de Castilla y León (Consejería de Educación, Valladolid, España; LE246A11-2 and LE029P17).

Conflicts of Interest: The authors declare no conflict of interest. Those funding our research had no role in the design of the study, in the collection, analyses, or interpretation of data, or in the writing of the manuscript.

References

1. Piironen, V.; Lindsay, D.G.; Miettinen, T.A.; Toivo, J.; Lampi, A.M. Plant sterols: Biosynthesis, biological function and their importance to human nutrition. *J. Sci. Food Agric.* **2000**, *80*, 939–966. [CrossRef]
2. Fernandes, P.; Cabral, J.M.S. Phytosterols: Applications and recovery methods. *Bioresour. Technol.* **2007**, *98*, 2335–2350. [CrossRef] [PubMed]
3. Wollam, J.; Antebi, A. Sterol regulation of metabolism, homeostasis, and development. *Annu. Rev. Biochem.* **2011**, *80*, 885–916. [CrossRef] [PubMed]
4. Kodner, R.B.; Pearson, A.; Summons, R.E.; Knoll, A.H. Sterols in the red and green algae: Quantification, phylogeny, and relevance for the interpretation of geologic steranes. *Geobiology* **2008**, *6*, 411–420. [CrossRef] [PubMed]
5. Bird, C.W.; Lynch, J.M.; Pirt, F.J.; Reid, W.W. Steroids and squalene in *Methylococcus capsulatus* grown on methane. *Nature* **1971**, *230*, 473–474. [CrossRef] [PubMed]
6. Patt, T.E.; Hanson, R.S. Intracytoplasmic membrane, phospholipid, and sterol content of *Methylobacterium organophilum* cells grown under different conditions. *J. Bacteriol.* **1978**, *134*, 636–644.
7. Kohl, W.; Gloe, A.; Reichenbach, H. Steroids from the myxobacterium *Nannocystis exedens*. *J. Gen. Microbiol.* **1983**, *129*, 1629–1635. [CrossRef]
8. Schouten, S.; Bowman, J.P.; Rijpstra, W.I.C.; Damsté, J.S.S. Sterols in a psychrophilic methanotroph, *Methylosphaera hansonii*. *FEMS Microbiol. Lett.* **2000**, *186*, 193–195. [CrossRef]
9. Bode, H.B.; Zeggel, B.; Silakowski, B.; Wenzel, S.C.; Reichenbach, H.; Müller, R. Steroid biosynthesis in prokaryotes: Identification of myxobacterial steroids and cloning of the first bacterial 2,3(S)-oxidosqualene cyclase from the mixobacterium *Stigmatella aurantiaca*. *Mol. Microbiol.* **2003**, *47*, 471–481. [CrossRef]

10. Pearson, A.; Budin, M.; Brocks, J.J. Phylogenetic and biochemical evidence for sterol synthesis in the bacterium *Gemmata obscuriglobus*. *Proc. Natl. Acad. Sci. USA* **2003**, *100*, 15352–15357. [CrossRef]
11. Lamb, D.C.; Jackson, C.J.; Warrilow, A.G.; Manning, N.J.; Kelly, D.E.; Kelly, S.L. Lanosterol biosynthesis in the prokaryote *Methylococcus capsulatus*. Insight into the evolution of sterol biosynthesis. *Mol. Biol. Evol.* **2007**, *24*, 1714–1721. [CrossRef] [PubMed]
12. Gawas, D.; García, R.; Huch, V.; Müller, R. A highly conjugated dihydroxylated C28 steroid from a myxobacterium. *J. Nat. Prod.* **2011**, *74*, 1281–1283. [CrossRef] [PubMed]
13. García, R.; Gemperlein, K.; Müller, R. *Minicystis rosea* gen. Nov., sp. Nov., a polyunsaturated fatty acid-rich and steroid-producing soil myxobacterium. *Int. J. Syst. Evol. Microbiol.* **2014**, *64*, 3733–3742. [CrossRef] [PubMed]
14. Banta, A.B.; Wei, J.H.; Welander, P.V. A distinc pathway for tetrahymanol synthesis in bacteria. *Proc. Natl. Acad. Sci. USA* **2015**, *112*, 13478–13483. [CrossRef] [PubMed]
15. Banta, A.B.; Wei, J.H.; Gill, C.C.C.; Giner, J.-L.; Welander, P.V. Synthesis of arborane triterpenols by a bacterial oxidosqualene cyclase. *Proc. Natl. Acad. Sci. USA* **2017**, *114*, 245–250. [CrossRef] [PubMed]
16. Lange, I.G.; Daxenberger, A.; Schiffer, B.; Witters, H.; Ibarreta, D.; Meyer, H.D.D. Sex hormones originating from different livestock production systems: Fate and potential disrupting activity in the environment. *Anal. Chim. Acta* **2002**, *473*, 27–37. [CrossRef]
17. Ridlon, J.M.; Kang, D.-J.; Hylemon, P.B. Bile salt biotransformations by human intestinal bacteria. *J. Lipid Res.* **2006**, *47*, 241–259. [CrossRef] [PubMed]
18. Froehner, S.; Martins, R.F.; Errera, M.R. Assessment of fecal sterols in Barigui River sediments in Curitiba, Brazil. *Environ. Monit. Assess.* **2009**, *157*, 591–600. [CrossRef]
19. Chang, H.S.; Choo, K.H.; Lee, B.; Choi, S.J. The methods for identification, analysis, and removal of endocrine disrupting compounds (EDCs) in water. *J. Hazard. Mater.* **2009**, *172*, 1–12. [CrossRef]
20. Santos, L.H.; Araújo, A.N.; Fachini, A.; Pena, A.; Delerue-Matos, C.; Montenegro, M.C. Ecotoxicological aspects related to the presence of pharmaceuticals in the aquatic environment. *J. Hazard. Mater.* **2010**, *175*, 45–95. [CrossRef]
21. Prost, K.; Birk, J.J.; Lehndorff, E.; Gerlach, R.; Amelung, W. Steroid biomarkers revisited-improved source identification of faecal remains in archaeological soil material. *PLoS ONE* **2017**, *12*, e0164882. [CrossRef] [PubMed]
22. Matić Bujagić, I.; Grujić, S.; Laušević, M.; Hofmann, T.; Micić, V. Emerging contaminants in sediment core from the Iron Gate I Reservoir on the Danube River. *Sci. Total Environ.* **2019**, *662*, 77–87. [CrossRef] [PubMed]
23. Gagné, F.; Blaise, C.; André, C. Occurrence of pharmaceutical products in a municipal effluent and toxicity to rainbow trout (*Oncorhynchus mykiss*) hepatocytes. *Ecotoxicol. Environ. Saf.* **2006**, *64*, 329–336. [CrossRef] [PubMed]
24. Crane, M.; Watts, C.; Boucard, T. Chronic aquatic environmental risks from exposure to human pharmaceuticals. *Sci. Total Environ.* **2006**, *367*, 23–41. [CrossRef] [PubMed]
25. Galli, R.; Braun, C. Integrative risk assessment of endocrine disruptors in Switzerland. *Chimia* **2008**, *62*, 417–423. [CrossRef]
26. Tong, W.-Y.; Dong, X. Microbial biotransformation: Recent developments on steroid drugs. *Recent Pat. Biotechnol.* **2009**, *3*, 141–153. [CrossRef] [PubMed]
27. Runnalls, T.J.; Margiotta-Casaluci, L.; Kugathas, S.; Sumpter, J.P. Pharmaceuticals in the aquatic environment: Steroids and anti-steroids as high priorities for research. *Hum. Ecol. Risk Assess.* **2010**, *16*, 1318–1338. [CrossRef]
28. Philipp, B.; Erdbrink, H.; Suter, M.J.; Schink, B. Degradation of and sensitivity to cholate in *Pseudomonas* sp. strain Chol1. *Arch. Microbiol.* **2006**, *185*, 192–201. [CrossRef]
29. Merino, E.; Barrientos, A.; Rodríguez, J.; Naharro, G.; Luengo, J.M.; Olivera, E.R. Isolation of cholesterol- and deoxycholate-degrading bacteria from soil samples: Evidence of a common pathway. *Appl. Microbiol. Biotechnol.* **2013**, *97*, 891–904. [CrossRef]
30. Liu, L.; Zhu, W.; Cao, Z.; Xu, B.; Wang, G.; Luo, M. High correlation between genotypes and phenotypes of environmental bacteria *Comamonas testosteroni* strains. *BMC Genom.* **2015**, *16*, 110. [CrossRef]
31. Holert, J.; Yücel, O.; Suvekbala, V.; Kulić, Z.; Möller, H.; Philipp, B. Evidence of distinct pathways for bacterial degradation of the steroid compound cholate suggests the potential for metabolic interactions by interspecies cross-feeding. *Environ. Microbiol.* **2014**, *16*, 1424–1440. [CrossRef] [PubMed]

32. Zhang, T.; Xiong, G.; Maser, E. Characterization of the steroid degrading bacterium S19-1 from the Baltic Sea at Kiel, Germany. *Chem. Biol. Interact.* **2011**, *191*, 83–88. [CrossRef] [PubMed]
33. Fernández, L.; Louvado, A.; Esteves, V.I.; Gomes, N.C.M.; Almeida, A.; Cunha, Â. Biodegradation of 17β-estradiol by bacteria isolated from deep sea sediments in aerobic and anaerobic media. *J. Hazard. Mater.* **2017**, *323*, 359–366. [CrossRef] [PubMed]
34. Holert, J.; Cardenas, E.; Bergstrand, L.H.; Zaikova, E.; Hahn, A.S.; Hallam, S.J.; Mohn, W.W. Metagenomes reveal global distribution of bacterial steroid catabolism in natural, engineered, and host environments. *mBio* **2018**, *9*, e02345-17. [CrossRef] [PubMed]
35. Pandey, A.K.; Sassetti, C.M. Mycobacterial persistence requires the utilization of host cholesterol. *Proc. Natl. Acad. Sci. USA* **2008**, *105*, 4376–4380. [CrossRef] [PubMed]
36. van der Geize, R.; Grommen, A.W.; Hessels, G.I.; Jacobs, A.A.; Dijkhuizen, L. The steroid catabolic pathway of the intracellular pathogen *Rhodococcus equi* is important for pathogenesis and a target for vaccine development. *PLoS Pathog.* **2011**, *7*, e1002181. [CrossRef]
37. Söhngen, N.L. Benzin, petroleum, paraffinöl und paraffin als kohlenstoff- und energiequelle für mikroben. *Zent. Bakteriol. Parasitenkd. Infekt.* **1913**, *27*, 595–609.
38. Tak, J.D. On bacteria decomposing cholesterol. *Antonie Leeuwenhoek* **1942**, *8*, 32–40. [CrossRef]
39. Arnaudi, C. *Flavobacterium dehydrogenans* (*Micrococcus dehydrogenans*) und seine Fähigkeit zur oxidation von steroiden sowie substanzen aus der sexualhormonreibe. *Zentr. Bakt. Parasitenk. II* **1942**, *105*, 352–366.
40. Turfitt, G.E. Microbiological agencies in the degradation of steroids: I. The cholesterol-decomposing organisms of soils. *J. Bacteriol.* **1944**, *47*, 487–493.
41. Turfitt, G.E. The microbiological degradation of steroids 4. Fission of the steroid molecule. *Biochem. J.* **1948**, *42*, 376–383. [CrossRef] [PubMed]
42. Horvath, J.; Kramli, A. Microbiological oxidation of cholesterol with *Azotobacter*. *Nature* **1947**, *160*, 639. [CrossRef] [PubMed]
43. Whitmarsh, J.M. Intermediates of microbiological metabolism of cholesterol. *Biochem. J.* **1964**, *90*, 23–24.
44. Brown, R.L.; Peterson, G.E. Cholesterol oxidation by soil Actinomycetes. *J. Gen. Microbiol.* **1966**, *45*, 441–450. [CrossRef]
45. Arima, K.; Nagasawa, M.; Bae, M.; Tamura, G. Microbial transformation of sterols. Part I. Decomposition of cholesterol by microorganisms. *Agric. Biol. Chem.* **1969**, *33*, 1636–1643.
46. Chipley, J.R.; Dreyfuss, M.S.; Smucker, R.A. Cholesterol metabolism by *Mycobacterium*. *Microbios* **1975**, *12*, 199–207. [PubMed]
47. Martin, C.K.A. Microbial transformation of β-sitosterol by *Nocardia* sp. M29. *Eur. J. Appl. Microbiol.* **1976**, *2*, 243–255. [CrossRef]
48. Ferreira, N.P.; Tracey, R.P. Numerical taxonomy of cholesterol-degrading soil bacteria. *J. Appl. Microbiol.* **1984**, *57*, 429–446. [CrossRef]
49. Mahato, S.B.; Banerjee, S.; Sahu, N.P. Metabolism of progesterone and testosterone by a *Bacillus* sp. *Steroids* **1984**, *43*, 545–558. [CrossRef]
50. Mahato, S.B.; Banerjee, S. Metabolism of 11-deoxycortisol by a *Bacillus* species. *J. Steroid Biochem.* **1986**, *25*, 995–999. [CrossRef]
51. Yoshimoto, T.; Nagai, F.; Fujimoto, J.; Watanabe, K.; Mizukoshi, H.; Makino, T.; Kimura, K.; Saino, H.; Sawada, H.; Omura, H. Degradation of strogens by *Rhodococcus zopfii* and *Rhodococcus equi* isolates from activated sludge in wastewater treatment plants. *Appl. Environ. Microbiol.* **2004**, *70*, 5283–5289. [CrossRef] [PubMed]
52. Fahrbach, M.; Kuever, J.; Meinke, R.; Kämpfer, P.; Hollender, J. *Denitratisoma oestradiolicum* gen. nov., sp. nov., a 17beta-oestradiol-degrading, denitrifying betaproteobacterium. *Int. J. Syst. Evol. Microbiol.* **2006**, *56*, 1547–1552. [CrossRef] [PubMed]
53. Drzyzga, O.; Navarro Llorens, J.M.; Fernández de Las Heras, L.; García Fernández, E.; Perera, J. *Gordonia cholesterolivorans* sp. nov., a cholesterol-degrading actinomycete isolated from sewage sludge. *Int. J. Syst. Evol. Microbiol.* **2009**, *59*, 1011–1015. [CrossRef] [PubMed]
54. Fernández de las Heras, L.; García Fernández, E.; Navarro Llorens, J.M.; Perera, J.; Drzyzga, O. Morphological, physiological, and molecular characterization of a newly isolated steroid-degrading Actinomycete, identified as *Rhodococcus ruber* strain Chol-4. *Curr. Microbiol.* **2009**, *59*, 548–553. [CrossRef] [PubMed]

55. Van der Geize, R.; Yam, K.; Heuser, T.; Wilbrink, M.H.; Hara, H.; Anderton, M.C.; Sim, E.; Dijkhuizen, L.; Davies, J.E.; Mohn, W.W.; et al. A gene cluster encoding cholesterol catabolism in a soil actinomycete provides insight into *Mycobacterium tuberculosis* survival in macrophages. *Proc. Natl. Acad. Sci. USA* **2007**, *104*, 1947–1952. [CrossRef] [PubMed]
56. Uhía, I.; Galán, B.; Morales, V.; García, J.L. Initial step in the catabolism of cholesterol by *Mycobacterium smegmatis* mc^2 155. *Environ. Microbiol.* **2011**, *13*, 943–959. [CrossRef]
57. Mohn, W.W.; Wilbrink, M.H.; Casabon, I.; Stewart, G.R.; Liu, J.; van der Geize, R.; Eltis, L.D. A gene cluster encoding cholate catabolism in *Rhodococcus* spp. *J. Bacteriol.* **2012**, *194*, 6712–6719. [CrossRef]
58. Fernández de las Heras, L.; Perera, J.; Navarro Llorens, J.M. Cholesterol to cholestenone oxidation by ChoG, the main extracellular oxidase of *Rhodococcus ruber* strain Chol-4. *J. Steroid Biochem. Mol. Biol.* **2013**, *139*, 33–44. [CrossRef]
59. Liu, Y.; Ge, F.; Chen, G.; Li, W.; Ma, P.; Zhang, G.; Zeng, L. *Gordonia neofelifaecis* sp. nov., a cholesterol side-chain-cleaving actinomycete isolated from the faeces of *Neofelis nebulosa*. *Int. J. Syst. Evol. Microbiol.* **2011**, *61*, 165–169. [CrossRef]
60. Horinouchi, M.; Hayashi, T.; Kudo, T. Steroid degradation in *Comamonas testosteroni*. *J. Steroid Biochem. Mol. Biol.* **2012**, *129*, 4–14. [CrossRef]
61. Holert, J.; Alam, I.; Larsen, M.; Antunes, A.; Bajic, V.B.; Stingl, U.; Philipp, B. Genome sequence of *Pseudomonas* sp. strain Chol1, a model organism for the degradation of bile acids and other steroid compounds. *Genome Announc.* **2013**, *1*, e00014-12. [CrossRef] [PubMed]
62. Barrientos, A.; Merino, E.; Casabon, I.; Rodríguez, J.; Crowe, A.M.; Holert, J.; Philipp, B.; Eltis, L.D.; Olivera, E.R.; Luengo, J.M. Functional analyses of three acyl-CoA synthetases involved in bile acid degradation in *Pseudomonas putida* DOC21. *Environ. Microbiol.* **2015**, *17*, 47–63. [CrossRef] [PubMed]
63. Tarlera, S.; Denner, E.B.M. *Sterolibacterium denitrificans* gen. nov., sp. nov., a novel cholesterol-oxidizing, denitrifying member of the β-*Proteobacteria*. *Int. J. Syst. Evol. Microbiol.* **2003**, *53*, 1085–1091. [PubMed]
64. Chiang, Y.R.; Ismail, W.; Heintz, D.; Schaeffer, C.; Van Dorsselaer, A.; Fuchs, G. Study of anoxic and oxic cholesterol metabolism by *Sterolibacterium denitrificans*. *J. Bacteriol.* **2008**, *190*, 905–914. [CrossRef] [PubMed]
65. Fahrbach, M.; Kuever, J.; Remesch, M.; Huber, B.E.; Kämpfer, P.; Dott, W.; Hollender, J. *Steroidobacter denitrificans* gen. nov., sp. nov., a steroidal hormone-degrading gammaproteobacterium. *Int. J. Syst. Evol. Microbiol.* **2008**, *58*, 2215–2223. [CrossRef] [PubMed]
66. Plésiat, P.; Nikaido, H. Outer membranes of gram-negative bacteria are permeable to steroid probes. *Mol. Microbiol.* **1992**, *6*, 1323–1333. [CrossRef] [PubMed]
67. Mohn, W.W.; van der Geize, R.; Stewart, G.R.; Okamoto, S.; Liu, J.; Dijkhuizen, L.; Eltis, L.D. The actinobacterial *mce4* locus encodes a steroid transporter. *J. Biol. Chem.* **2008**, *283*, 35368–35374. [CrossRef] [PubMed]
68. Van der Geize, R.; de Jong, W.; Hessels, G.I.; Grommen, A.W.F.; Jacobs, A.A.C.; Dijkhuizen, L. A novel method to generate unmarked gene deletions in the intracellular pathogen *Rhodococcus equi* using 5-fluorocytosine conditional lethality. *Nucleic Acids Res.* **2008**, *36*, e151. [CrossRef]
69. Drzyzga, O.; Fernández de las Heras, L.; Morales, V.; Navarro Llorens, J.M.; Perera, J. Cholesterol degradation by *Gordonia cholesterolivorans*. *Appl. Environ. Microbiol.* **2011**, *77*, 4802–4810. [CrossRef]
70. Klepp, L.I.; Forrellad, M.A.; Osella, A.V.; Blanco, F.C.; Stella, E.J.; Bianco, M.V.; de la Paz Santangelo, M.; Sassetti, C.; Jackson, M.; Cataldi, A.A.; et al. Impact of the deletion of the six *mce* operons in *Mycobacterium smegmatis*. *Microbes Infect.* **2012**, *14*, 590–599. [CrossRef]
71. Cole, S.T.; Brosch, R.; Parkhill, J.; Garnier, T.; Churcher, C.; Harris, D.; Gordon, S.V.; Eiglmeier, K.; Gas, S.; Barry, C.E., 3rd; et al. Deciphering the biology of *Mycobacterium tuberculosis* from the complete genome sequence. *Nature* **1998**, *393*, 537–544. [CrossRef] [PubMed]
72. García-Fernández, J.; Papavinasasundaram, K.; Galán, B.; Sassetti, C.M.; García, J.L. Molecular and functional analysis of the *mce4* operon in *Mycobacterium smegmatis*. *Environ. Microbiol.* **2017**, *19*, 3689–3699. [CrossRef] [PubMed]
73. Kendall, S.L.; Withers, M.; Soffair, C.N.; Moreland, N.J.; Gurcha, S.; Sidders, B.; Frita, R.; ten Bokum, A.; Besra, G.S.; Lott, J.S.; et al. A highly conserved transcriptional repressor controls a large regulon involved in lipid degradation in *Mycobacterium smegmatis* and *Mycobacterium tuberculosis*. *Mol. Microbiol.* **2007**, *65*, 684–699. [CrossRef] [PubMed]
74. Casali, N.; Riley, L.W. A phylogenomic analysis of the Actinomycetales *mce* operons. *BMC Genom.* **2007**, *8*, 60. [CrossRef] [PubMed]

75. Joshi, S.M.; Pandey, A.K.; Capite, N.; Fortune, S.M.; Rubin, E.J.; Sassetti, C.M. Characterization of mycobacterial virulence genes through genetic interaction mapping. *Proc. Natl. Acad. Sci. USA* **2006**, *103*, 11760–11765. [CrossRef] [PubMed]
76. Nazarova, E.V.; Montague, C.R.; La, T.; Wilburn, K.M.; Sukumar, N.; Lee, W.; Caldwell, S.; Russell, D.G.; VanderVen, B.C. Rv3723/LucA coordinates fatty acid and cholesterol uptake in *Mycobacterium tuberculosis*. *eLife* **2017**, *6*, e26969. [CrossRef] [PubMed]
77. Sassetti, C.M.; Rubin, E.J. Genetic requirements for mycobacterial survival during infection. *Proc. Natl. Acad. Sci. USA* **2003**, *100*, 12989–12994. [CrossRef] [PubMed]
78. Rosas-Magallanes, V.; Stadthagen-Gomez, G.; Rauzier, J.; Barreiro, L.B.; Tailleux, L.; Boudou, F.; Griffin, R.; Nigou, J.; Jackson, M.; Gicquel, B.; et al. Signature-tagged transposon mutagenesis identifies novel *Mycobacterium tuberculosis* genes involved in the parasitism of human macrophages. *Infect. Immun.* **2007**, *75*, 504–507. [CrossRef] [PubMed]
79. Song, H.; Sandie, R.; Wang, Y.; Andrade-Navarro, M.A.; Niederweis, M. Identification of outer membrane proteins of *Mycobacterium tuberculosis*. *Tuberculosis* **2008**, *88*, 526–544. [CrossRef] [PubMed]
80. Feltcher, M.E.; Gunawardena, H.P.; Zulauf, K.E.; Malik, S.; Griffin, J.E.; Sassetti, C.M.; Chen, X.; Braunstein, M. Label-free Quantitative Proteomics reveals a role for the *Mycobacterium tuberculosis* SecA2 pathway in exporting solute binding proteins and Mce transporters to the cell wall. *Mol. Cell. Proteom.* **2015**, *14*, 1501–1516. [CrossRef] [PubMed]
81. Perkowski, E.F.; Miller, B.K.; McCann, J.R.; Sullivan, J.T.; Malik, S.; Allen, I.C.; Godfrey, V.; Hayden, J.D.; Braunstein, M. An orphaned Mce-associated membrane protein of *Mycobacterium tuberculosis* is a virulence factor that stabilizes Mce transporters. *Mol. Microbiol.* **2016**, *100*, 90–107. [CrossRef] [PubMed]
82. Somalinga, V.; Mohn, W.W. *Rhodococcus jostii* porin A (RjpA) functions in cholate uptake. *Appl. Environ. Microbiol.* **2013**, *79*, 6191–6193. [CrossRef] [PubMed]
83. Haußmann, U.; Wolters, D.A.; Fränzel, B.; Eltis, L.D.; Poetsch, A. Physiological adaptation of the *Rhodococcus jostii* RHA1 membrane proteome to steroids as growth substrates. *J. Proteome Res.* **2013**, *12*, 1188–1198. [CrossRef] [PubMed]
84. Swain, K.; Casabon, I.; Eltis, L.D.; Mohn, W.W. Two transporters essential for reassimilation of novel cholate metabolites by *Rhodococcus jostii* RHA1. *J. Bacteriol.* **2012**, *194*, 6720–6727. [CrossRef] [PubMed]
85. Birkenmaier, A.; Holert, J.; Erdbrink, H.; Moeller, H.M.; Friemel, A.; Schoenenberger, R.; Suter, M.J.; Klebensberger, J.; Philipp, B. Biochemical and genetic investigation of initial reactions in aerobic degradation of the bile acid cholate in *Pseudomonas* sp. strain Chol1. *J. Bacteriol.* **2007**, *189*, 7165–7173. [CrossRef] [PubMed]
86. Lin, C.W.; Wang, P.H.; Ismail, W.; Tsai, Y.W.; El Nayal, A.; Yang, C.Y.; Yang, F.C.; Wang, C.H.; Chiang, Y.R. Substrate uptake and subcellular compartmentation of anoxic cholesterol catabolism in *Sterolibacterium denitrificans*. *J. Biol. Chem.* **2015**, *290*, 1155–1169. [CrossRef] [PubMed]
87. Holert, J.; Jagmann, N.; Philipp, B. The essential function of genes for a hydratase and an aldehyde dehydrogenase for growth of *Pseudomonas* sp. strain Chol1 with the steroid compound cholate indicates an aldolytic reaction step for deacetylation of the side chain. *J. Bacteriol.* **2013**, *195*, 3371–3380. [CrossRef]
88. Petrusma, M.; Dijkhuizen, L.; van der Geize, R. *Rhodococcus rhodochrous* DSM 43269 3-ketosteroid 9α-hydroxylase, a two-component iron-sulfur-containing monooxygenase with subtle steroid substrate specificity. *Appl. Environ. Microbiol.* **2009**, *75*, 5300–5307. [CrossRef]
89. Griffin, J.E.; Gawronski, J.D.; Dejesus, M.A.; Ioerger, T.R.; Akerley, B.J.; Sassetti, C.M. High-resolution phenotypic profiling defines genes essential for mycobacterial growth and cholesterol catabolism. *PLoS Pathog.* **2011**, *7*, e1002251. [CrossRef]
90. Thomas, S.T.; VanderVen, B.C.; Sherman, D.R.; Russell, D.G.; Sampson, N.S. Pathway profiling in *Mycobacterium tuberculosis*: Elucidation of cholesterol-derived catabolite and enzymes that catalyze its metabolism. *J. Biol. Chem.* **2011**, *286*, 43668–43678. [CrossRef]
91. Uhía, I.; Galán, B.; Kendall, S.L.; Stoker, N.G.; García, J.L. Cholesterol metabolism in *Mycobacterium smegmatis*. *Environ. Microbiol. Rep.* **2012**, *4*, 168–182. [CrossRef] [PubMed]
92. Li, W.; Ge, F.; Zhang, Q.; Ren, Y.; Yuan, J.; He, J.; Li, W.; Chen, G.; Zhang, G.; Zhuang, Y.; et al. Identification of gene expression profiles in the actinomycete *Gordonia neofelifaecis* grown with different steroids. *Genome* **2014**, *57*, 345–353. [CrossRef] [PubMed]

93. Bergstrand, L.H.; Cardenas, E.; Holert, J.; Van Hamme, J.D.; Mohn, W.W. Delineation of steroid-degrading microorganisms through comparative genomic analysis. *mBio* **2016**, *7*, e00166-16. [CrossRef] [PubMed]
94. Rosłoniec, K.Z.; Wilbrink, M.H.; Capyk, J.K.; Mohn, W.W.; Ostendorf, M.; van der Geize, R.; Dijkhuizen, L.; Eltis, L.D. Cytochrome P450 125 (CYP125) catalyses C26-hydroxylation to initiate sterol side-chain degradation in *Rhodococcus jostii* RHA1. *Mol. Microbiol.* **2009**, *74*, 1031–1043. [CrossRef] [PubMed]
95. Kreit, J. Microbial catabolism of sterols: Focus on the enzymes that transform the sterol 3β-hydroxy-5-en into 3-keto-4-en. *FEMS Microbiol. Lett.* **2017**, *364*, fnx007. [CrossRef] [PubMed]
96. Li, J.; Vrielink, A.; Brick, P.; Blow, D.M. Crystal structure of cholesterol oxidase complexed with a steroid substrate: Implications for flavin adenine dinucleotide dependent alcohol oxidases. *Biochemistry* **1993**, *32*, 11507–11515. [CrossRef]
97. Coulombe, R.; Yue, K.Q.; Ghisla, S.; Vrielink, A. Oxygen access to the active site of cholesterol oxidase through a narrow channel is gated by an Arg-Glu pair. *J. Biol. Chem.* **2001**, *276*, 30435–30441. [CrossRef]
98. Horinouchi, S.; Ishizuka, H.; Beppu, T. Cloning, nucleotide sequence, and transcriptional analysis of the NAD(P)-dependent cholesterol dehydrogenase gene from a *Nocardia* sp. and its hyperexpression in *Streptomyces* spp. *Appl. Environ. Microbiol.* **1991**, *57*, 1386–1393.
99. Yang, X.; Dubnau, E.; Smith, I.; Sampson, N.S. Rv1106c from *Mycobacterium tuberculosis* is a 3β-hydroxysteroid dehydrogenase. *Biochemistry* **2007**, *46*, 9058–9067. [CrossRef]
100. Yang, X.; Gao, J.; Smith, I.; Dubnau, E.; Sampsonn, N.S. Cholesterol is not an essential source of nutrition for *Mycobacterium tuberculosis* during infection. *J. Bacteriol.* **2011**, *193*, 1473–1476. [CrossRef]
101. Brzostek, A.; Rumijowska-Galewicz, A.; Dziadek, B.; Wojcik, E.A.; Dziadek, J. ChoD and HsdD can be dispensable for cholesterol degradation in mycobacteria. *J. Steroid Biochem. Mol. Biol.* **2013**, *134*, 1–7. [CrossRef] [PubMed]
102. Hoffmann, F.; Maser, E. Carbonyl reductases and pluripotent hydroxysteroid dehydrogenases of the short-chain dehydrogenase/reductase superfamily. *Drug Metab. Rev.* **2007**, *39*, 87–144. [CrossRef] [PubMed]
103. Marcus, P.I.; Talalay, P. Induction and purification of α- and β-hydroxysteroid dehydrogenases. *J. Biol. Chem.* **1956**, *218*, 661–674. [PubMed]
104. Oppermann, U.C.T.; Maser, E. Characterization of a 3 α-hydroxysteroid dehydrogenase/carbonyl reductase from the gram-negative bacterium *Comamonas testosteroni*. *Eur. J. Biochem.* **1996**, *241*, 744–749. [CrossRef] [PubMed]
105. Oppermann, U.C.T.; Belai, I.; Maser, E. Antibiotic resistance and enhanced insecticide catabolism as consequences of steroid induction in the Gram-negative bacterium *Comamonas testosteroni*. *J. Steroid Biochem. Mol. Biol.* **1996**, *58*, 217–223. [CrossRef]
106. Möbus, E.; Maser, E. Molecular cloning, overexpression, and characterization of steroid-inducible 3α-hydroxysteroid dehydrogenase/carbonyl reductase from *Comamonas testosteroni*. A novel member of the short-chain dehydrogenase/reductase superfamily. *J. Biol. Chem.* **1998**, *273*, 30888–30896. [CrossRef]
107. Horinouchi, M.; Kurita, T.; Hayashi, T.; Kudo, T. Steroid degradation genes in *Comamonas testosteroni* TA441: Isolation of genes encoding a Δ4(5)-isomerase and 3α- and 3β-dehydrogenases and evidence for a 100 kb steroid degradation gene hot spot. *J. Steroid Biochem. Mol. Biol.* **2010**, *122*, 253–263. [CrossRef]
108. Holert, J.; Yücel, O.; Jagmann, N.; Prestel, A.; Möller, H.M.; Philipp, B. Identification of bypass reactions leading to the formation of one central steroid degradation intermediate in metabolism of different bile salts in *Pseudomonas* sp. strain Chol1. *Environ. Microbiol.* **2016**, *18*, 3373–3389. [CrossRef]
109. Shi, C.J.; Tai, H.H.; Tsong, Y.Y. The mechanism of microbial conversion of cholesterol into 17-keto steroids. *J. Am. Chem. Soc.* **1967**, *89*, 1957–1958.
110. Shi, C.J.; Tai, H.H.; Tsong, Y.Y.; Lee, S.S.; Coombe, R.G. Mechanisms of steroid oxidation by microorganisms 0.14. Pathway of cholesterol side-chain degradation. *Biochemistry* **1968**, *7*, 808–818.
111. Reddy, J.K.; Hashimoto, T. Peroxisomal β-oxidation and peroxisome proliferator activated receptor α: An adaptive metabolic system. *Annu. Rev. Nutr.* **2001**, *21*, 193–230. [CrossRef] [PubMed]
112. Capyk, J.K.; Kalscheuer, R.; Stewart, G.R.; Liu, J.; Kwon, H.; Zhao, R.; Okamoto, S.; Jacobs, W.R., Jr.; Eltis, L.D.; Mohn, W.W. Mycobacterial cytochrome p450 125 (cyp125) catalyzes the terminal hydroxylation of C27 steroids. *J. Biol. Chem.* **2009**, *284*, 35534–35542. [CrossRef] [PubMed]
113. Ouellet, H.; Guan, S.; Johnston, J.B.; Chow, E.D.; Kells, P.M.; Burlingame, A.L.; Cox, J.S.; Podust, L.M.; de Montellano, P.R. *Mycobacterium tuberculosis* CYP125A1, a steroid C27 monooxygenase that detoxifies intracellularly generated cholest-4-en-3-one. *Mol. Microbiol.* **2010**, *77*, 730–742. [CrossRef] [PubMed]

114. McLean, K.J.; Lafite, P.; Levy, C.; Cheesman, M.R.; Mast, N.; Pikuleva, I.A.; Leys, D.; Munro, A.W. The structure of *Mycobacterium tuberculosis* CYP125: Molecular basis for cholesterol binding in a P450 needed for host infection. *J. Biol. Chem.* **2009**, *284*, 35524–35533. [CrossRef] [PubMed]
115. Ouellet, H.; Johnston, J.B.; de Montellano, P.R. Cholesterol catabolism as a therapeutic target in *Mycobacterium tuberculosis*. *Trends Microbiol.* **2011**, *19*, 530–539. [CrossRef]
116. Johnston, J.B.; Ouellet, H.; Ortiz de Montellano, P.R. Functional redundancy of steroid C26-monooxygenase activity in *Mycobacterium tuberculosis* revealed by biochemical and genetic analyses. *J. Biol. Chem.* **2010**, *285*, 36352–36360. [CrossRef]
117. Frank, D.J.; Madrona, Y.; Ortiz de Montellano, P.R. Cholesterol ester oxidation by mycobacterial cytochrome P450. *J. Biol. Chem.* **2014**, *289*, 30417–30425. [CrossRef]
118. Wilbrink, M.H.; Petrusma, M.; Dijkhuizen, L.; van der Geize, R. FadD19 of *Rhodococcus rhodochrous* DSM43269, a steroid-coenzyme A ligase essential for degradation of C-24 branched sterol side chains. *Appl. Environ. Microbiol.* **2011**, *77*, 4455–4464. [CrossRef]
119. Casabon, I.; Swain, K.; Crowe, A.M.; Eltis, L.D.; Mohn, W.W. Actinobacterial acyl coenzyme A synthetases involved in steroid side-chain catabolism. *J. Bacteriol.* **2014**, *196*, 579–587. [CrossRef]
120. Thomas, S.T.; Sampson, N.S. *Mycobacterium tuberculosis* utilizes a unique heterotetrameric structure for dehydrogenation of the cholesterol side chain. *Biochemistry* **2013**, *52*, 2895–2904. [CrossRef]
121. Yang, M.; Lu, R.; Guja, K.E.; Wipperman, M.F.; St Clair, J.R.; Bonds, A.C.; Garcia-Diaz, M.; Sampson, N.S. Unraveling cholesterol catabolism in *Mycobacterium tuberculosis*: ChsE4-ChsE5 $\alpha 2\beta 2$ acyl-CoA dehydrogenase initiates β-oxidation of 3-oxo-cholest-4-en-26-oyl CoA. *ACS Infect. Dis.* **2015**, *1*, 110–125. [CrossRef] [PubMed]
122. Kim, J.J.; Battaile, K.P. Burning fat: The structural basis of fatty acid β-oxidation. *Curr. Opin. Struct. Biol.* **2002**, *12*, 721–728. [CrossRef]
123. Yang, M.; Guja, K.E.; Thomas, S.T.; Garcia-Diaz, M.; Sampson, N.S. A distinct MaoC-like enoyl-CoA hydratase architecture mediates cholesterol catabolism in *Mycobacterium tuberculosis*. *ACS Chem. Biol.* **2014**, *9*, 2632–2645. [CrossRef] [PubMed]
124. Griffin, J.E.; Pandey, A.K.; Gilmore, S.A.; Mizrahi, V.; McKinney, J.D.; Bertozzi, C.R.; Sassetti, C.M. Cholesterol catabolism by *Mycobacterium tuberculosis* requires transcriptional and metabolic adaptations. *Chem. Biol.* **2012**, *19*, 218–227. [CrossRef] [PubMed]
125. Wipperman, M.F.; Yang, M.; Thomas, S.T.; Sampson, N.S. Shrinking the FadE proteome of *Mycobacterium tuberculosis*: Insights into cholesterol metabolism through identification of an $\alpha 2\beta 2$ heterotetrameric acyl coenzyme A dehydrogenase family. *J. Bacteriol.* **2013**, *195*, 4331–4341. [CrossRef] [PubMed]
126. Nesbitt, N.M.; Yang, X.; Fontán, P.; Kolesnikova, I.; Smith, I.; Sampson, N.S.; Dubnau, E. A thiolase of *Mycobacterium tuberculosis* is required for virulence and production of androstenedione and androstadienedione from cholesterol. *Infect. Immun.* **2010**, *78*, 275–282. [CrossRef] [PubMed]
127. Schaefer, C.M.; Lu, R.; Nesbitt, N.M.; Schiebel, J.; Sampson, N.S.; Kisker, C. FadA5 a thiolase from *Mycobacterium tuberculosis*: A steroid-binding pocket reveals the potential for drug development against tuberculosis. *Structure* **2015**, *23*, 21–33. [CrossRef]
128. Lu, R.; Schaefer, C.M.; Nesbitt, N.M.; Kuper, J.; Kisker, C.; Sampson, N.S. Catabolism of the cholesterol side chain in *Mycobacterium tuberculosis* is controlled by a Redox-Sensitive Thiol Switch. *ACS Infect. Dis.* **2017**, *3*, 666–675. [CrossRef]
129. Anbazhagan, P.; Harijan, R.K.; Kiema, T.R.; Janardan, N.; Murthy, M.R.; Michels, P.A.; Juffer, A.H.; Wierenga, R.K. Phylogenetic relationships and classification of thiolases and thiolase-like proteins of *Mycobacterium tuberculosis* and *Mycobacterium smegmatis*. *Tuberculosis* **2014**, *94*, 405–412. [CrossRef]
130. Xu, L.Q.; Liu, Y.J.; Yao, K.; Liu, H.H.; Tao, X.Y.; Wang, F.Q.; Wei, D.Z. Unraveling and engineering the production of 23,24-bisnorcholenic steroids in sterol metabolism. *Sci. Rep.* **2016**, *6*, 21928. [CrossRef]
131. Wilbrink, M.H.; van der Geize, R.; Dijkhuizen, L. Molecular characterization of *lpt3* and *lpt4*, essential for C24-branched chain sterol-side-chain degradation in *Rhodococcus rhodochrous* DSM 43269. *Microbiology* **2012**, *158*, 3054–3062. [CrossRef] [PubMed]
132. Gilbert, S.; Hood, L.; Seah, S.Y.K. Characterization of an aldolase involved in cholesterol side chain degradation in *Mycobacterium tuberculosis*. *J. Bacteriol.* **2018**, *200*, e00512-17. [CrossRef] [PubMed]
133. Holert, J.; Kulić, Ž.; Yücel, O.; Suvekbala, V.; Suter, M.J.; Möller, H.M.; Philipp, B. Degradation of the acyl side chain of the steroid compound cholate in *Pseudomonas* sp. strain Chol1 proceeds via an aldehyde intermediate. *J. Bacteriol.* **2013**, *195*, 585–595. [CrossRef] [PubMed]

134. Genti-Raimondi, S.; Tolmasky, M.E.; Patrito, L.C.; Flury, A.; Actis, L.A. Molecular cloning and expression of the β-hydroxysteroid dehydrogenase gene from *Pseudomonas testosteroni*. *Gene* **1991**, *105*, 43–49. [CrossRef]
135. Schultz, R.M.; Groman, E.V.; Engel, L.L. 3(17)β-Hydroxysteroid dehydrogenase of *Pseudomonas testosteroni*. A convenient purification and demonstration of multiple molecular forms. *J. Biol. Chem.* **1977**, *252*, 3775–3783. [PubMed]
136. Minard, P.; Legoy, M.D.; Thomas, D. 3β,17β-hydroxysteroid dehydrogenase of *Pseudomonas testosteroni*. Kinetic evidence for the bifunctional activity at a common catalytic site. *FEBS Lett.* **1985**, *188*, 85–90. [CrossRef]
137. Yu, Y.; Liu, C.; Wang, B.; Li, Y.; Zhang, H. Characterization of 3,17β-hydroxysteroid dehydrogenase in *Comamonas testosteroni*. *Chem. Biol. Interact.* **2015**, *234*, 221–228. [CrossRef]
138. Itagaki, E.; Matushita, H.; Hatta, T. Steroid transhydrogenase activity of 3-ketosteroid-Δ1-dehydrogenase from *Nocardia corallina*. *J. Biochem.* **1990**, *108*, 122–127. [CrossRef]
139. Rohman, A.; van Oosterwijk, N.; Thunnissen, A.M.W.H.; Dijkstra, B.W. Crystal structure and site-directed mutagenesis of 3-ketosteroid Δ1-dehydrogenase from *Rhodococcus erythropolis* SQ1 explain its catalytic mechanism. *J. Biol. Chem.* **2013**, *288*, 35559–35568. [CrossRef]
140. Horinouchi, M.; Hayashi, T.; Yamamoto, T.; Kudo, T. A new bacterial steroid degradation gene cluster in *Comamonas testosteroni* TA441 which consists of aromatic-compound degradation genes for seco-steroids and 3-ketosteroid dehydrogenase genes. *Appl. Environ. Microbiol.* **2003**, *69*, 4421–4430. [CrossRef]
141. Olivera, E.R.; de la Torre, M.; Barrientos, A.; Luengo, J.M. Steroid catabolism in bacteria: Genetic and functional analyses of *stdH* and *stdJ* in *Pseudomonas putida* DOC21. *Can. J. Biotechnol.* **2018**, *2*, 88–99. [CrossRef]
142. Brzostek, A.; Pawelczyk, J.; Rumijowska-Galewicz, A.; Dziadek, B.; Dziadek, J. *Mycobacterium tuberculosis* is able to accumulate and utilize cholesterol. *J. Bacteriol.* **2009**, *191*, 6584–6591. [CrossRef] [PubMed]
143. Knol, J.; Bodewits, K.; Hessels, G.I.; Dijkhuizen, L.; van der Geize, R. 3-Keto-5α-steroid Δ(1)-dehydrogenase from *Rhodococcus erythropolis* SQ1 and its orthologue in *Mycobacterium tuberculosis* H37Rv are highly specific enzymes that function in cholesterol catabolism. *Biochem. J.* **2008**, *410*, 339–346. [CrossRef] [PubMed]
144. Van der Geize, R.; Hessels, G.I.; Dijkhuizen, L. Molecular and functional characterization of the *kstD2* gene of *Rhodococcus erythropolis* SQ1 encoding a second 3-ketosteroid Δ(1)-dehydrogenase isoenzyme. *Microbiology* **2002**, *148*, 3285–3292. [CrossRef] [PubMed]
145. van der Geize, R.; Hessels, G.I.; van Gerwen, R.; van der Meijden, P.; Dijkhuizen, L. Unmarked gene deletion mutagenesis of *kstD*, encoding 3-ketosteroid Δ1-dehydrogenase, in *Rhodococcus erythropolis* SQ1 using *sacB* as counter-selectable marker. *FEMS Microbiol. Lett.* **2001**, *205*, 197–202. [CrossRef]
146. Brzostek, A.; Sliwiński, T.; Rumijowska-Galewicz, A.; Korycka-Machała, M.; Dziadek, J. Identification and targeted disruption of the gene encoding the main 3-ketosteroid dehydrogenase in *Mycobacterium smegmatis*. *Microbiology* **2005**, *151*, 2393–2402. [CrossRef]
147. Florin, C.; Kohler, M.; Grandguillot, M.; Plesiat, P. *Comamonas testosteroni* 3-ketosteroid-Δ4(5α)-dehydrogenase: Gene and protein characterization. *J. Bacteriol.* **1996**, *178*, 3322–3330. [CrossRef]
148. van Oosterwijk, N.; Knol, J.; Dijkhuizen, L.; van der Geize, R.; Dijkstra, B.W. Structure and catalytic mechanism of 3-ketosteroid-Delta4-(5α)-dehydrogenase from *Rhodococcus jostii* RHA1 genome. *J. Biol. Chem.* **2012**, *287*, 30975–30983. [CrossRef]
149. van der Geize, R.; Hessels, G.I.; van Gerwen, R.; van der Meijden, P.; Dijkhuizen, L. Molecular and functional characterization of *kshA* and *kshB*, encoding two components of 3-ketosteroid 9alpha-hydroxylase, a class IA monooxygenase, in *Rhodococcus erythropolis* strain SQ1. *Mol. Microbiol.* **2002**, *45*, 1007–1018. [CrossRef]
150. van der Geize, R.; Hessels, G.I.; Nienhuis-Kuiper, M.; Dijkhuizen, L. Characterization of a second *Rhodococcus erythropolis* SQ1 3-ketosteroid 9alpha-hydroxylase activity comprising a terminal oxygenase homologue, KshA2, active with oxygenase-reductase component KshB. *Appl. Environ. Microbiol.* **2008**, *74*, 7197–7203. [CrossRef]
151. Capyk, J.K.; D'Angelo, I.; Strynadka, N.C.; Eltis, L.D. Characterization of 3-ketosteroid 9{α}-hydroxylase, a Rieske oxygenase in the cholesterol degradation pathway of *Mycobacterium tuberculosis*. *J. Biol. Chem.* **2009**, *284*, 9937–9946. [CrossRef] [PubMed]
152. Capyk, J.K.; Casabon, I.; Gruninger, R.; Strynadka, N.C.; Eltis, L.D. Activity of 3-ketosteroid 9α-hydroxylase (KshAB) indicates cholesterol side chain and ring degradation occur simultaneously in *Mycobacterium tuberculosis*. *J. Biol. Chem.* **2011**, *286*, 40717–40724. [CrossRef] [PubMed]

153. García, J.L.; Uhía, I.; Galán, B. Catabolism and biotechnological applications of cholesterol degrading bacteria. *Microb. Biotechnol.* **2012**, *5*, 679–699. [CrossRef] [PubMed]
154. Donova, M.; Gulevskaya, S.; Dovbnya, D.; Puntus, I. *Mycobacterium* sp. mutant strain producing 9α-hydroxy androstenedione from sitosterol. *Appl. Microbiol. Biotechnol.* **2005**, *67*, 671–678. [CrossRef] [PubMed]
155. Guevara, G.; Fernández de las Heras, L.; Perera, J.; Navarro Llorens, J.M. Functional differentiation of 3-ketosteroid Δ^1-dehydrogenase isoenzymes in *Rhodococcus ruber* strain Chol-4. *Microb. Cell Fact.* **2017**, *16*, 42. [CrossRef] [PubMed]
156. Yao, K.; Xu, L.Q.; Wang, F.Q.; Wei, D.Z. Characterization and engineering of 3-ketosteroid Δ^1-dehydrogenase and 3-ketosteroid 9α-hydroxylase in *Mycobacterium neoaurum* ATCC 25795 to produce 9a-hydroxy-4-androstene-3,17-dione through the catabolism of sterols. *Metab. Eng.* **2014**, *24*, 181–191. [CrossRef] [PubMed]
157. Hu, Y.; van der Geize, R.; Besra, G.S.; Gurcha, A.; Liu, M.; Rohde, M.; Singh, M.; Coares, A. 3-Ketosteroid 9α-hydroxylase is an essential factor in the pathogenesis of *Mycobacterium tuberculosis*. *Mol. Microbiol.* **2010**, *75*, 107–121. [CrossRef]
158. Bragin, E.Y.; Shtratnikova, V.Y.; Dovbnya, D.V.; Schelkunov, M.I.; Pekov, Y.A.; Malakho, S.G.; Egoroba, O.V.; Ivashina, T.V.; Sokolov, S.L.; Ashapkin, V.V.; et al. Comparative analysis of genes encoding key steroid core oxidation enzymes in fast-growing *Mycobacterium* spp. Strains. *J. Steroid Biochem. Mol. Biol.* **2013**, *138*, 41–53. [CrossRef]
159. Petrusma, M.; Hessels, G.; Dijkhuizen, L.; van der Geize, R. Multiplicity of 3-ketosteroid-9α-hydroxylase enzymes in *Rhodococcus rhodochrous* DSM43269 for specific degradation of different classes of steroids. *J. Bacteriol.* **2011**, *193*, 3931–3940. [CrossRef]
160. Horinouchi, M.; Hayashi, T.; Kudo, T. The genes encoding the hydroxylase of 3-hydroxy-9,10-secoandrosta-1,3,5(10)-triene-9,17-dione in steroid degradation in *Comamonas testosteroni* TA441. *J. Steroid. Biochem. Mol. Biol.* **2004**, *92*, 143–154. [CrossRef]
161. Dresen, C.; Lin, L.Y.; D'Angelo, I.; Tocheva, E.I.; Strynadka, N.; Eltis, L.D. A flavin-dependent monooxygenase from *Mycobacterium tuberculosis* involved in cholesterol catabolism. *J. Biol. Chem.* **2010**, *285*, 22264–22275. [CrossRef] [PubMed]
162. Horinouchi, M.; Yamamoto, T.; Taguchi, K.; Arai, H.; Kudo, T. Meta-cleavage enzyme gene *tesB* is necessary for testosterone degradation in *Comamonas testosteroni* TA441. *Microbiology* **2001**, *147*, 3367–3375. [CrossRef] [PubMed]
163. Yam, K.C.; D'Angelo, I.; Kalscheuer, R.; Zhu, H.; Wang, J.X.; Snieckus, V.; Ly, L.H.; Converse, P.J.; Jacobs, W.R. Jr.; Strynadka, N.; et al. Studies of a ring-cleaving dioxygenase illuminate the role of cholesterol metabolism in the pathogenesis of *Mycobacterium tuberculosis*. *PLoS Pathog.* **2009**, *5*, e1000344. [CrossRef] [PubMed]
164. Horinouchi, M.; Hayashi, T.; Koshino, H.; Yamamoto, T.; Kudo, T. Gene encoding the hydrolase for the product of the meta-cleavage reaction in testosterone degradation by *Comamonas testosteroni*. *Appl. Environ. Microbiol.* **2003**, *69*, 2139–2152. [CrossRef] [PubMed]
165. Lack, N.; Lowe, E.D.; Liu, J.; Eltis, L.D.; Noble, M.E.; Sim, E.; Westwood, I.M. Structure of HsaD, a steroid-degrading hydrolase, from *Mycobacterium tuberculosis*. *Acta Crystallogr. Sect. F Struct. Biol. Cryst. Commun.* **2008**, *64*, 2–7. [CrossRef] [PubMed]
166. Lack, N.A.; Yam, K.C.; Lowe, E.D.; Horsman, G.P.; Owen, R.L.; Sim, E.; Eltis, L.D. Characterization of a carbon-carbon hydrolase from *Mycobacterium tuberculosis* involved in cholesterol metabolism. *J. Biol. Chem.* **2010**, *285*, 434–443. [CrossRef] [PubMed]
167. Horinouchi, M.; Hayashi, T.; Koshino, H.; Kurita, T.; Kudo, T. Identification of 9,17-dioxo-1,2,3,4,10,19-hexanorandrostan-5-oic acid, 4-hydroxy-2-oxohexanoic acid, and 2-hydroxyhexa-2,4-dienoic acid and related enzymes involved in testosterone degradation in *Comamonas testosteroni* TA441. *Appl. Environ. Microbiol.* **2005**, *71*, 5275–5281. [CrossRef] [PubMed]
168. Carere, J.; McKenna, S.E.; Kimber, M.S.; Seah, S.Y. characterization of an aldolase_dehydrogenase complex from the cholesterol degrading pathway of *Mycobacterium tuberculosis*. *Biochemistry* **2013**, *52*, 3502–3511. [CrossRef]
169. Horinouchi, M.; Hayashi, T.; Koshino, H.; Kudo, T. ORF18 disrupted mutant of *Comamonas testosteroni* TA441 accumulates significant amounts of 9,17-dioxo-1,2,3,4,10,19-hexanorandrostan-5-oic acid and its derivatives after incubation with steroids. *J. Steroid Biochem. Mol. Biol.* **2006**, *101*, 78–84. [CrossRef]

170. Casabon, I.; Crowe, A.M.; Liu, J.; Eltis, L.D. FadD3 is an acyl-CoA synthetase that initiates catabolism of cholesterol rings C and D in actinobacteria. *Mol. Microbiol.* **2013**, *87*, 269–283. [CrossRef]

171. Crowe, A.M.; Casabon, I.; Brown, K.L.; Liu, J.; Lian, J.; Rogalski, J.C.; Hurst, T.E.; Snieckus, V.; Foster, L.J.; Eltis, L.D. Catabolism of the last two steroid rings in *Mycobacterium tuberculosis* and other bacteria. *mBio* **2017**, *8*, e00321-17. [CrossRef] [PubMed]

172. Horinouchi, M.; Hayashi, T.; Koshino, H.; Kudo, T. Identification of 9alpha-hydroxy-17-oxo-1,2,3,4,10,19-hexanorandrostan-5-oic acid in steroid degradation by *Comamonas testosteroni* TA441 and its conversion to the corresponding 6-en-5-oyl Coenzyme A (CoA) involving open reading frame 28 (ORF28)- and ORF30-encoded acyl-CoA dehydrogenases. *J. Bacteriol.* **2014**, *196*, 3598–3608.

173. Horinouchi, M.; Hayashi, T.; Koshino, H.; Malon, M.; Hirota, H.; Kudo, T. Identification of 9α-hydroxy-17-oxo-1,2,3,4,10,19-hexanorandrost-6-en-5-oic acid and β-oxidation products of the C-17 side chain in cholic acid degradation by *Comamonas testosteroni* TA441. *J. Steroid Biochem. Biol. Mol.* **2014**, *143*, 306–322. [CrossRef] [PubMed]

174. Horinouchi, M.; Koshino, H.; Malon, M.; Hirota, H.; Hayashi, T. Steroid degradation in *Comamonas testosteroni* TA441: Identification of metabolites and the genes involved in the reactions necessary before D-ring cleavage. *Appl. Environ. Microbiol.* **2018**, *84*, e01324-18. [CrossRef] [PubMed]

175. Crowe, A.M.; Workman, S.D.; Watanabe, N.; Worrall, L.J.; Strynadka, N.C.J.; Eltis, L.D. IpdAB, a virulence factor in *Mycobacterium tuberculosis*, is a cholesterol ring-cleaving hydrolase. *Proc. Natl. Acad. Sci. USA* **2018**, *115*, E3378–E3387. [CrossRef] [PubMed]

176. Horinouchi, M.; Koshino, H.; Malon, M.; Hirota, H.; Hayashi, T. Identification of 9-oxo-1,2,3,4,5,6,10,19-octanor-13,17-secoandrost-8(14)-ene-7,17-dioic acid as a metabolite of steroid degradation in *Comamonas testosteroni* TA441 and the genes involved in the conversion. *J. Steroid Biochem. Mol. Biol.* **2019**, *185*, 268–276. [CrossRef]

177. Horinouchi, M.; Malon, M.; Hirota, H.; Hayashi, T. Identification of 4-methyl-5-oxo-octane-1,8-dioic acid and the derivatives as metabolites of steroidal C,D-ring degradation in *Comamonas testosteroni* TA441. *J. Steroid Biochem. Mol. Biol.* **2019**, *185*, 277–286. [CrossRef]

178. Harwood, C.S.; Parales, R.E. The β-ketoadipate pathway and the biology of self-identity. *Annu. Rev. Microbiol.* **1996**, *50*, 553–590. [CrossRef]

179. Chen, Y.L.; Wang, C.H.; Yang, F.C.; Ismail, W.; Wang, P.H.; Shih, C.J.; Wu, Y.C.; Chiang, Y.R. Identification of *Comamonas testosteroni* as an androgen degrader in sewage. *Sci. Rep.* **2016**, *6*, 35386. [CrossRef]

180. Kendall, S.L.; Burgess, P.; Balhana, R.; Withers, M.; Ten Bokum, A.; Lott, J.S.; Gao, C.; Uhía-Castro, I.; Stocker, N.G. Cholesterol utilization in mycobacteria is controlled by two TetR-type transcriptional regulators: KstR and KstR2. *Microbiology* **2010**, *156*, 1362–1371. [CrossRef]

181. Ho, N.A.T.; Dawes, S.S.; Crowe, A.M.; Casabon, I.; Gao, C.; Kendall, S.L.; Baker, E.N.; Eltis, L.D.; Lott, J.S. The structure of the transcriptional repressor KstR in complex with CoA thioester cholesterol metabolites sheds light on the regulation of cholesterol catabolism in *Mycobacterium tuberculosis*. *J. Biol. Chem.* **2016**, *291*, 7256–7266. [CrossRef] [PubMed]

182. García-Fernández, E.; Medrano, F.J.; Galán, B.; García, J.L. Deciphering the transcriptional regulation of cholesterol catabolic pathway in mycobacteria: Identification of the inducer of KstR repressor. *J. Biol. Chem.* **2014**, *289*, 17576–17588. [CrossRef] [PubMed]

183. Casabon, I.; Zhu, S.-H.; Otani, H.; Liu, J.; Mohn, W.W.; Eltis, L.D. Regulation of KstR2 regulon of *Mycobacterium tuberculosis* by a cholesterol catabolite. *Mol. Microbiol.* **2013**, *89*, 1201–1212. [CrossRef] [PubMed]

184. Crowe, A.M.; Stogios, P.J.; Casabon, I.; Evdokimova, E.; Savchenko, A.; Eltis, L.D. Structural and functional characterization of a ketosteroid transcriptional regulator of *Mycobacterium tuberculosis*. *J. Biol. Chem.* **2015**, *290*, 872–882. [CrossRef] [PubMed]

185. Horinouchi, M.; Kurita, T.; Yamamoto, T.; Hatori, E.; Hayasi, T.; Kudo, T. Steroid degradation gene cluster of *Comamonas testosteroni* consisting of 18 putative genes from *meta*-cleavage enzyme gene *tesB* to regulator gene *tesR*. *Biochem. Biophys. Res. Commun.* **2004**, *324*, 597–604. [CrossRef]

186. Pruneda-Paz, J.L.; Linares, M.; Cabrera, J.E.; Genti-Raimondi, S. TeiR, a LuxR-type transcriptional factor required for testosterone degradation in *Comamonas testosteroni*. *J. Bacteriol.* **2004**, *186*, 1430–1437. [CrossRef]

187. Göhler, A.; Xiong, G.; Paulsen, S.; Trentmann, G.; Maser, E. Testosterone-inducible regulator is a kinase that drives steroid sensing and metabolism in *Comamonas testosteroni*. *J. Biol. Chem.* **2008**, *283*, 17380–17390. [CrossRef]

188. Linares, M.; Pruneda-Paz, J.L.; Reyna, L.; Genti-Raimondi, S. Regulation of testosterone degradation in *Comamonas testosteroni*. *J. Steroid Biochem. Mol. Biol.* **2008**, *112*, 145–150. [CrossRef]
189. Xiong, G.; Maser, E. Regulation of the steroid-inducible 3α-hydroxysteroid dehydrogenase/carbonyl reductase gene in *Comamonas testosteroni*. *J. Biol. Chem.* **2001**, *276*, 9961–9970. [CrossRef]
190. Xiong, G.; Martin, H.J.; Maser, E. Identification and characterization of a novel translational repressor of the steroid-inducible 3α-hydroxysteroid dehydrogenase/carbonyl reductase in *Comamonas testosteroni*. *J. Biol. Chem.* **2003**, *278*, 47400–47407. [CrossRef]
191. Gong, W.; Xiong, G.; Maser, E. Identification and characterization of the LysR-type transcriptional regulator HsdR for steroid-inducible expression of the 3a-hydroxysteroid dehydrogenase/carbonyl reductase gene in *Comamonas testosteroni*. *Appl. Environ. Microbiol.* **2011**, *78*, 941–950. [CrossRef] [PubMed]
192. Gong, W.; Xiong, G.; Maser, E. Oligomerization and negative autoregulation of the LysR-type transcriptional regulator HsdR from *Comamonas testosteroni*. *J. Steroid Biochem. Mol. Biol.* **2012**, *132*, 203–211. [CrossRef] [PubMed]
193. Li, M.; Xiong, G.; Maser, E. A novel transcriptional repressor PhaR for the steroid-inducible expression of the 3,17β-hydroxysteroid dehydrogenase gene in *Comamonas testosteroni* ATCC11996. *Chem. Biol. Interact.* **2013**, *202*, 116–125. [CrossRef] [PubMed]
194. Ji, Y.; Pan, T.; Zhang, Y.; Xiong, G.; Yu, Y. Functional analysis of a novel repressor LuxR in *Comamonas testosteroni*. *Chem. Biol. Interact.* **2017**, *276*, 113–120. [CrossRef] [PubMed]
195. Pan, T.; Huang, P.; Xiong, G.; Maser, E. Isolation and characterization of a repressor TetR for 3,17β-HSD expressional regulation in *Comamonas testosteroni*. *Chem. Biol. Interact.* **2015**, *234*, 205–212. [CrossRef] [PubMed]
196. Wu, Y.; Huang, P.; Xiong, G.; Maser, E. Identification and isolation of a regulator protein for 3,17β-HSD expression regulation in *Comamonas testosteroni*. *Chem. Biol. Interact.* **2015**, *234*, 197–204. [CrossRef]
197. Yu, C.P.; Roh, H.; Chu, K.H. 17β-Estradiol-degrading bacteria isolated from activated sludge. *Environ. Sci. Technol.* **2007**, *41*, 486–492. [CrossRef]
198. Chen, Y.L.; Yu, C.P.; Lee, T.H.; Goh, K.S.; Chu, K.H.; Wang, P.H.; Ismail, W.; Shih, C.J.; Chiang, Y.R. Biochemical mechanisms and catabolic enzymes involved in bacterial estrogen degradation pathways. *Cell Chem. Biol.* **2017**, *24*, 712–724. [CrossRef]
199. Roh, H.; Chu, K.H. A 17beta-estradiol-utilizing bacterium, *Sphingomonas* strain KC8: Part I—Characterization and abundance in wastewater treatment plants. *Environ. Sci. Technol.* **2010**, *44*, 4943–4950. [CrossRef]
200. Liang, R.; Liu, H.; Liu, J. Genome sequence of *Pseudomonas putida* strain SJTE-1, a bacterium capable of degrading estrogens and persistent organic pollutants. *J. Bacteriol.* **2012**, *194*, 4781–4782. [CrossRef]
201. Wang, P.; Zheng, D.; Peng, W.; Wang, Y.; Wang, X.; Xiong, W.; Liang, R. Characterization of 17β-hydroxysteroid dehydrogenase and regulators involved in estrogen degradation in *Pseudomonas putida* SJTE-1. *Appl. Microbiol. Biotechnol.* **2019**, *103*, 2413–2425. [CrossRef] [PubMed]
202. Wu, K.; Lee, T.H.; Chen, Y.L.; Wang, Y.S.; Wang, P.H.; Yu, C.P.; Chu, K.H.; Chiang, Y.R. Metabolites involved in aerobic degradation of the A and B rings of estrogen. *Appl. Environ. Microbiol.* **2019**, *85*, e02223-18. [CrossRef] [PubMed]
203. Qin, D.; Me, C.; Hu, A.; Zhang, F.; Hu, H.; Yu, C.P. *Altererythrobacter estronivorus* sp. nov., an estrogen-degrading strain isolated from Yundang Lagoon of Xiamen City in China. *Curr. Microbiol.* **2016**, *72*, 634–640. [CrossRef] [PubMed]
204. Fujii, K.; Satomi, M.; Morita, N.; Motomura, T.; Tanaka, T.; Kikuchi, S. *Novosphingobium tardaugens* sp. nov., an oestradiol-degrading bacterium isolated from activated sludge of a sewage treatment plant in Tokyo. *Int. J. Syst. Evol. Microbiol.* **2003**, *53*, 47–52. [CrossRef] [PubMed]
205. Chen, Y.L.; Fu, H.-Y.; Lee, T.-H.; Shih, C.-J.; Huang, L.; Wang, Y.-S.; Ismail, W.; Chiang, Y.-R. Estrogen degraders and estrogen degradation pathway identified in an activated sludge. *Appl. Environ. Microbiol.* **2018**, *84*, e00001-18. [CrossRef] [PubMed]
206. Kurisu, F.; Ogura, M.; Saitoh, S.; Yamazoe, A.; Yagi, O. Degradation of natural estrogen and identification of the metabolites produced by soil isolates of *Rhodococcus* sp. and *Sphingomonas* sp. *J. Biosci. Bioeng.* **2010**, *109*, 576–582. [CrossRef] [PubMed]
207. Nakai, S.; Yamamura, A.; Tanaka, S.; Shi, J.; Nishikawa, M.; Nakashimada, Y.; Hosomi, M. Pathway of 17β-estradiol degradation by *Nitrosomonas europaea* and reduction in 17β-estradiol-derived estrogenic activity. *Environ. Chem. Lett.* **2011**, *9*, 1–6. [CrossRef]

208. Lee, H.B.; Liu, D. Degradation of 17β-estradiol and its metabolites by sewage bacteria. *Water Air Soil Pollut.* **2002**, *134*, 353–368. [CrossRef]
209. Harder, J.; Probian, C. Anaerobic mineralization of cholesterol by a novel type of denitrifying bacterium. *Arch. Microbiol.* **1997**, *167*, 269–274. [CrossRef]
210. Wang, P.-H.; Leu, Y.-L.; Ismail, W.; Tang, S.-L.; Tsai, C.-Y.; Chen, H.-J.; Kao, A.-T.; Chiang, Y.-R. Anaerobic and aerobic cleavage of the steroid core ring structure by *Steroidobacter denitrificans*. *J. Lipid Res.* **2013**, *54*, 1493–1504. [CrossRef]
211. Wang, P.-H.; Yu, C.-P.; Lee, T.-H.; Lin, C.-W.; Ismail, W.; Wey, S.-P.; Kuo, A.-T.; Chiang, Y.-R. Anoxic androgen degradation by the denitrifying bacterium *Sterolibacterium denitrificans* via the 2,3-seco pathway. *Appl. Environ. Microbiol.* **2014**, *80*, 3442–3452. [CrossRef] [PubMed]
212. Yang, F.C.; Chen, Y.L.; Tang, S.L.; Yu, C.P.; Wang, P.H.; Ismail, W.; Wang, C.H.; Ding, J.Y.; Yang, C.Y.; Yang, C.Y.; et al. Integrated multi-omics analyses reveal the biochemical mechanisms and phylogenetic relevance of anaerobic androgen biodegradation in the environment. *ISME J.* **2016**, *10*, 1967–1983. [CrossRef] [PubMed]
213. Chiang, Y.R.; Ismail, W.; Müller, M.; Fuchs, G. Initial steps in the anoxic metabolism of cholesterol by the denitrifying *Sterolibacterium denitrificans*. *J. Biol. Chem.* **2007**, *282*, 13240–13249. [CrossRef] [PubMed]
214. Dermer, J.; Fuchs, G. Molybdoenzyme that catalyzes the anaerobic hydroxylation of a tertiary carbon atom in the side chain of cholesterol. *J. Biol. Chem.* **2012**, *287*, 36905–36916. [CrossRef] [PubMed]
215. Rugor, A.; Wójcik-Augustyn, A.; Niedzialkowska, E.; Mordalski, S.; Staroń, J.; Bojarski, A.; Szaleniec, M. Reaction mechanism of sterol hydroxylation by steroid C25 dehydrogenase–Homology model, reactivity and isoenzymatic diversity. *J. Inorg. Biochem.* **2017**, *173*, 28–43. [CrossRef] [PubMed]
216. Warnke, M.; Jacoby, C.; Jung, T.; Agne, M.; Mergelsberg, M.; Starke, R.; Jehmlich, N.; von Bergen, M.; Richnow, H.H.; Brüls, T.; et al. A patchwork pathway for oxygenase-independent degradation of side chain containing steroids. *Environ. Microbiol.* **2017**, *19*, 4684–4699. [CrossRef] [PubMed]
217. Jacoby, C.; Eipper, J.; Warnke, M.; Tiedt, O.; Mergelsberg, M.; Stärk, H.J.; Daus, B.; Martín-Moldes, Z.; Zamarro, M.T.; Díaz, E.; et al. Four Molybdenum-dependent steroid C-25 hydroxylases: Heterologous overproduction, role in steroid degradation, and application for 25-hydroxyvitamin D_3 synthesis. *mBio* **2018**, *9*, e00694-18. [CrossRef] [PubMed]
218. Wang, P.-H.; Lee, T.H.; Ismail, W.; Tsai, C.Y.; Lin, C.W.; Tsai, Y.W.; Chiang, Y.R. An oxygenase-independent cholesterol catabolic pathway operates under oxic conditions. *PLoS ONE* **2013**, *8*, e66675. [CrossRef] [PubMed]
219. Warnke, M.; Jung, T.; Jacoby, C.; Agne, M.; Feller, F.M.; Philipp, B.; Seiche, W.; Breit, B.; Boll, M. Functional characterization of three specific acyl-coenzyme A synthetases involved in anaerobic cholesterol degradation in *Sterolibacterium denitrificans* Chol1S. *Appl. Environ. Microbiol.* **2018**, *84*, e02721-17. [CrossRef] [PubMed]
220. Chiang, Y.R.; Ismail, W.; Gallien, S.; Heintz, D.; Van Dorsselaer, A.; Fuchs, G. Cholest-4-en-3-one-Δ1-dehydrogenase, a flavoprotein catalyzing the second step in anoxic cholesterol metabolism. *Appl. Environ. Microbiol.* **2008**, *74*, 107–113. [CrossRef] [PubMed]
221. Chiang, Y.R.; Fang, J.Y.; Ismail, W.; Wang, P.H. Initial steps in anoxic testosterone degradation by *Steroidobacter denitrificans*. *Microbiology* **2010**, *156*, 2253–2259. [CrossRef] [PubMed]
222. Fahrbach, M.; Krauss, M.; Preiss, A.; Kohler, H.P.; Hollender, J. Anaerobic testosterone degradation in *Steroidobacter denitrificans*—Identification of transformation products. *Environ. Pollut.* **2010**, *158*, 2572–2581. [CrossRef] [PubMed]
223. Leu, Y.L.; Wang, P.H.; Shiao, M.S.; Ismail, W.; Chiang, Y.R. A novel testosterone catabolic pathway in bacteria. *J. Bacteriol.* **2011**, *193*, 4447–4455. [CrossRef] [PubMed]

© 2019 by the authors. Licensee MDPI, Basel, Switzerland. This article is an open access article distributed under the terms and conditions of the Creative Commons Attribution (CC BY) license (http://creativecommons.org/licenses/by/4.0/).

genes

MDPI

Article

Testosterone Degradative Pathway of *Novosphingobium tardaugens*

Juan Ibero, Beatriz Galán, Eduardo Díaz and José L. García *

Department of Microbial and Plant Biotechnology, Centro de Investigaciones Biológicas, Agencia Estatal Consejo Superior de Investigaciones Científicas. Ramiro de Maeztu 9, 28040 Madrid, Spain; jicaballero@cib.csic.es (J.I.); bgalan@cib.csic.es (B.G.); ediaz@cib.csic.es (E.D.)
* Correspondence: jlgarcia@cib.csic.es

Received: 4 September 2019; Accepted: 28 October 2019; Published: 31 October 2019

Abstract: In this work, we have shown that *Novosphingobium tardaugens* NBRC 16725 (strain ARI-1), a bacterial strain that was isolated due to its capacity to mineralize the estrogenic endocrine compound 17β-estradiol, is also able to mineralize testosterone, the androgenic endocrine compound. Using in silico analysis, we predicted a new putative steroid degradation (SD) gene cluster in strain ARI-1, which resembles genes involved in testosterone degradation in *Comamonas testosteroni* and other testosterone degrading bacteria like Actinobacteria (like *Rhodococcus* and *Mycobacteria* genera) although with significant differences in gene organization. A whole transcriptomic analysis of *N. tardaugens* revealed that testosterone produces a limited induction of the genes of the SD cluster that show a high basal expression in its absence. The 3β/17β-hydroxysteroid dehydrogenase involved in the first metabolic step of testosterone degradation was identified by using genetic and biochemical approaches. The construction of knockout mutant strains in the genes of the SD cluster together with in silico analyses suggests the existence of gene redundancy in the genome of *N. tardaugens*. This work will expand the knowledge about the metabolic pathways and biotransformation capabilities of a Gram-negative bacterium that could become a new model system in the bacterial steroid degradation field.

Keywords: testosterone; steroid; catabolism; transcriptomic; biodegradation

1. Introduction

Endocrine disruptors (EDCs) are chemicals that interfere with the endocrine system and produce adverse effects in both humans and wildlife. Numerous studies have reported the feminization and/or masculinization of freshwater wildlife exposed to estrogens and/or androgens in polluted rivers [1]. Natural estrogens and androgens enter the environment through the excretions of humans, domestic or farm animals, and wildlife. 17β-estradiol (E2) (estrogen) and testosterone (TES) (androgen), are the most ubiquitously sexual hormones found as pollutants in soil and water systems [2–5]. These compounds contaminate the waste water treatment plant effluents and occur at low concentration (ng/L to μg/L) [5,6]. Microbial degradation is a crucial mechanism to eliminate steroid hormones from contaminated systems and the persistence and fate of TES and E2 have been studied previously [3,5] showing that only 6% of E2 and 63% of TES could be mineralized in native soils under aerobic conditions. Similar results were obtained in water treatment plants [4]. This indicates that either a limited number of organisms can mineralize E2 [7–14] (Table S1) or that this compound is mineralized at lower rates compared to TES [15].

Although some bacteria are able to partially degrade/transform E2, only few of them, isolated mainly during the last 20 years, have been described to be able to completely metabolize E2 and use it as a sole carbon and energy source, either aerobically or anaerobically [15] (Table S1). In some cases, these bacteria need to grow in a rich medium to metabolize E2 (e.g., *Vibrio* sp. strain H5 [16] and

Buttiauxella sp. strain S19-1 [17]). In other cases, a defined mixed culture of co-degraders are required to eliminate this compound (e.g., *Achromobacter xylosoxidans* and *Ralstonia* sp.) [8]. Nevertheless, there are only few bacteria described so far that can catabolize both E2 and TES [16–23] (Table S1).

N. tardaugens NBRC 16725 (strain ARI-1) is a Gram-negative, aerobic, rod-shaped and non-motile α-Proteobacterium isolated from a sewage treatment plant in Tokyo [7,24]. It was isolated due to its capacity to mineralize E2 and it has been used when immobilized in alginate to remove estrogens from sewage and cow dung [25]. Nevertheless, nobody has described so far its capacity to degrade TES or other androgens.

The complete oxic mineralization of TES has been studied in detail in *Comamonas testosteroni* [26,27] (Figure 1). The aerobic catabolism of TES is initiated by dehydrogenation of the 17β-hydroxyl group to produce androst-4-en-3,17-dione (AD), which undergoes a further dehydrogenation to form androsta-1,4-diene-3,17-dione (ADD). The subsequent cleavage of the core ring system is catalysed by several oxygenases that utilize oxygen as co-substrate [26,28] (Figure 1).

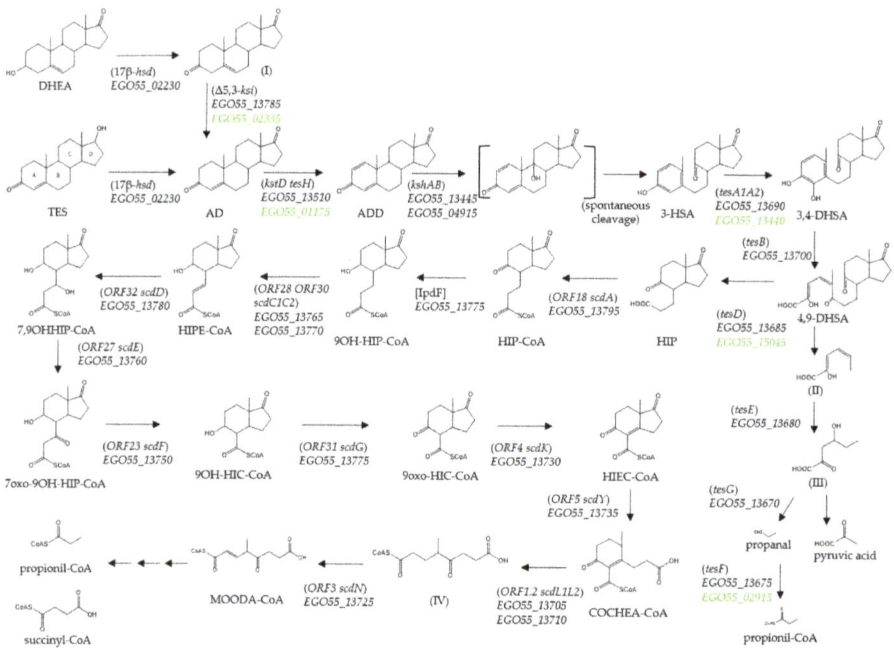

Figure 1. Proposed TES degradation pathway in *N. tardaugens*. Compound names are shown or indicated with an abbreviation: (I) androst-5-ene-3,17-dione; (3-HSA) 3-hydroxy-9,10-secoandrosta-1,3,5(10)-triene-9,17-dione; (3,4-DHSA) 3,4-dihydroxy-9,10-secoandrosta-1,3,5(10)-triene-9,17-dione; (4,9-DHSA) 4,5-9,10-diseco-3-hydroxy-5,9,17-trioxoandrosta-1(10),2-diene-4-oic acid; (II) 2-hydroxyhexa-2,4-dienoate; (III) 4-hydroxy-2-oxohexanoate; (HIP) 9,17-dioxo-1,2,3,4,10,19-hexanorandrostan-5-oic acid CoA ester; (HIPE-CoA) 9-hydroxy-17-oxo-1,2,3,4,10,19-hexanorandrost-6-en-5-oic acid; (9OH-HIC-CoA) 9α-hydroxy-17-oxo-1,2,3,4,5,6,10,19-octanorandrostan-7-oic acid, (HIEC-CoA) 9,17-dioxo-1,2,3,4,5,6,10,19-octanorandrost-8(14)-en-7-oic acid; (COCHEA-CoA) 9-oxo-1,2,3,4,5,6,10,19-octanor-13,17-secoandrost-8(14)-ene-7,17-dioic acid; (IV) 4-methyl-5-oxo-octane-1,8-dioic acid; and (MOODA-CoA) 4-methyl-5-oxo-oct-2-ene-1,8-dioic acid. Enzyme names are: (3,17β-*hsd*) 3,17β-hydroxysteroid dehydrogenase; (Δ5,3-*ksi*) 3-ketosteroid Δ4(5)-isomerase; (*kstD*) 3-ketosteroid-delta1-dehydrogenase; (*kshAB*) 3-ketosteroid 9alpha-hydroxylase; (*tesA1A2*) 3-HSA hydroxylase; (*tesB*) 3,4-DHSA 4,5-dioxygenase; (*tesE*) 2-hydroxyhexa-2,4-dienoate hydratase; (*tesG*) 4-hydroxy-2-oxovalerate aldolase; (*tesF*) propionaldehyde

dehydrogenase; (*tesD*) 4,9-DHSA hydrolase; (*ORF18 scdA*) HIP-CoA ligase; [*IpdF*] 5-oxo HIC-CoA oxidase; (*ORF28 ORF30 scdC1C2*) acyl-CoA dehydrogenase; (*ORF32 scdD*) enoyl-CoA hydratase; (*ORF27 scdE*) dehydrogenase; (*ORF23 scdF*) CoA acetyl transferase; (*ORF31 scdG*) hydroxylacyl dehydrogenase; (*ORF4 scdK*) acyl-CoA dehydrogenase; (*ORF5 scdY*) enoyl-CoA hydratase; (*ORF1,2 scdL1L2*) β -ketoacyl-CoA-transferase; and (*ORF3 scdN*) CoA-hydratase. The proposed catabolic genes from *N. tardaugens* are indicated in italics by their locus code (*EGO55_XXXX*), in green other possible candidates are considered. The homologous genes from model bacteria *C. testosteroni* TA441 and *M. tuberculosis* H37Rv are shown in brackets and square brackets, respectively.

In this work, we have analysed the growth of *N. tardaugens* NBRC 16725 (strain ARI-1) with TES and other C-19 steroids as the sole carbon and energy sources. The complete genome sequence of this bacterium has been recently reported [29] allowing us to predict a SD gene cluster presumably involved in the catabolism of TES and other C-19 steroids. We have determined the expression of the catabolic genes by using a transcriptomic approach. The enzyme 17β-hydroxysteroid dehydrogenase, encoded outside of the predicted SD cluster, involved in the first step of TES catabolism in *N. tardaugens* ARI-1, has been identified by biochemical analyses and a metabolic pathway for the degradation of TES in this strain has been proposed.

2. Materials and Methods

2.1. Chemicals

Testosterone (TES), 17β-estradiol (E2), estrone (E1), pyruvate, chloroform, n-hexane, ethyl acetate, sulphuric acid and acetonitrile were purchased from Merck KGaA (Darmstadt, Germany). Androst-4-en-3,17-dione (AD) and androsta-1,4-diene-3,17-dione (ADD) were purchased from TCI EUROPE (Boereveldseweg, Belgium). TES-17-acetate (TES-Ac) was purchased from CYMIT QUÍMICA S.L. (Barcelona, Spain). Trans-dehydroandrosterone (DHEA) and pregnenolone (PREG) were purchased from Fluka (Switzerland). Randomly methylated β-cyclodextrin (TRMB-T Randomly Methylated BCD) (CDX) was purchased from Cyclodex (Alachua, USA). Other chemicals and reagents were purchased from Merck KGaA (Darmstadt, Germany).

2.2. Strains and Growth Media

Bacterial strains and plasmids used in this study are listed in Table S2. *N. tardaugens* NBRC 16725 (strain ARI-1) was purchased from the Leibniz-Institut DSMZ type culture collection. This strain and its mutants were cultured at 30 °C in an orbital shaker at 200 rpm. Nutrient broth (NB) (Difco) was used as rich medium to grow this strain. Minimal medium M63 [KH_2PO_4 (136 g/L), $(NH_4)_2SO_4$ (20 g/L), $FeSO_4{\cdot}7H_2O$ (5 mg/L), pH 7.0] was supplemented with 0.39 mM $CaCl_2$, 1 mM $MgSO_4$ and the appropriate carbon source concentration (we used a carbon equimolar concentration for each substrate tested). Steroids and pyruvate stock solutions were prepared in PBS buffer and 70 mM CDX so the final carbon concentration in the culture was 36 mM in 13.33 mM CDX. *Escherichia coli* DH10B, BL21 (DE3) and HB101 strains were grown at 37 °C in an orbital shaker at 200 rpm in lysogeny broth (LB) medium [30]. The appropriate antibiotics, i.e., chloramphenicol (34 mg/mL), kanamycin (50 mg/mL) or rifampicin (50 mg/mL) were added when needed.

2.3. DNA Manipulation

Molecular biology and DNA manipulations where performed as described elsewhere [31]. *N. tardaugens* genomic DNA was extracted as described before [29]. Plasmid DNA was purified using High Pure Plasmid Isolation Kit (Roche). DNA fragments where purified with QIAquick PCR Purification Kit (Qiagen) or QIAquick Gel Extraction Kit (Qiagen). *E. coli* cells were transformed using the RbCl method or by electroporation using a Gene Pulser (Bio-Rad) [32]. DNA amplification was performed in a Mastercycler Gradient (Eppendorf) using the oligonucleotides listed in Table S3, which were purchased from Merck KGaA (Darmstadt, Germany). Phusion High-Fidelity DNA

Polymerase (New England Biolabs) was used for cloning amplifications and Taq DNA polymerase (Biotools) for screening. All PCR products were checked by agarose gel electrophoresis and those aimed for cloning were confirmed by DNA sequencing by Secugen S.L. (Spain). Digestion of DNA fragments was done using restriction enzymes (New England Biolab) and ligation was performed with Instant Sticky-end Ligase Master Mix (New England Biolabs).

2.4. Gene Expression Analyses

2.4.1. RNA Extraction

Total RNA of *N. tardaugens* cells was extracted from cultures grown in minimal medium with 20 mM CDX and TES or pyruvate as carbon sources. Cells where harvested in mid exponential phase (OD_{600} 0.6) and stored at −80 °C. Pellets where thawed and cells were lysed in 400 µL TE buffer (10 mM Tris-HCl, 1 mM EDTA, pH 7.5) containing lysozyme (50 mg/mL) following three freezing–thawing cycles. High Pure Isolation Kit (Roche), followed by DNA-free DNA Removal Kit (Invitrogen) treatment, was used to obtained pure RNA. Purity and concentration were measured in a ND1000 spectrophotometer (Nanodrop Technologies).

2.4.2. RNA-seq

RNA-seq was done in Macrogen Korea. Total RNA integrity was checked using an Agilent Technologies 2100 Bioanalyzer. Ribosomal RNA was removed from the total RNA with Ribo-Zero rRNA Removal Kit to later construct a 100 bp paired-end library using TruSeq RNA Sample Prep Kit v2 that was quality-checked in an Agilent Technologies 2100 Bioanalyzer using a DNA 1000 chip. Library sequencing was performed in a HiSeq 3000 4000 (Illumina) using TruSeq 3000 4000 SBS Kit v3 as reagent. Bioinformatic analysis was performed by the Bioinformatics and Biostatistics service of the Centre for Biological Research (CIB-CSIC). Raw read data quality was checked using FastQC and trimmed with Trimmomatic. Trimmed reads were mapped against the genome sequence of *N. tardaugens* (accession number CP034179) using Bowtie2 and expression quantification was done using HTSeq-count. An average of 50 million raw sequencing reads (approximately 6.7 billion base pairs; average 1300× genome coverage per sample) were generated from samples from two independent experiments in the presence of pyruvate or TES, each with three biological replicates. After trimming the raw sequence reads, an average 48.7 million high-quality clean reads were mapped to the *N. tardaugens* reference genome and between 98.4% and 82.5% were uniquely mapped (Table S4). Differential expression analysis was done using Deseq2 and GO-term enrichment analysis was performed with GOSeq. The dissimilarity matrix shown in the heatmap was obtained with the euclidean distance and the cluster analysis was performed with the Ward's minimum variance method. Bioinformatic analysis software was used with default settings. Raw read data obtained from the three replicates of the transcriptome of the strain grown in pyruvate and TES have been deposited in the Sequence Read Archive (SRA) database of the National Centre for Biotechnology Information (NCBI) under accession numbers SRR9027780, SRR9027781, SRR9027779 (Bioproject PRJNA541800) and SRR9027897, SRR9027898, SRR9027896 (Bioproject PRJNA541801), respectively.

2.5. Isolation of a Rifampicin Resistant Phenotype of *N. tardaugens*

N. tardaugens wt strain, sensitive to rifampicin, was cultured in NB medium supplemented with rifampicin up to stationary phase (≈48 h). Cells where then plated on NB plates supplemented with the antibiotic and one single colony was picked, grown and used as *N. tardaugens* Rf^r, suitable for conjugation.

2.6. Construction of *N. tardaugens* Knockout Strains

The knockout strains were constructed by double homologous recombination using the suicide vector pK18mobsacB [33]. *N. tardaugens* genomic DNA was used as template to amplify two fragments of ≈700 bp containing the upstream and downstream regions of the gene to delete. The fragments were digested with the appropriate restriction enzymes and cloned in the unique sites of the plasmid. The ligation product was transformed into *E. coli* DH10B competent cells and once recombinant candidates were PCR-checked, the cloned region was confirmed by sequencing. The plasmids were transformed by triparental conjugation [34] into *N. tardaugens* Rf^r as recipient strain using *E. coli* HB101 (pRK600) [35] as helper and *E. coli* DH10B, harbouring the corresponding vector, as donor. The strains resulting of plasmid integration in this first recombination event were selected in NB agar plates containing kanamycin and rifampicin and screened by PCR. Selected candidates were grown up to stationary phase (≈48 h) in NB medium and then plated in NB supplemented with 5% sucrose. The clones that are resistant to sucrose and sensitive to kanamycin were checked by PCR and the amplicon was sequenced to confirm the second cross-over event.

2.7. Heterologous Production of the Putative 3β/17β-Hydroxysteroid Dehydrogenase from N. tardaugens

To overproduce the putative 3β/17β-hydroxysteroid dehydrogenase (3β/17β-HSD), genes *EGO55_02235-EGO55_02230* (*hsd70-hsd60*) were cloned together or separately into the expression vector pET-29a. The DNA fragments containing *EGO55_02235-EGO55_02230*, *EGO55_02235* and *EGO55_02230* were amplified with primers 5′NdeIhsdTandemEcolif-3′BamHIhsdTandemEcolir, 5′NdeIhsd70SUBf-3′XhoIhsd70SUBr, 5′NdeIhsd60SUBf-3′XhoIhsd60SUBr, respectively (Table S3), using *N. tardaugens* genomic DNA as template. The DNA fragments were digested with the corresponding restriction enzymes *Nde*I-*Bam*HI, *Nde*I-*Xho*I and *Nde*I-*Xho*I respectively, and then ligated into the pET29a vector yielding pET29Hsd70-Hsd60, pET29Hsd70 and pET29Hsd60 plasmids, respectively. Electrocompetent cells of *E. coli* BL21 (DE3) were transformed with plasmids pET29Hsd70-Hsd60, pET29Hsd70 and pET29Hsd60. The resulting strains *E. coli* BL21 (DE3) (pET29Hsd70-Hsd60), *E. coli* BL21 (DE3) (pET29Hsd70) and *E. coli* BL21 (DE3) (pET29Hsd60) were grown in 50 mL LB containing kanamycin up to an OD_{600} of 0.5–0.8 and gene expression was induced with 0.2 mM IPTG (isopropyl-β-D-thiogalactopyranoside). After 3 h of induction, the cells were harvested, washed with 0.85% saline solution and resuspended in 20 mM Tris-HCl (pH 8.0). Cells were lysed by sonication using a Branson sonicator applying three cycles of 30-s bursts at maximum power alternated with 30 s cooling in ice. Soluble fraction was separated by centrifugation in a Sorvall Linx 6000 SS-34 rotor (15 min at 4 °C and 14000 rpm) and protein concentration was calculated using the Bradford method [36]. The overproduction of Hsd70 and Hsd60 proteins in the soluble fraction of the crude extract was checked by SDS-polyacrylamide gel electrophoresis (Figure S1).

2.8. Enzymatic Assay of 3β/17β-HSD Activity

Enzymatic assays of 3β/17β-HSD activity were performed in 500 µL total reaction volume. Reaction mixture consisted of 0.5 mg protein of the crude extracts obtained from the recombinant strains, 500 µM of the steroidal substrate dissolved in CDX (3.5 mM final concentration of CDX), 1 mM NAD+ and 50 mM sodium phosphate buffer (pH 7.0). Reaction assays were started by adding NAD+ and stopped by adding 2 volumes of chloroform.

2.9. Organic Phase Extraction and Thin Layer Chromatography (TLC) analysis

The presence of steroidal compounds in culture media and enzymatic assay mixtures was determined after organic solvent extraction by TLC analysis. Two volumes of chloroform were added and the mixture was vortexed for 30 s and centrifuged for 1 min at 13,000 rpm in an Eppendorf microcentrifuge. The organic phase was extracted and dried. The dried sample was dissolved in 100 µL of acetonitrile and analysed by thin layer chromatography (TLC). For TLC analysis, 10 µL of the standards and the samples dissolved in acetonitrile were spotted in silica gel plates (TLC Silicagel 60 F_{254}, Merck Millipore) and n-hexane: ethyl acetate (10:4 v/v) was used as developing system. Steroid products were visualized by UV and revealed by spraying 20% (v/v) sulphuric acid and heating at 120 °C.

2.10. In Silico Analyses

Gene product prediction was done with Rapid Annotations using Subsystems Technology (RAST) [37]. Homologous genes search in different bacteria was performed by using the Standard Protein Basic Local Alignment Search Tool (BLASTp) [38].

3. Results

3.1. Catabolism of C-19 Compounds in N. tardaugens

Although the ability of *N. tardaugens* strain ARI-1 to grow in E2 as a sole carbon and energy source has been described [24], nothing was known regarding its capacity to mineralize other C-19 steroids, such as TES, AD, ADD and DHEA. Figure 2a shows that *N. tardaugens* is able to mineralize these compounds. Moreover, we were unable to detect by TLC any of these steroids after 18 h cultivation (39 h in case of DHEA) (Figure 2b). Interestingly, strain ARI-1 was also able to grow in the xenobiotic steroid compound TES-Ac generating TES as transient intermediate (Figure 2a,b). Taken together all these results reveal that *N. tardaugens* contains an efficient aerobic degradative pathway for C-19 steroids.

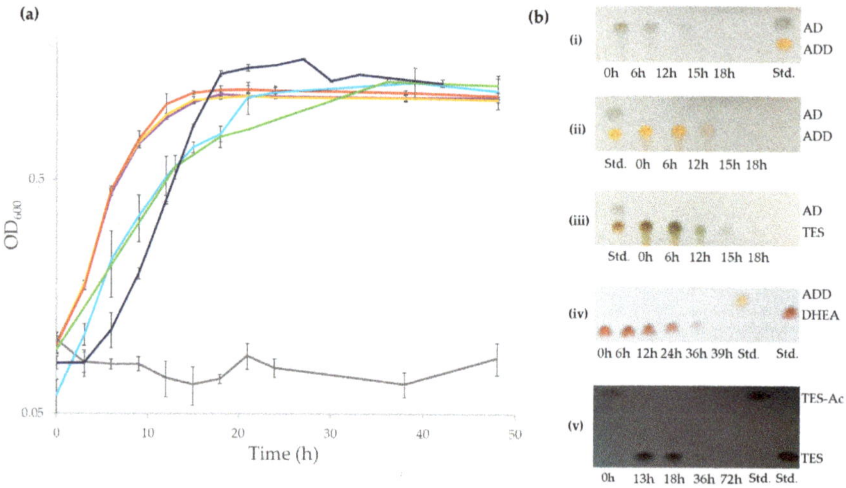

Figure 2. (**a**) Bacterial growth (Log_2OD600) of *N. tardaugens* NBRC 16725 when cultured in M63 minimal medium containing 1.89 mM AD (purple), 1.89 mM ADD (yellow), 1.89 mM TES (red), 1.89 mM DHEA (light blue), 1.71 mM TES-Ac (green), 2 mM E2 (dark blue) and 13.33 mM CDX (grey) and (**b**) TLC analysis of the organic extraction of the culture of *N. tardaugens* along time growing in: (**i**) AD, (**ii**) ADD, (**iii**) TES, (**iv**) DHEA and (**v**) TES-Ac. The AD, ADD, TES, DHEA and TES-Ac standards (Std.) (1 mM) are also shown.

3.2. In Silico Identification of N. tardaugens Genes for Catabolism of C-19 Compounds

In silico analysis of the assembled *N. tardaugens* genome (4,358,096 bp) [29] using as query the amino acid sequences of the coding genes responsible for TES degradation in *C. testosteroni* TA441 revealed the existence of a putative steroid degradation gene cluster covering a 26.4 kb region (*EGO55_13795-EGO55_13670* (Figure 3). The predicted SD cluster is organized in two regions that are transcribed divergently, i.e., *EGO55_13690-EGO55_13670* and *EGO55_13695-EGO55_13795* (Figure 3b). In contrast to *C. testosteroni*, where the TES degradative genes are organized in two different clusters (Figure 3a), *N. tardaugens* shows a more compact gene organization.

Remarkably, the four genes essential for the three initial steps of C-19 steroid catabolism, i.e., the *hsd* gene encoding the 3β/17β-hydroxysteroid dehydrogenase (3β/17β-HSD), the *kstD* gene encoding the ketosteroid dehydrogenase, and the *kshA* and *kshB* genes encoding the oxidase and reductase subunits of the 9α-ketosteroid hydroxylase, respectively (Figure 1), are not contained within this degradation cluster. Interestingly, several genes homologous to *kstD*, *kshA* and *hsd* from *C. testosteroni* have been found distributed along the genome of *N. tardaugens* (Table S5). However, only one gene, *EGO55_04915*, encoded a protein showing identity (33%) with the reductase subunit of the ketosteroid hydroxylase *ORF17* (*kshB*) of *C. testosteroni*. Finally, a TeiR-like regulatory protein which is a positive regulator that induces the expression of the TES cluster in *C. testosteroni* [39], is not present in the *N. tardaugens* genome. Furthermore, the absence of other putative regulatory genes located near the SD cluster suggest either the absence of a specific regulation by TES or at least a different transcriptional regulation in this bacterium.

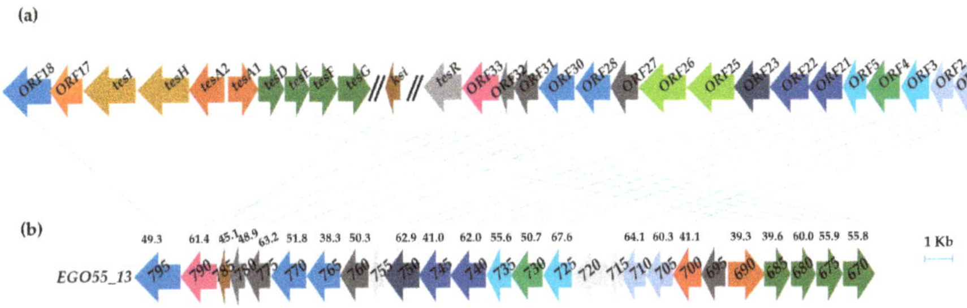

Figure 3. (**a**) *Comamonas testosteroni* TA441 mega-cluster of steroid degradation (accession number LC010134). (**b**) Putative steroid degradation gene cluster of *N. tardaugens* NBRC 16725 (accesion number CP034179). Genes encoding same funtion are pictured in the same color and connected with a line. Percent identities (BLASTp) of the gene products are pictured compared with those of *C. testosteroni*.

We have tried to demonstrate the involvement of the SD cluster in the degradation of C-19 compounds by constructing two *N. tardaugens* knockout mutant strains in two representative genes of the SD cluster, i.e., the *EGO55_13795* (*fadD3*) and *EGO55_13685* (*tesD*) genes. When both mutants were grown in TES as the sole carbon and energy sources their growth was not impaired (Figure S2). Interestingly, the analysis of the *N. tardaugens* genome revealed the existence of several homologous genes of the SD cluster located along the chromosome (Table S5) suggesting that some reactions involved in C19 steroid degradation could be replaced by such homologous genes. Thus, the analysis of the *N. tardaugens* genome revealed the existence of 9 *fadD3* and 3 *tesD* homologous genes (Table S5) that would explain the observed growth phenotype of the *fadD3* and *tesD* mutant strains.

3.3. Whole Transcriptomic Analysis of *N. tardaugens* Grown in Testosterone

To determine the expression of the genes involved in the degradation of C-19 steroids we performed RNA-seq analyses in *N. tardaugens* cultured using pyruvate (control condition) or TES as carbon sources. Differential expression analysis yielded 2046 differentially expressed genes (DEGs) (from 3980 total genes in genome), where 863 were up regulated and 1183 were down regulated in TES condition compared to pyruvate (being FC 2 or −2, respectively, the cut-off value) (Table S6), showing a noticeable contrast in differential expression pattern (Figure 4). The functional classification of DEGs in different GO terms is shown in Figure 5. The highest level of up regulation (fold change (FC) > 10) was observed in 49 genes (Table S6) but other 111 genes were notably up regulated (5 < FC < 10) (Table S6).

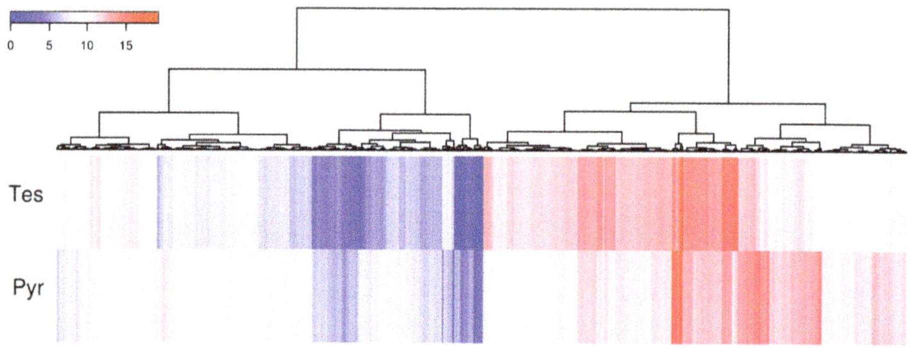

Figure 4. Heatmap diagram of cluster analysis showing the log2 mean normalize expression in each experimental growth condition for those genes where a FC > 2 and FC < −2 was observed.

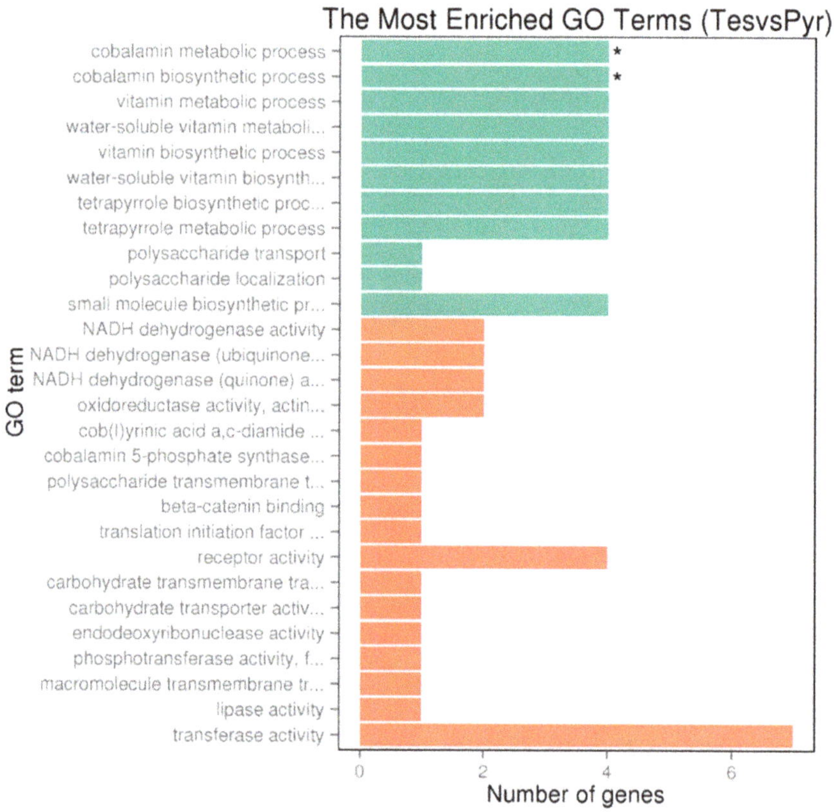

Figure 5. GO enrichment bar chart of upregulated genes (DEG) representing the number of DEGs enriched in biological process, cellular component and molecular function. Colors represent different GO types: biological process (green) and molecular function (orange). The term with a star "*" is significantly enriched term (corrected $p_{\text{value}} < 0.05$).

The data obtained from the transcriptomic analysis show a slight differential induction of the SD cluster in the presence of TES with respect to the pyruvate control condition (Table S6). Nevertheless, it is important to notice that the basal expression level of these genes in pyruvate condition is already high when compared to that of housekeeping genes. The level of induction of the genes included in the *EGO55_13695-EGO55_13795* genes is slightly higher than those of the *EGO55_13690-EGO55_13670* genes (Table S6). Several genes showed a fourfold increase in expression levels: *EGO55_13735*, *EGO55_13740*, *EGO55_13745* and *EGO55_13750*. FC values of previously identified genes homologous to *kstD* and *kshA* from *C. testosteroni* TA441 (Table S5), allowed us to propose *EGO55_13510* (3.5-fold induction) and *EGO55_13445* (7.8-fold induction), as the KstD and KshA involved in TES degradation pathway in *N. tardaugens*, respectively (Figure 1). These *kstD* and *kshA* genes are located approximately 30 and 50 kb away from the SD cluster, respectively. Additionally, closed to the predicted kshA coding gene (*EGO55_13445*) there is a predicted TesA2 coding gene (*EGO55_13440*) that is induced 7.2-fold, which is a higher value than the one located within the SD cluster. Also the gene *EGO55_15045*, encoding a putative TesD homolog, showed higher induction than the corresponding gene in the SD cluster. These suggest that *EGO55_13440* and *EGO55_15045* genes could be involved in TES degradation in *N. tardaugens*.

Interestingly, *EGO55_01995*, *EGO55_02005*, *EGO55_02015* and *EGO55_02020* genes, encoding a putative propionyl-CoA carboxylase biotin-containing subunit, a methylmalonyl-CoA mutase, a methylmalonyl-CoA epimerase, and a propionyl-CoA carboxylase carboxyl transferase subunit, respectively, are found among the most upregulated genes in TES condition (Table S6). This result allowed us to identify a cluster (*EGO55_01990-EGO55_02025*) containing genes showing a high level of identity to those experimentally described as responsible of the methylmalonyl degradation pathway for propionic acid degradation [40–44] (Table S7). Furthermore, this cluster also contains a putative biotin synthase (*EGO55_02000*) that is upregulated, in agreement with the requirement of B7 cofactor for propionyl-CoA carboxylase activity [45] needed to metabolize propionyl-CoA, presumably generated in TES degradation (Figure 1).

As shown in Table S6, up to 19 genes annotated as involved in B12 cofactor biosynthesis pathway are significantly upregulated in TES condition. Figure 5 also shows that the highest number of upregulated genes in TES condition belong to cobalamin biosynthetic and metabolic processes, water-soluble vitamin biosynthesis, vitamin biosynthetic process and vitamin and water-soluble vitamin metabolic processes GO terms. High expression levels of genes involved in cobalamin synthesis pathway correlates with the requirement of this cofactor by the methylmalonyl-CoA mutase, which is upregulated in TES condition as indicated above (Table S6).

3.4. Identification of the Initial Biochemical Step of TES Degradation Pathway in N. tardaugens

In *C. testosteroni*, the degradation of TES starts by the dehydrogenation of the 17β-hydroxyl group to render AD. This step is catalysed by a short chain dehydrogenase 3β/17β-HSD, a tetrameric NAD(H)-dependent reversible enzyme [46]. Comparative gene analyses yielded up to 16 proteins homologous to the 3β/17β-HSD of *C. testosteroni* in *N. tardaugens* genome (Table S5). Due to this gene redundancy, we looked at the induction fold of those genes in the presence of TES. Among the 16 genes the *EGO55_02235* was slightly induced 2-fold and it is located in tandem with *EGO55_02230* encoding also a putative 3β/17β-HSD. These isoenzymes, Hsd60 and Hsd70, show 80% amino acid sequence identity. They form a putative four-gene operon together with the *EGO55_02225* gene coding for a putative esterase and the *EGO55_02240* gene annotated as a permease of the major facilitator superfamily. To prove the involvement of Hsd60 and Hsd70 in TES catabolism, a double knock out mutant was produced, where *EGO_02230* and *EGO_02235* genes were deleted. The mutant strain was grown in TES as sole carbon and energy source showing that its growth was not impaired (Figure S2). This result is not surprising given the number of homologous genes found in *N. tardaugens* genome (Table S5) that could be replacing the enzymatic activity of the deleted genes.

To further determine the putative role of Hsd60 and Hsd70 in TES metabolism the *EGO55_02230* and *EGO55_02235* genes were cloned in the pET29a vector and the resulting plasmid, named pET29Hsd70-Hsd60, was transformed in *E. coli* BL21 (DE3) cells to overproduce both enzymes (Figure S1). Enzymatic assays using crude extracts from *E. coli* BL21 (DE3) (pET29Hsd70-Hsd60) cells revealed a 17β-HSD reaction converting TES to AD (Figure 6a) and E2 to E1 (Figure 6b). Same enzymatic assays using DHEA and PREG as substrates showed that the catalytic activity present in the crude extract is able to transform them (according to their mobility in TLC) into the expected keto compounds, i.e., Δ5 androstadione and isoprogesterone, respectively (Figure 6c,d). The crude extract showed a lower activity when PREG (Figure 6d) was used as substrate, suggesting that the C-17 chain of PREG impairs the recognition of substrate. In this sense, no activity was detected when cholesterol with a long C-17 side chain was used as substrate.

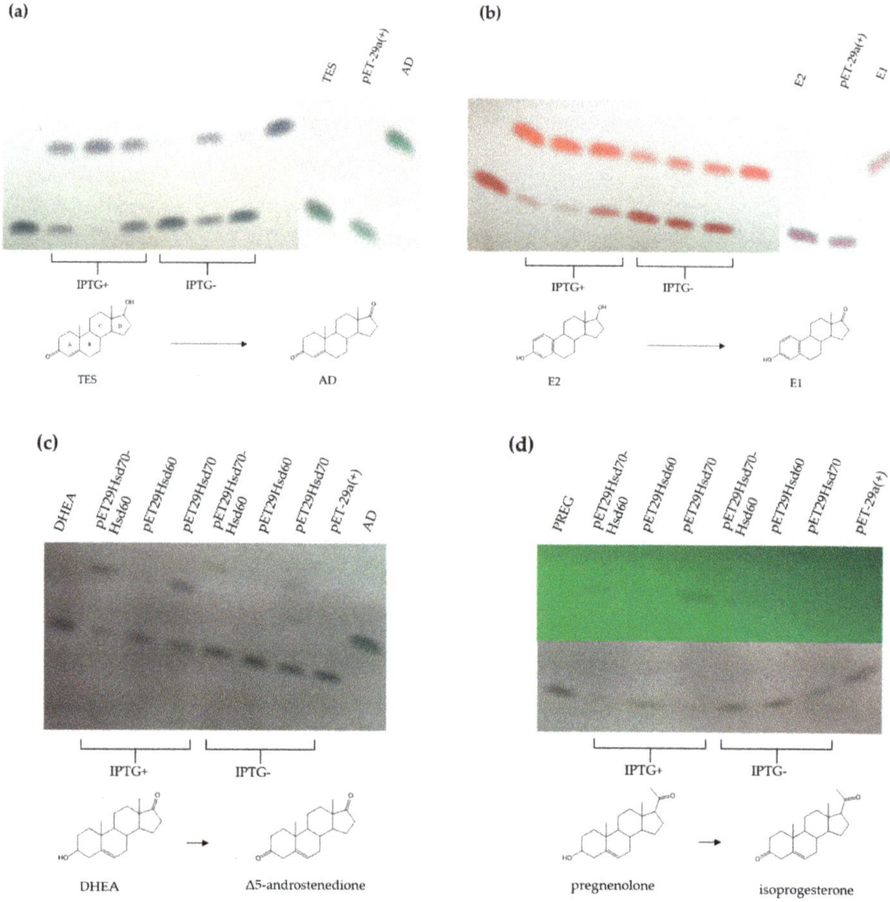

Figure 6. TLC analysis of the enzymatic reaction of crude extracts from *E. coli* BL21(DE3) cells harbouring pETHsd70-Hsd60, pETHsd70 and pETHsd60 transforming (**a**) TES, (**b**) E2, (**c**) DHEA and (**d**) PREG. In (**d**) the lower panel shows reduced products revealed with UV light. Standards of AD, TES, E2, E1, DHEA and PREG were also added. The use of inductor (IPTG) for overexpression is indicated. The molecular structure of the steroidal compounds involved in the reductive reactions are represented.

Once the 3β/17β-HSD activity was determined within these cell extracts, the genes encoding Hsd70 and Hsd60 were cloned separately into the pET29a vector to explore their individual catalytic abilities. The resulting plasmids, pETHsd70 and pET29Hsd60, were transformed in *E. coli* BL21 (DE3) cells. Enzymatic assays using crude extracts showed, interestingly, that Hsd60 enzyme catalyzes more efficiently the dehydrogenation of C17-OH (Figure 6a,b), whereas Hsd70 catalyzes more efficiently the dehydrogenation of C3-OH (Figure 6c,d).

4. Discussion

N. tardaugens, ARI-1 strain was isolated based on its ability to metabolize E2 [24]. Here we show that in addition this strain has a prominent capability to metabolize other steroids. Particularly, we have demonstrated that this strain is able to degrade steroids that can also be considered as toxic EDCs, such as TES, TES-Ac, DHEA and other C-19 steroids, like AD and ADD (Figure 2).

The recent reduction of the number of contigs in a previous work [29] allowed us to identify the genes that could be responsible for the degradation of C-19 steroids in *N. tardaugens*. A comparison with the genes described for other C-19 degradation pathways, particularly with those described for the degradation of TES in *C. testosteroni*, was carried out. A significant finding was that ARI-1 contains most of the predicted genes in a very compact cluster compared with *C. testosteroni* that has at least two main clusters to encode the complete pathway (Figure 3) [26]. The same cluster structure as that of *N. tardaugens* has been described in other estrogen degrader strains such as *Sphingomonas* sp. KC8 strain and *Altererythrobacter estronivorus* [18]. In the particular case of the KC8 strain, those genes are being expressed in the E2 and TES grown cells, suggesting that strain KC8 uses the same gene products to degrade the C/D rings of both C18 and C19 compounds [18]. Since many of the genes of the SD cluster putatively responsible for CD-ring degradation of TES (C19) in *N. tardaugens* do not have any other homolog in the genome, it seems very likely that this cluster is also responsible for the degradation of C18 (E2) compounds. This is in accordance with recent work where a great level of conservation of predicted CD-ring degradation genes was found among different genera of steroid-metabolizing proteobacteria [47,48].

In *N. tardaugens*, only few genes required for C-19 degradation are located outside the SD cluster. We speculate that the genes *EGO55_02230* (*hsd*), *EGO55_13510* (*kstD*) or *EGO55_13445* (*kshA*), that are required to metabolize C19 compounds, but that are not included in the SD cluster, might be involved in the aerobic degradation of other steroid compounds, so its location outside the cluster suggests that they could be playing a more global role.

It is interesting that *kshB* (*EGO55_04915*) (*ORF17* in *C. testosteroni*) is not present in the SD cluster and no homologous gene to *tesA1* is found in the genome of *N. tardaugens*. KshB and TesA1 are the flavin reductase components of the flavin-dependent two-component monooxygenases KshA and TesA2, respectively. It has been demonstrated that these type of reductase components act in trans and are not highly specific and does not require a particular interaction [49,50]. However, it cannot be discarded that the reduced flavin required by KshA and TesA2 should be provided by unspecific reductases in *N. tardaugens*.

There are some genes included within the SD cluster of *N. tardaugens* that are not found in the TES cluster of *C. testosteroni*. This is the case of *EGO55_13755*, *EGO55_13720* and *EGO_13715*, which encode a nuclear transport factor 2 family protein, a lipid transfer protein and a benzoylsuccinyl-CoA thiolase, respectively. The precise role of these enzymes in the SD cluster remains unknown. On the contrary, there are some genes located within the TES cluster of *C. testosteroni* that have been located outside of the SD cluster. This is the case of *EGO55_13615* gene that is homologous to tesI gene from *C. testosteroni* encoding a ketosteroid Δ4-dehydrogenase, involved in epiandrosterone degradation [51]. In addition, the *ORF25* and *ORF26* genes of TES cluster encoding a 6-aminohexanoate-cyclic-dimer hydrolase are homologous to the *EGO55_02680* and *EGO55_04710* of *N. tardaugens*, respectively, that are located far from the SD cluster. These genes are also absent in the SD cluster of other estrogen-degrading strains like *Sphingomonas* sp. KC8 and *Altererythrobacter estronivorus* [18]. The role of these genes in TES catabolism in *C. testosteroni* is not clear since their disruption does not impair TES degradation [26].

Remarkably, we have detected a large number of genes dispersed in the genome of *N. tardaugens* that are homologous to those contained in the SD cluster and are expressed at significant levels in the presence of TES. This is the case of *EGO55_15045* (*tesD* homologous), *EGO55_02915* (*tesF* homologous), *EGO55_02335* (*ksi* homologous) and *EGO55_13440* (*tesA2* homologous), *EGO55_09255* and *EGO55_20150* (*ORF23* homologous), and *EGO55_01175* (*kstD* homologous) (Figure 1, Table S5). Therefore, we cannot discard the possibility that these genes could be involved in the metabolism of C-19 compounds in *N. tardaugens* enhancing the versatility and robustness of steroid metabolism.

In spite of the fact that the expression of other C-19 clusters, as occurs in *C. testosteroni*, are regulated by a specific regulator, in ARI-1 the data obtained from the transcriptomic analyses showed only a slight differential expression of the genes from the SD cluster in the presence of TES (Table S6). The catabolic enzymes for TES degradation described so far were not constitutively expressed but, rather, were significantly induced by their respective substrates [39,52]. For instance, in *C. testosteroni* the LuxR-type transcription activator TeiR regulates the transcription of genes involved in the initial enzymatic steps of TES degradation [39]. Moreover, the *teiR* deletion mutant was not able to use TES as a carbon source. TesR from *C. testosteroni* TA441, almost identical to that of TeiR from *C. testosteroni* ATCC11996, was shown to be necessary for induction of the steroid degradation gene clusters, *tesB* to *tesR*, *tesA1* to *tesG*, and *tesA2* to *ORF18* in *C. testosteroni* TA441 [53]. Interestingly, it has been reported that *tesR*-like regulatory genes are only present in *C. testosteroni* strains but are not found in other testosterone-degrading bacteria [26]. The in silico analysis showed that neither a *teiR* homologue nor other putative regulatory genes are present in the vicinity of the SD cluster in ARI-1. This observation is consistent with the transcriptomic analysis revealing a significant basal expression of the SD cluster in the absence of TES, which might be explained because this pathway could be fundamental for the survival of the ARI-1 strain in the specific niche where it was isolated.

There are other genes that appear to be induced by pleiotropic induction effects in the presence of TES and which are not directly related to the degradative C-19 pathway but rather to central metabolic processes (e.g., stress processes, requirements of cofactor synthesis, etc.). For instance, the production of propionyl-CoA as a presumable metabolite of TES degradation or the requirements of CoA to mineralize the TES rings can promote the generation of multiple stresses and pleiotropic differential expression signals when compared with the transcriptome of pyruvate metabolism [54,55]. According to the metabolic scheme shown in Figure 1, TES would be converted into succinyl-CoA (1 mol), pyruvate (1 mol), acetyl-CoA (3 mol) and propionyl-CoA (2 mol). The analysis of the genome revealed that *N. tardaugens* does not have a methylcitrate cycle and thus, it metabolizes propionyl-CoA by the methylmalonyl-CoA pathway to synthesize succinyl-CoA. The finding that the methyl-malonyl-CoA pathway is highly upregulated under TES growth conditions, is a solid evidence that propionyl-CoA is formed.

3β/17β-HSDs are essential enzymes in the biosynthesis of all classes of mammalian steroids. They catalyze the interconversion of alcohol and carbonyl functions stereospecifically in defined positions using oxidized or reduced NAD(H) or NADP(H) as co-substrates [56]. It is also well known that 3β/17β-HSDs are involved in the first catabolic step in TES degradation and, particularly, the 3β/17β-HSD of *C. testosteroni* has been extensively studied [20,46,57,58]. It catalyzes the reversible reduction/dehydrogenation of the oxo/β-hydroxy groups at positions C3 and C17 of steroids, including hormones and isobile acids. The dual positional specificity of this 3β/17β-HSD has been explained after resolving its 3D structure [46]. Kinetic studies revealed an ordered mechanism and suggested a single catalytic site accommodating both the 3β and 17β activities [46]. *N. tardaugens* has 16 genes that might encode putative 3β/17β-HSD homologous to the enzyme described in *C. testosteroni* (Table S5). The transcriptomic data obtained in the presence of TES allowed us to postulate specific candidates that could be involved in the metabolism of TES. Furthermore, after exploring the genetic environment of the 16 candidates, we identified the *EGO55_02235*-*EGO55_02230* tandem genes that code for two putative 3β/17β-HSD isoenzymes (Hsd60 and Hsd70) as part of a putative four-genes operon together with the *EGO55_02225* gene coding for a putative esterase and the *EGO55_02240* gene encoding a putative permease of the major facilitator superfamily. Taking into account all these reasons we cloned and tested the activity of the two putative HSDs of this operon demonstrating that they have a differential activity on C-19 steroids (Figure 6). Our results showed that isoenzyme Hsd60 is more efficient performing the C17-OH dehydrogenation than Hsd70 (Figure 6a,b), which in turn performed more efficiently the dehydrogenation of C3-OH (Figure 6c,d). The substrate specificity of Hsd60 makes this enzyme a very interesting candidate for the development of biocatalysts for TES production using mycobacterial strains that accumulate AD as described [59]. The alignment of the two proteins showed a variable region at the carboxyl-terminal domain (196–230 residues) that only share 42% amino acid sequence identity, in contrast with the 80% of the whole polypeptide chain. It has been described that the steroid binding pocket in 3α, 20β-HSD from Streptomyces hydrogenans is formed by the carboxyl-terminal 60 residues [60]. Interestingly, the sequence alignment of 3α, 20β-HSD with Hsd60 and Hsd70 showed that the variable region detected for both isoenzymes align perfectly with the sequence responsible for steroid binding at 3α, 20β-HSD (Figure S3). The lower similarity of this region is consistent with the variable substrate specificity observed between the two isoenzymes.

It is well known that many steroids are currently used as pharmaceutical drugs in their esterified forms. The esterification of natural or synthetic androgens or anabolic steroids renders metabolic resistant prohormones with improved oral bioavailability, increased lipophilicity, and extended elimination half-life. This is for instance the case of TES that can be esterified with acetate (TES-Ac) that was first described in 1936 and was one of the first androgen esters to be synthesized and used as an anabolic steroid. To introduce steroid esters in the C-19 degradation pathway they should be firstly hydrolyzed by an esterase. On the other hand, the uptake of these lipophilic xenobiotic compounds might require a specific transport system to cross the two membranes of a Gram-negative bacterium. Then, the production of an esterase together with two HSD enzymes and the assistance of a transport protein appear to constitute an efficient system to handle the degradation of these pharmaceutical compounds that frequently contaminate the municipal waste waters from where *N. tardaugens* was isolated [7]. In agreement with this hypothesis, we have demonstrated that *N. tardaugens* is able to catabolize TES-Ac (Figure 2).

The fact that *N. tardaugens* can degrade both C3-OH and C17-OH steroids (Figure 2) supports the existence of both HSD activities. This finding does not rule out that some of the other HSD homologous enzymes encoded in the genome of *N. tardaugens* can fulfil these activities and additional experiments should be carried out to demonstrate a specific implication of this operon in the metabolism of steroids. Nevertheless, this result opens a new and interesting scenario to study in more detail the degradation of xenobiotic steroid esters of EDCs.

Several examples of bacteria capable of using both cholesterol and C-19 steroids as the only carbon and energy source have been described (e.g., *Rhodococcus erythropolis* strain SQ1 [61], *Rhodococcus ruber* strain Chol-4 [62], *Rhodococcus rhodochrous* DSM 43269 [63], *Gordonia neofelifaecis* NRRL B-59395 [64]). Using the genes of the SD cluster of *N. tardaugens* as a template, we screened the presence of homologous genes in other steroid metabolizing bacteria (Table S8). Apart from *C. testosteroni* TA441, we observed homologous genes in *Sphingomonas* sp. KC8, *Pseudomonas* sp. Chol1, *Rhodococcus jostii* RHA1, *Mycobacterium tuberculosis* H37Rv, *Mycobacterium smegmatis* mc^{2}155, *Altererythrobacter estronivorus* MH-B5 and *Sterolibacterium denitrificans* Chol-1S(T) genomes. In spite of their large phylogenetic distances, a significant identity was observed with some Actinobacteria, e.g., RHA1, H37Rv and mc^{2}155 strains (Table S8). This suggests a great level of conservation of key degradative enzymes among bacteria adapted to metabolize different steroidal compound.

Genetic manipulation of the ARI-1 strain will pave the way to unravel in more detail not only TES degradative pathway but also the other pathways involved in the degradation of E2 and several steroids that are used as carbon and energy sources by this bacterium. The data presented will enable to build upon the knowledge on metabolic pathways and the biotransformation capabilities of this

Gram-negative bacterium that could become a new, model system in the steroid field. Moreover, *N. tardaugens* NBRC 16725 (strain ARI-1) might cover a wide spectrum of steroid biotransformation reactions and their improvement may lead to promising alternative biotechnological processes.

Supplementary Materials: The following are available online at http://www.mdpi.com/2073-4425/10/11/871/s1, Figure S1: 12.5% SDS-polyacrylamide gel electrophoresis of the overproduction of Hsd70-Hsd60, Hsd60 and Hsd70 proteins in the soluble fraction of the crude extract of *E. coli* BL21(DE3) strains.Twenty-five µg of total protein of each sample where loaded, Figure S2: Bacterial growth (Log_2OD_{600}) of *N. tardaugens* NBRC 16725 (red), *N. tardaugens ΔfadD3* (yellow), *N. tardaugens ΔtesD* (purple) and *N. tardaugens Δhsd* (green) strains when cultured in M63 minimal medium containing 1.89 mM TES, Figure S3: Protein alignment of the 3α,20β-HSD from *Streptomycens hydrogenans* with HSD60 and HSD70. The key amino acids of the active site are in red. The variable sequence is shown in a red box. This sequence is involved in substrate binding in the 3α, 20β-HSD from *S. hydrogenans*, Table S1: Isolated bacteria able to degrade E2 and/or TES, Table S2: Bacterial strains and plasmids used in this study, Table S3: Primers used in this study, Table S4: Statistical results of whole genome sequencing and mapping of all transcripts, Table S5: Genes found in *N. tardaugens* genome homologous to genes involved in testosterone degradation in *C. testosteroni* TA441. Percentage of identity (ID %) and fold change (FC) increase in expression levels when *N. tardaugens* grows in testosterone are shown, Table S6: Gene expression analysis (RNA-seq) of *N. tardaugens* grown in testosterone condition compared to pyruvate. Genes located in the SD cluster (blue) and those involved in methylmalonyl-CoA pathway (green) and cofactor B12 biosynthesis pathway (orange) are highlighted, Table S7: Methylmalonyl-CoA degradation cluster in *N. tardaugens* NBRC 16725. Genes homologous to those described as involved in the pathway are highlighted in green, Table S8: Steroid degradation genes in the putative testosterone degradation pathway of *N. tardaugens* (accession CP034179). Homologous genes found in the genomes of *C. testosteroni* TA441 (accession LC010134), *Sphingomonas* sp. KC8 (accession CP016306), *Pseudomonas* sp. Chol1 (accession AMSL00000000), *R. jostii* RHA1 (accession CP000431), *M. tuberculosis* H37Rv (accession AL123456.3), *M. smegmatis* mc2 155 (accession CP000480), *A. estronivorus* MH-B5 (accession NZ_JRQQ00000000) and *S. denitrificans* Chol (accession LT837803) are listed and the percentage identity is shown. A cut-off value of 39 % identity was used.

Author Contributions: J.I. carried out molecular genetic studies, in silico analysis, fermentation experiments and helped to draft the manuscript. E.D. and J.L.G. conceived the study. J.L.G. and B.G. coordinated the study and drafted the manuscript. All authors read, reviewed and approved the final manuscript.

Funding: This research was funded by Ramón Areces Foundation.

Acknowledgments: The bioinformatics support of the Bioinformatics and Biostatistics Service of CIB particularly Guillermo Padilla Alonso is greatly appreciated. The technical work of A. Valencia is greatly appreciated. We acknowledge support of the publication fee by the CSIC Open Access Publication Support Initiative through its Unit of Information Resources for Research (URICI). We also acknowledge the financial support provided by the Ramón Areces Foundation.

Conflicts of Interest: The authors declare no conflicts of interest.

References

1. Milla, S.; Depiereux, S.; Kestemont, P. The effects of estrogenic and androgenic endocrine disruptors on the immune system of fish: A review. *Ecotoxicology* **2011**, *20*, 305–319. [CrossRef] [PubMed]
2. Chen, Y.L.; Wang, C.H.; Yang, F.C.; Ismail, W.; Wang, P.H.; Shih, C.J.; Wu, Y.C.; Chiang, Y.R. Identification of *Comamonas testosteroni* as an androgen degrader in sewage. *Sci. Rep.* **2016**, *6*, 1–13. [CrossRef] [PubMed]
3. Fan, Z.; Casey, F.X.M.; Hakk, H.; Larsen, G.L. Persistence and fate of 17β-estradiol and testosterone in agricultural soils. *Chemosphere* **2007**, *67*, 886–895. [CrossRef] [PubMed]
4. Bradley, P.M.; Barber, L.B.; Chapelle, F.H.; Gray, J.L.; Kolpin, D.W.; McMahon, P.B. Biodegradation of 17β-estradiol, estrone and testosterone in stream sediments. *Environ. Sci. Technol.* **2009**, *43*, 1902–1910. [CrossRef]
5. Fan, Z.; Wu, S.; Chang, H.; Hu, J. Behaviors of glucocorticoids, androgens and progestogens in a municipal sewage treatment plant: Comparison to estrogens. *Environ. Sci. Technol.* **2011**, *45*, 2725–2733. [CrossRef]
6. Jarošová, B.; Erseková, A.; Hilscherová, K.; Loos, R.; Gawlik, B.M.; Giesy, J.P.; Bláha, L. Europe-Wide survey of estrogenicity in wastewater treatment plant effluents: The need for the effect-based monitoring. *Environ. Sci. Pollut. Res.* **2014**, *21*, 10970–10982. [CrossRef]
7. Fujii, K.; Kikuchi, S.; Satomi, M.; Ushio-Sata, N.; Morita, N. Degradation of 17β-estradiol by a gram-negative bacterium isolated from activated sludge in a sewage treatment plant in Tokyo, Japan. *Appl. Environ. Microbiol.* **2002**, *68*, 2057–2060. [CrossRef]

8. Weber, S.; Leuschner, P.; Kämpfer, P.; Dott, W.; Hollender, J. Degradation of estradiol and ethinyl estradiol by activated sludge and by a defined mixed culture. *Appl. Microbiol. Biotechnol.* **2005**, *67*, 106–112. [CrossRef]
9. Haiyan, R.; Shulan, J.; ud din Ahmad, N.; Dao, W.; Chengwu, C. Degradation characteristics and metabolic pathway of 17alpha-ethynylestraidiol by *Sphingobacterium* sp. JCR5. *Chemosphere* **2007**, *66*, 340–346. [CrossRef]
10. Jiang, L.; Yang, J.; Chen, J. Isolation and characteristics of 17β-estradiol-degrading *Bacillus* spp. strains from activated sludge. *Biodegradation* **2010**, *21*, 729–736. [CrossRef]
11. Kurisu, F.; Ogura, M.; Saitoh, S.; Yamazoe, A.; Yagi, O. Degradation of natural estrogen and identification of the metabolites produced by soil isolates of *Rhodococcus* sp. and *Sphingomonas* sp. *J. Biosci. Bioeng.* **2010**, *109*, 576–582. [CrossRef] [PubMed]
12. Zeng, Q.; Li, Y.; Gu, G.; Zhao, J.; Zhang, C.; Luan, J. Sorption and biodegradation of 17β-estradiol by acclimated aerobic activated sludge and isolation of the bacterial strain. *Environ. Eng. Sci.* **2009**, *26*, 783–790. [CrossRef]
13. Chen, Y.L.; Fu, H.Y.; Lee, T.H.; Shih, C.J.; Huang, L.; Wang, Y.S.; Ismail, W.; Chiang, Y.R. Estrogen degraders and estrogen degradation pathway identified in an activated sludge. *Appl. Environ. Microbiol.* **2018**, *84*, e00001–e00018. [CrossRef] [PubMed]
14. Fahrbach, M.; Kuever, J.; Meinke, R.; Kampfer, P.; Hollender, J. *Denitratisoma oestradiolicum* gen. nov., sp. nov., a 17beta-oestradiol-degrading, denitrifying betaproteobacterium. *Int. J. Syst. Evol. Microbiol.* **2006**, *56*, 1547–1552. [CrossRef]
15. Yu, C.P.; Deeb, R.A.; Chu, K.H. Microbial degradation of steroidal estrogens. *Chemosphere* **2013**, *91*, 1225–1235. [CrossRef]
16. Sang, Y.; Xiong, G.; Maser, E. Identification of a new steroid degrading bacterial strain H5 from the Baltic Sea and isolation of two estradiol inducible genes. *J. Steroid Biochem. Mol. Biol.* **2012**, *129*, 22–30. [CrossRef]
17. Zhang, T.; Xiong, G.; Maser, E. Characterization of the steroid degrading bacterium S19-1 from the Baltic Sea at Kiel, Germany. *Chem. Biol. Interact.* **2011**, *191*, 83–88. [CrossRef]
18. Chen, Y.L.; Yu, C.P.; Lee, T.H.; Goh, K.S.; Chu, K.H.; Wang, P.H.; Ismail, W.; Shih, C.J.; Chiang, Y.R. Biochemical mechanisms and catabolic enzymes involved in bacterial estrogen degradation pathways. *Cell Chem. Biol.* **2017**, *24*, 712–724. [CrossRef]
19. Li, S.; Liu, J.; Sun, M.; Ling, W.; Zhu, X. Isolation, characterization, and degradation performance of the 17β-estradiol-degrading bacterium *Novosphingobium* sp. E2S. *Int. J. Environ. Res. Public Health* **2017**, *14*, 115. [CrossRef]
20. Payne, D.W.; Talalay, P. Isolation of novel microbial 3α-, 3β-, and 17β-hydroxysteroid dehydrogenases. Purification, characterization, and analytical applications of a 17β-hydroxysteroid dehydrogenase from an *Alcaligenes* sp. *J. Biol. Chem.* **1985**, *260*, 13648–13655.
21. Hashimoto, T.; Onda, K.; Morita, T.; Luxmy, B.S.; Tada, K.; Miya, A.; Murakami, T. Contribution of the estrogen-degrading bacterium *Novosphingobium* sp. strain JEM-1 to estrogen removal in wastewater treatment. *J. Environ. Eng.* **2009**, *136*, 890–896. [CrossRef]
22. Yoshimoto, T.; Nagai, F.; Fujimoto, J.; Watanabe, K.; Mizukoshi, H.; Makino, T.; Kimura, K.; Saino, H.; Sawada, H.; Omura, H. Degradation of estrogens by *Rhodococcus zopfii* and *Rhodococcus equi* isolates from activated sludge in wastewater treatment plants. *Appl. Environ. Microbiol.* **2004**, *70*, 5283–5289. [CrossRef] [PubMed]
23. Fahrbach, M.; Kuever, J.; Remesch, M.; Huber, B.E.; Kämpfer, P.; Dott, W.; Hollender, J. *Steroidobacter denitrificans* gen. nov., sp. nov., a steroidal hormone-degrading gammaproteobacterium. *Int. J. Syst. Evol. Microbiol.* **2008**, *58*, 2215–2223. [CrossRef] [PubMed]
24. Fujii, K.; Satomi, M.; Morita, N.; Motomura, T.; Tanaka, T.; Kikuchi, S. *Novosphingobium tardaugens* sp. nov., an oestradiol-degrading bacterium isolated from activated sludge of a sewage treatment plant in Tokyo. *Int. J. Syst. Evol. Microbiol.* **2003**, *53*, 47–52. [CrossRef] [PubMed]
25. Liu, J.; Li, S.; Li, X.; Gao, Y.; Ling, W. Removal of estrone, 17β-estradiol, and estriol from sewage and cow dung by immobilized *Novosphingobium* sp. ARI-1. *Environ. Technol.* **2018**, *39*, 2423–2433. [CrossRef]
26. Horinouchi, M.; Hayashi, T.; Kudo, T. Steroid degradation in *Comamonas testosteroni*. *J. Steroid Biochem. Mol. Biol.* **2012**, *129*, 4–14. [CrossRef]
27. Horinouchi, M.; Koshino, H.; Malon, M.; Hirota, H.; Hayashi, T. Steroid degradation in *Comamonas testosteroni* TA441: Identification of the entire β-oxidation cycle of the cleaved B-ring. *Appl. Environ. Microbiol.* **2019**, *85*, e01204–e01219. [CrossRef]

28. Horinouchi, M.; Hayashi, T.; Yamamoto, T.; Kudo, T. A new bacterial steroid degradation gene cluster in *Comamonas testosteroni* TA441 which consists of aromatic-compound degradation genes for seco-steroids and 3-ketosteroid dehydrogenase genes. *Appl. Environ. Microbiol.* **2003**, *69*, 4421–4430. [CrossRef]
29. Ibero, J.; Sanz, D.; Galán, B.; Díaz, E.; García, J.L. High-Quality whole-genome sequence of an estradiol-degrading strain, *Novosphingobium tardaugens* NBRC 16725. *Microb. Resour. Annoucements* **2019**, *8*, 17–19. [CrossRef]
30. Bertani, G. Studies on lysogenesis. I. The mode of phage liberation by lysogenic *Escherichia coli*. *J. Bacteriol.* **1951**, *62*, 293–300.
31. Sambrook, J.; Russel, D. *Molecular Cloning: A Laboratory Manual*; Cold Spring Harbor Laboratory Press: New York, NY, USA, 2001.
32. Wirth, R.; Friesenegger, A.; Fiedler, S. Transformation of various species of gram-negative bacteria belonging to 11 different genera by electroporation. *Mol. Gen. Genet. MGG* **1989**, *216*, 175–177. [CrossRef] [PubMed]
33. Schafer, A.; Tauch, A.; Jager, W.; Kalinowski, J.; Thierbach, G.; Puhler, A. Small mobilizable multi-purpose cloning vectors derived from the *Escherichia coli* plasmids pK18 and pK19: Selection of defined deletions in the chromosome of *Corynebacterium glutamicum*. *Gene* **1994**, *145*, 69–73. [CrossRef]
34. Herrero, M.; de Lorenzo, V.; Timmis, K.N. Transposon vectors containing non-antibiotic resistance selection markers for cloning and stable chromosomal insertion of foreign genes in gram-negative bacteria. *J. Bacteriol.* **1990**, *172*, 6557–6567. [CrossRef] [PubMed]
35. Sharma, S.B.; Signer, E.R. Temporal and spatial regulation of the symbiotic genes of *Rhizobium meliloti* in planta revealed by transposon Tn5-gusA. *Genes Dev.* **1990**, *4*, 344–356. [CrossRef] [PubMed]
36. Bradford, M.M. A rapid and sensitive method for the quantitation of microgram quantities of protein utilizing the principle of protein-dye binding. *Anal. Biochem.* **1976**, *72*, 248–254. [CrossRef]
37. Aziz, R.K.; Bartels, D.; Best, A.; DeJongh, M.; Disz, T.; Edwards, R.A.; Formsma, K.; Gerdes, S.; Glass, E.M.; Kubal, M.; et al. The RAST Server: Rapid annotations using subsystems technology. *BMC Genom.* **2008**, *9*, 1–15. [CrossRef]
38. Altschul, S.F.; Gish, W.; Miller, W.; Myers, E.W.; Lipman, D.J. Basic local alignment search tool. *J. Mol. Biol.* **1990**, *215*, 403–410. [CrossRef]
39. Pruneda-Paz, J.L.; Linares, M.; Cabrera, J.E.; Genti-Raimondi, S. TeiR, a LuxR-Type transcription factor required for testosterone degradation in *Comamonas testosteroni*. *J. Bacteriol.* **2004**, *186*, 1430–1437. [CrossRef]
40. Hou, J.; Xiang, H.; Han, J. Propionyl coenzyme A (propionyl-CoA) carboxylase in *Haloferax mediterranei*: Indispensability for propionyl-CoA assimilation and impacts on global metabolism. *Appl. Environ. Microbiol.* **2015**, *81*, 794–804. [CrossRef]
41. Sanyal, I.; Flint, D.H.; Lee, S.L. Biosynthesis of pimeloyl-CoA, a biotin precursor in *Escherichia coli*, follows a modified fatty acid synthesis pathway: 13C-labeling studies. *J. Am. Chem. Soc.* **1994**, *116*, 2637–2638. [CrossRef]
42. Haller, T.; Buckel, T.; Retey, J.; Gerlt, J.A. Discovering new enzymes and metabolic pathways: Conversion of succinate to propionate by *Escherichia coli*. *Biochemistry* **2000**, *39*, 4622–4629. [CrossRef] [PubMed]
43. Bobik, T.A.; Rasche, M.E. Purification and partial characterization of the *Pyrococcus horikoshii* methylmalonyl-CoA epimerase. *Appl. Microbiol. Biotechnol.* **2004**, *63*, 682–685. [CrossRef] [PubMed]
44. Carter, M.S.; Alber, B.E. Transcriptional regulation by the short-chain fatty acyl coenzyme A regulator (ScfR) PccR controls propionyl coenzyme A assimilation by *Rhodobacter sphaeroides*. *J. Bacteriol.* **2015**, *197*, 3048–3056. [CrossRef] [PubMed]
45. Tong, L. Structure and function of biotin-dependent carboxylases. *Cell. Mol. Life Sci.* **2013**, *70*, 863–891. [CrossRef]
46. Benach, J.; Filling, C.; Oppermann, U.C.T.; Roversi, P.; Bricogne, G.; Berndt, K.D.; Jörnvall, H.; Ladenstein, R. Structure of bacterial 3β/17β-hydroxysteroid dehydrogenase at 1.2 Å resolution: A model for multiple steroid recognition. *Biochemistry* **2002**, *41*, 14659–14668. [CrossRef]
47. Van Hamme, J.D.; Bergstrand, L.H.; Mohn, W.W.; Cardenas, E.; Holert, J. Delineation of steroid-degrading microorganisms through comparative genomic analysis. *MBio* **2016**, *7*, 1–13.
48. Crowe, A.M.; Casabon, I.; Brown, K.L.; Liu, J.; Lian, J.; Rogalski, J.C.; Hurst, T.E.; Snieckus, V.; Foster, L.J.; Eltis, L.D. Catabolism of the last two steroid rings in *Mycobacterium tuberculosis* and other bacteria. *MBio* **2017**, *8*, e00321-17. [CrossRef]

49. Valton, J.; Filisetti, L.; Fontecave, M.; Niviere, V. A two-component flavin-dependent monooxygenase involved in actinorhodin biosynthesis in *Streptomyces coelicolor*. *J. Biol. Chem.* **2004**, *279*, 44362–44369. [CrossRef]
50. Galan, B.; Diaz, E.; Prieto, M.A.; Garcia, J.L. Functional analysis of the small component of the 4-hydroxyphenylacetate 3-monooxygenase of *Escherichia coli* W: A prototype of a new Flavin:NAD(P)H reductase subfamily. *J. Bacteriol.* **2000**, *182*, 627–636. [CrossRef]
51. Grandguillot, M.; Florin, C.; Köhler, T.; Plesiat, P. *Comamonas testosteroni* 3-ketosteroid-delta 4(5 α)-dehydrogenase: Gene and protein characterization. *J. Bacteriol.* **1996**, *178*, 3322–3330.
52. Göhler, A.; Xiong, G.; Paulsen, S.; Trentmann, G.; Maser, E. Testosterone-Inducible regulator is a kinase that drives steroid sensing and metabolism in *Comamonas testosteroni*. *J. Biol. Chem.* **2008**, *283*, 17380–17390. [CrossRef] [PubMed]
53. Kudo, T.; Kurita, T.; Hatori, E.; Yamamoto, T.; Horinouchi, M.; Hayashi, T. Steroid degradation gene cluster of *Comamonas testosteroni* consisting of 18 putative genes from meta-cleavage enzyme gene tesB to regulator gene tesR. *Biochem. Biophys. Res. Commun.* **2004**, *324*, 597–604.
54. Griffin, J.E.; Pandey, A.K.; Gilmore, S.A.; Mizrahi, V.; Mckinney, J.D.; Bertozzi, C.R.; Sassetti, C.M. Cholesterol catabolism by *Mycobacterium tuberculosis* requires transcriptional and metabolic adaptations. *Chem. Biol.* **2012**, *19*, 218–227. [CrossRef] [PubMed]
55. Lee, W.; Vanderven, B.C.; Fahey, R.J.; Russell, D.G. Intracellular *Mycobacterium tuberculosis* exploits host-derived fatty acids to limit metabolic stress. *J. Biol. Chem.* **2013**, *288*, 6788–6800. [CrossRef] [PubMed]
56. Straus, J.F. The synthesis and metabolism of steroid hormones. *Yen Jaffe's Reprod. Endocrinol.* **2009**, *5*, 79–104.
57. Benach, J.; Knapp, S.; Oppermann, U.C.T.; Hagglund, O.; Jornvall, H.; Ladenstein, R. Crystallization and crystal packing of recombinant 3 (or 17) β-hydroxpsteroid dehydrogenase from *Comamonas testosteroni* ATTC 11996. *Eur. J. Biochem.* **1996**, *236*, 144–148. [CrossRef]
58. Oppermann, U.C.T.; Filling, C.; Berndt, K.D.; Persson, B.; Benach, J.; Ladenstein, R.; Jörnvall, H. Active site directed mutagenesis of 3β/17β-hydroxysteroid dehydrogenase establishes differential effects on short-chain dehydrogenase/reductase reactions. *Biochemistry* **1997**, *36*, 34–40. [CrossRef]
59. Fernández-Cabezón, L.; Galán, B.; García, J.L. Engineering *Mycobacterium smegmatis* for testosterone production. *Microb. Biotechnol.* **2017**, *10*, 151–161. [CrossRef]
60. Ghosh, D.; Wawrzak, Z.; Weeks, C.M.; Duax, W.L.; Erman, M. The refined three-dimensional structure of 3α,20β-hydroxysteroid dehydrogenase and possible roles of the residues conserved in short-chain dehydrogenases. *Structure* **1994**, *2*, 629–640. [CrossRef]
61. Van Der Geize, R.; Hessels, G.I.; van Gerwen, R.; Vrijbloed, J.W.; van Der Meijden, P.; Dijkhuizen, L. Targeted disruption of the *kstD* gene encoding a 3-ketosteroid delta(1)-dehydrogenase isoenzyme of *Rhodococcus erythropolis* strain SQ1. *Appl. Environ. Microbiol.* **2000**, *66*, 2029–2036. [CrossRef]
62. Petrusma, M.; Hessels, G.; Dijkhuizen, L.; van der Geize, R. Multiplicity of 3-ketosteroid-9alpha-hydroxylase enzymes in *Rhodococcus rhodochrous* DSM43269 for specific degradation of different classes of steroids. *J. Bacteriol.* **2011**, *193*, 3931–3940. [CrossRef] [PubMed]
63. Fernandez de las Heras, L.; van der Geize, R.; Drzyzga, O.; Perera, J.; Maria Navarro Llorens, J. Molecular characterization of three 3-ketosteroid-delta(1)-dehydrogenase isoenzymes of *Rhodococcus ruber* strain Chol-4. *J. Steroid Biochem. Mol. Biol.* **2012**, *132*, 271–281. [CrossRef] [PubMed]
64. Li, W.; Ge, F.; Zhang, Q.; Ren, Y.; Yuan, J.; He, J.; Li, W.; Chen, G.; Zhang, G.; Zhuang, Y.; et al. Identification of gene expression profiles in the actinomycete *Gordonia neofelifaecis* grown with different steroids. *Genome* **2014**, *57*, 345–353. [CrossRef] [PubMed]

© 2019 by the authors. Licensee MDPI, Basel, Switzerland. This article is an open access article distributed under the terms and conditions of the Creative Commons Attribution (CC BY) license (http://creativecommons.org/licenses/by/4.0/).

genes

MDPI

Article

A Case of Adaptive Laboratory Evolution (ALE): Biodegradation of Furfural by *Pseudomonas pseudoalcaligenes* CECT 5344

M. Isabel Igeño [1,2,†], Daniel Macias [1,†] and Rafael Blasco [1,2,*]

1 Departamento de Bioquímica y Biología Molecular y Genética, Facultad de Veterinaria, Universidad de Extremadura, 10003 Caceres, Spain

2 Meat and Meat Products Research Institute (IProCar), BioMic Research Group, Universidad de Extremadura, 10003 Caceres, Spain

* Correspondence: rblasco@unex.es; Tel.: +34-927-257-161

† These authors have contributed equally to the development of this work.

Received: 30 April 2019; Accepted: 27 June 2019; Published: 29 June 2019

Abstract: *Pseudomonas pseudoalcaligenes* CECT 5344 is a bacterium able to assimilate cyanide as a nitrogen source at alkaline pH. Genome sequencing of this strain allowed the detection of genes related to the utilization of furfurals as a carbon and energy source. Furfural and 5-(hydroxymethyl) furfural (HMF) are byproducts of sugars production during the hydrolysis of lignocellulosic biomass. Since they inhibit the yeast fermentation to obtain bioethanol from sugars, the biodegradation of these compounds has attracted certain scientific interest. *P. pseudoalcaligenes* was able to use furfuryl alcohol, furfural and furoic acid as carbon sources, but after a lag period of several days. Once adapted, the evolved strain (R1D) did not show any more prolonged lag phases. The transcriptomic analysis (RNA-seq) of R1D revealed a non-conservative punctual mutation (L261R) in BN5_2307, a member of the AraC family of activators, modifying the charge of the HTH region of the protein. The inactivation of the mutated gene in the evolved strain by double recombination reverted to the original phenotype. Although the bacterium did not assimilate HMF, it transformed it into value-added building blocks for the chemical industry. These results could be used to improve the production of cost-effective second-generation biofuels from agricultural wastes.

Keywords: *Pseudomonas*; bioethanol; furfural; ALE; AraC

1. Introduction

The development of renewable resources is one of the key actions to palliate climate change, which is largely a consequence of the world's dependence on petrol. On the other hand, contamination of the environment is an inevitable consequence of human development. These are global problems that need international agreements [1]. Biotechnology can offer solutions to these challenges, such as the production of bioethanol as a substitute to gasoline [2]. Biotechnology can also offer alternatives to the physical-chemical treatment of contaminating compounds, either by avoiding their production, or by mitigating their impact once it has occurred. The biodegradation of pollutants is, in general, a technology that has good social acceptance [3]. *Pseudomonas pseudoalcaligenes* CECT 5344 was isolated from sludge of Guadalquivir River, and it is able to use cyanide as the only source of nitrogen [4]. Cyanide is an extremely toxic compound used in the synthesis of organic compounds such as nitriles, plastics, paints, adhesives, cosmetics, etc., while mining activities and the jewellery industry are the main source of cyanurated wastes [5–8]. This strain tolerates an unusually high concentration of cyanide (up to 30 mM) [4], but it requires a suitable carbon source for growing. The sequencing of the genome of *P. pseudoalcaligenes* CECT 5344 has made it possible to predict which carbon sources can be used by this bacterium, such as the assimilation of furanic compounds [9]. Furfurals are aromatic natural compounds formed by the dehydration of sugars during the thermochemical pre-treatment of the lignocellulosic materials to release fermentable sugars. The production of biofuels from lignocellulosic residues, which is part of the so-called second-generation biofuels, constitutes a viable option for reducing the greenhouse effect and for providing an alternative to fossil fuels [10,11]. There are different pre-treatment technologies of lignocellulosic residues. One of the parameters that has to be taken into account to optimize the process is avoiding the formation of potentially inhibitory compounds to the posterior yeast fermentation process [12]. From the food technology perspective, furfurals are potential carcinogenic compounds used as a marker of honey adulteration, generated by acid-catalyzed dehydration of carbohydrates of food-containing sugars [13]. In any case, furfural (F), or fufuralaldehyde, and 5-hydroxymethyl furfural (HMF) are natural products that can be eliminated by using the capacity of some microorganisms to metabolize them [14–22]. Other furanic derivatives are furoic acid (FA) and furfuryl alcohol (FFA), all of them with the common thread of having an aromatic furan ring. The variety of furanic compounds degrading species is limited mostly to Gram-negative aerobic bacteria and some Gram positives [17], with a few exceptions including fungi [14]. In the first degradation route currently proposed, furfural is oxidized to 2-furoic acid (FA), which is subsequently transformed into 2-oxoglutarate, a Krebs cycle intermediate [23]. The complete metabolic pathway for the assimilation of F and HMF, as well as the genetic of the process, was first described in the soil isolate *Cupriavidus basilensis* HMF14 [24] (Figure 1). In this strain, the *hmfABCDE* gene cluster is responsible for the assimilation of furoic acid. The first reaction in the pathway is catalysed by the 2-furoyl-CoA synthetase (HmfD), producing 2-furoyl-CoA from 2-furoic acid. The conversion of 2-furoyl-CoA is into 5-hydroxy-2-furoyl-CoA in *C. basilensis*; HMF14 is catalysed by the molybdenum-dependent 2-furoyl-CoA dehydrogenase (HmfABC). The final steps of the proposed furoic-acid metabolic pathway consist of the transformation of 5-oxo-2-furoyl-CoA into 2-oxoglutarate. No gene has been assigned to the hydrolysis of the lactone, whereas *hmfE* has been proposed to encode a specific 2-oxoglutaroyl-CoA thioesterase [24] (Figure 1). *P. pseudoalcaligenes* contains an *hmfABCDE* gene cluster homologous to the gene cluster shown to be essential for the assimilation of furfural in *C. basilensis* HMF14 (Figure 1). Concretely, the amino acid sequence of HmfA from *C. basilensis* HMF14 (GenBank ADE20399.1) is 64% identical to the homologous protein of *P. pseudoalcaligenes* (BN5_2298, 76% positives). The % identity/% similarity for the rest of the proteins are: 59%/72%, 77%/83%, 61%/75% and 80%/88%, for HmfB (GenBank ADE20400.1), HmfC (GenBank ADE20401.1), HmfD (GenBank ADE20402.1) and HmfE (GenBank ADE20403.1), respectively. Moreover, this locus also contains downstream *hmfABCDE*, a gene (*benE*) belonging to the Major Facilitator Superfamily (MSF)-family of transporters and two separate genes homologous to genes related to the assimilation of furfural in *Pseudomonas putida* Fu1 [9,25] (Figure 1). One of them belongs to the AraC-family of regulators. AraC from *P. putida* Fu1 (GenBank ACA09742.1) is 75% identical (88% similar) to its orthologous gene product in *P. pseudoalcaligenes* (BN5_2307, Figure 1). The other upstream gene (*PsfD*) codes for a putative conserved protein usually annotated as maturation factor for molybdenum containing dehydrogenases, like xanthine and CO dehydrogenases [9,25]. In that respect, the furoyl-CoA dehydrogenase was proposed to be a molydo-protein [26]. PsfD protein from *P. putida* Fu1 (GenBank ACA09741.1) is 81% identical (89% similar) to PsfD form *P. pseudoalcaligenes* (BN5_2306, Figure 1). To date, the architecture of this operon presented in Figure 1B has not been described in this context. Here we show that this operon is functional after an adaptation period ending up with the selection of a punctual mutant in the *araC*-type regulator. Therefore, the locus described here seems to be a hybrid furfural assimilating system containing horizontally transferred genes homologous to the catalytic genes for the assimilation of FA described in *C. basilensis* [24] and the regulatory and accessory genes described in *P. putida* Fu1 [25].

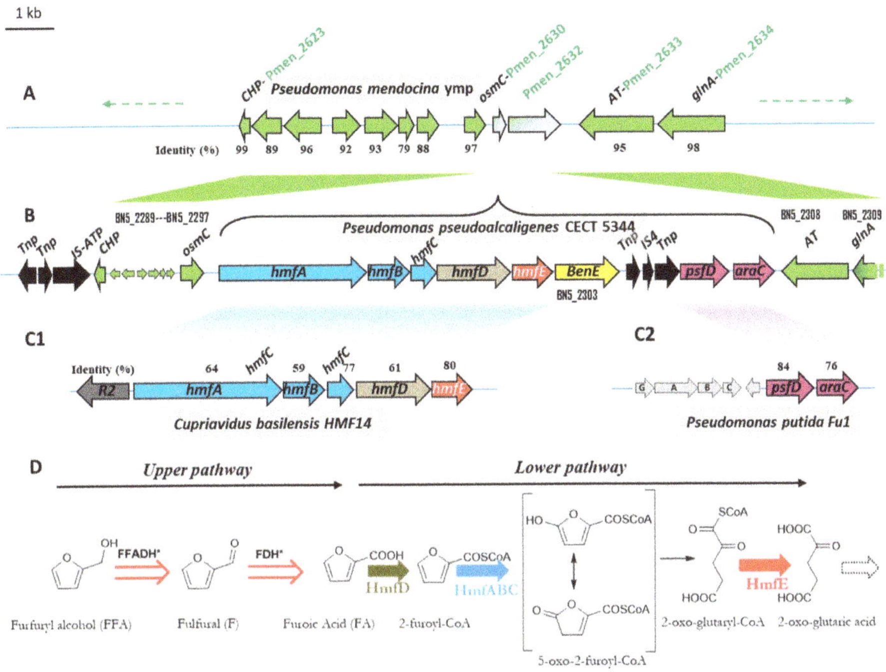

Figure 1. Scheme of the genetic organization of the *hmf* operon (**A–C**) (adapted from Ref. [9]) and predicted metabolic pathway for the assimilation of furfuryl alcohol in *Pseudomonas pseudoalcaligenes* (**D**). The *hmf* locus in *P. pseudoalcaligenes* CECT 5344 (delimited by a curly bracket, **B**) is located between BN5_2297 (*osmC*) and BN5_2308 (*AT*). The corresponding homologous genes in *Pseudomonas mendocina* ymp (*Pmen_2630-2633*) are consecutive in its genome (**A**) and the syntney of the homologous genes is conserved both upstream and downstream the *hmf* locus (green arrows). The *hmf* locus contains genes homologous to that described in the context of furfural degradation in *Cupriavidus basilensis* [24] (**C1**) and in *Pesudomonas putida* [25] (**C2**). The black arrows (panel **B**) represent genes involved in the transposition of mobile genetic elements. (**D**) Proposed pathway for the assimilation of FFA in *P. pseudoalcaligenes* R1. FFADH and FDH are furfuryl alcohol and furfural dehydrogenase, respectively. * FFADH and FDH could be the same enzyme, whose coding genes are unknown in *P. pseudoalcaligenes*. Both the oxidation of FFA and HMF in *C. basilensis* (not shown) converge in FA constituting the upper pathway. The lower pathway, encoded by the *hmfABCDE* operon (**C1**), is a series of reactions transforming FA into 2-oxo-glutaric acid. HmfA is a furoyl-CoA synthetase and HmfABC a furoyl CoA dehydrogenase. The transformation of 5-oxo-2-furoyl-CoA into 2-oxo-glutaril-CoA has no assigned gene and can be a spontaneous reaction or could be catalysed by an unspecific lactone hydrolase. Finally, HmfE was proposed to be the thioesterase rendering 2-oxo-glutaric acid from 2-oxo-glutaryl CoA [24].

2. Materials and Methods

2.1. Bacterial Strains, Media and Growth Conditions

Strain R1 is a spontaneous mutant of *P. pseudoalcaligenes* CECT5344 [27] resistant to rifampicin (40 µg/mL). *P. pseudoalcaligenes* CECT5344 strain R1D was obtained after four serial transfers of *P. pseudoalcaligenes* CECT5344 R1 to a M9 medium with furfural (10 mM) as the sole carbon source. A dilution 1:100 of the previously grown preculture was used as inoculum. For the rest of the growth curves, unless otherwise stated, the inoculum was the equivalent of an overnight culture diluted 1:10 into fresh medium. Bacterial growth was monitored by measuring the absorbance at 600 nm. Cells were grown in either minimal medium (M9) adjusted to pH 8.5 [4] or in LB medium [28] adjusted to pH 8.5. Cell cultures were prepared in Erlenmeyer flasks filled with 1/10 (v/v) of their nominal volume in order to ensure aerobic conditions and incubated on a rotatory shaker at 190 rpm and 30 °C. For minimal medium, ammonium chloride (5 mM) was used as the nitrogen source and 4 g/L of sodium acetate, furfural (10 mM), furfuryl alcohol (10 mM) or furoic acid (10 mM) were added as the sole carbon source. *Escherichia coli* XL1 *blue* MRF′ cells (Stratagene, Agilent Technologies, Santa Clara, CA, USA) were grown aerobically at 37 °C in complex LB medium [28] with ampicillin (100 μg/mL). Where appropriate, the following compounds were added to the media: X-Gal (5-bromo-4-chloro-3-indolyl-β-d-galactopyranoside, 0.2 mM, Appli-Chem (Barcelona, Spain), IPTG (isopropyl-β-d-1-thiogalactopyranoside, 0.5 mM, Sigma-Aldrich (St. Louis, MO, USA). Electrocompetent cells were prepared by the growth of cultures up to an optical density at 600 nm (OD_{600}) of 0.35 and centrifugation for 20 min at 1000× *g*, followed by three successive washes (4 °C) in 1:1, 1:2, 1:50, and 1:500 volumes of 10% glycerol. The last solution, which yields the stock of electrocompetent cells, contained yeast extract (0.125%) and tryptone (0.25%). A mixture of 50 μL of cells (2×10^{10} to 3×10^{10} CFU/mL) and 1 to 10 ng of DNA was electroporated in 2-mm cuvettes with a Bio-Rad Gene Pulser II apparatus, (Bio-rad, Hercules, CA, USA) operated at 2.5 kV, 25 μF, and 200 Ω (4- to 5-ms time constants).

2.2. Preparation of Cell-free Extracts

The cells of *P. pseudoalcaligenes* CECT 5344 R1D grown with furfuryl alcohol as a carbon source (500 mL culture) were collected by centrifugation at the end of the logarithmic phase and resuspended in 50 mM Tris-HCl (pH 8), containing a complete protease inhibitor cocktail, as recommended by the supplier (Roche, Penzberg, Germany) and glycerol (10%). The cells were disrupted by two passages through a French pressure cell operated at 130 MPa and the cell debris removed by centrifugation at 18,000× *g* for 15 min.

2.3. Enzymatic Assays

Furfural dehydrogenase (FDH) and furfuryl alcohol dehydrogenase (FFADH) were assayed as previously described for *C. basilensis* [24], but optimizing the assay for *P. pseudoalcaligenes* (pH and temperature). FFADH (E.C.1.1.1.–) was assayed spectophotometrically at 65 °C and pH 9.5 (50 mM Tris/phosphate/carbonate) following the increment of absorbance at 340 nm due to the production of NADH ($\varepsilon = 6220\ M^{-1} \cdot cm^{-1}$). The reaction mixture contained NAD^+ (1.5 mM), FFA (5 mM) and the appropriate amount of cell-free extract (50–100 μL cell-free extract, 0.5–1 mg protein, approximately) in a final volume of 1 mL. FDH (E.C. 1.2.3.1) was measured spectrophotometrically at 65 °C and pH 6.5 (50 mM Tris/phosphate/carbonate) following the increment of absorbance at 522 nm due to the reduction of the artificial electron acceptor DCPIP ($\varepsilon = 21{,}000\ M^{-1} \cdot cm^{-1}$). The reaction mixture contained, in a final volume of 1 mL, 0.33 mM PMS, 0.1 mM DCPIP, 5 mM F, and the appropriate amount of enzyme (50–100 μL cell-free extract, 0.5–1 mg protein, approximately).

2.4. Chromatographic Separation of FDH and FFADH

Cell-free extract from FFA-grown cells was loaded into an anion exchange chromatography (mono Q 5/50 GL, GE Healthcare (Chicago, IL, USA) attached to an Akta Purifier, GE. All the chormatographies were carried out at 4 °C. The column was equilibrated in buffer A (Tris/HCl 50 mM pH 8, 2 mM DTT and glycerol (2%)). The cell-free extract (2 mL) was loaded into the column at a flow rate of 1 mL/min. The unbound protein was washed with 12.5 mL of buffer A. Then, it was applied at a gradient of 1 mL from 0 to 0.1 M NaCl, always in the same buffer, which was maintained as isocratic during 5 mL, followed by a 20 mL gradient from 0.1 to 0.5 M NaCl and 0.5 to 1 M during 10 mL. Finally, the column was regenerated with 5 mL of buffer A containing 2 M NaCl. The fractions (0.5 mL) were analyzed for the presence FFADH and FDH activities.

2.5. Analytical Methods

2.5.1. HPLC

The concentration of furanic intermediates was determined as follows. The supernatants of the culture media were obtained by centrifugation of 1.0 mL of bacterial culture in a microfuge at maximal speed (14,000 rpm) for 10 minutes at room temperature. The supernatants were further filter-sterilized using a 0.2-µm filtration unit and stored at −20 °C. Concentrations of furan derivatives were determined from sample supernatants by high-performance liquid chromatography (HPLC) on a HPLC System Gold (Beckman) system. The column used was an Anion Exchange ION-300 (ICE-99-9850) (300 × 7.8 mm, Transgenomic, Omaha, NE, USA) operated at 65 °C. As eluent H_2SO_4 5 mN was used at a flow of 0.6 mL·min^{-1}. Furfural and furoic acid were detected at 278 nm and furfuryl alcohol at 217 nm.

Glucose concentration was colorimetrically measured by using a commercially available enzymatic test based on the glucose/peroxidase activities (Biosystems, Barcelona, Spain).

Protein concentration was determined by the Bradford procedure [29].

2.5.2. DNA Manipulation

DNA was sequenced using services provided by Sistemas Genómicos (Valencia, Spain). Genomic DNA was purified from *P. pseudoalcaligenes* CECT5344 cells grown in liquid LB medium at pH 8.5 using the GNome DNA isolation kit (QBIOgene). Plasmid DNA was purified with the Genopure plasmid Midi Kit (Roche) from *E. coli* XL1 *blue* MRF′ cells grown in liquid cultures of LB media supplemented with the antibiotic used for selection. The *E. coli* XL1 *blue* MRF′ strain was used to clone recombinant DNA, performing restriction enzyme digestion and ligation as recommended by the manufacturers (Fermentas and Promega, respectively). Plasmids were introduced into *E. coli* XL1 *blue* MRF′ and *P. pseudoalcaligenes* CECT5344 cells by electroporation, as described previously [27]. Mutagenesis of *edd* gene (BN5_3048): To amplify by PCR a section of the *edd* ORF based on the DNA sequence of the *P. pseudoalcaligenes* CECT5344 (GenBank accession no. JN408065), two couples of sets of specific primers flanking the *edd* gene were designed (edd_9U /$edd_{730}L$ and $edd_{1140}U$ /$edd_{1737}L$, Table 1). A BamHI restriction site was incorporated into $edd_{730}L$ and $edd_{1140}U$ primers. The amplified fragments were cloned separately into pGEM-T Easy vector (Promega) generating the p*edd*1 and the p*edd*2 plasmids. The p*edd*2 plasmid was linearized with ApaI and BamHI. The plasmid p*edd*1 was digested with the same enzymes and the resulting fragment was gel-purified and subcloned into p*edd*2 to generate p*edd*1-2, thus obtaining a PGEM-TE derivative plasmid with two internal sequences of *edd* gene separated by a BamHI restriction site. On the other hand, a 1.0 kb BamHI fragment from the pMS255 plasmid [30] containing the gentamicin resistance gene (*aacC1*) was cloned into BamHI-digested PGEM-TE plasmid containing the two internal *edd* fragments (Figure S1). Finally, the resulting suicide plasmid was transferred to *P. pseudoalcaligenes* CECT5344 R1D by electroporation. The mutants were selected on gentamicin (10 µg/mL, Sigma-Aldrich) and mutant strains resulting from double homologous recombination were isolated. The insertion of the gentamicin resistance gene and the loss of the plasmid backbone were confirmed by PCR. Mutagenesis *ara*C gene (BN5_2307): Based on the DNA sequence of the *P. pseudoalcaligenes* CECT5344 (GenBank accession no. JN408065), one couple of specific primers ($araC_{157}U$/$araC_{823}L$, Table 1) flanking the *ara*C gene was designed for amplification of a section of the *ara*C ORF by PCR. PCR was performed and the amplified fragment that contained a KpnI restriction site at position 542 was cloned into pGEM-T Easy vector (Promega), thus obtaining a PGEM-TE derivative plasmid with one internal sequence of *ara*C. The plasmid was digested with KpnI and ligated to a 1.0-kb KpnI fragment containing the gentamicin-resistance gene (*aac*C1) from the pMS255 plasmid [30]. (Figure S2). The resulting plasmid was transferred to *P. pseudoalcaligenes* CECT5344 R1D by electroporation. The mutants were selected on gentamicin (10 µg/mL, Sigma-Aldrich), and mutant strains resulting from double-homologous recombination were isolated. The insertion of the gentamicin resistance gene and the loss of the plasmid backbone were confirmed by PCR.

Table 1. Primers utilized in this work.

Name	Sequence (5′→3′)	Description
edd_9U	CCGCGTTGTTGAAGTGACCGA	Amplification of *edd* gene
$edd_{730}L$	GCCCGGATCCATGAAGGAAGC	Amplification of *edd* gene (BamHI).
$edd_{1140}U$	CTACACCCGGGATCCCTTCCT	Amplification of *edd* gene (BamHI).
$edd_{1737}L$	CGCATGAAGGCGAACAACTCG	Amplification of *edd* gene
$araC_{157}U$	CCGGGGCCCGACCGCAATGTG	Amplification of *araC* gene (ApaI)
$araC_{823}L$	GGGCCCGTCCACTACCCGCTG	Amplification of *araC gene* (ApaI)

2.5.3. PCR Reaction Conditions

PCR samples were prepared with the following components: 0.5 ng/μL DNA, 1.0 μM of each primer, 2.0 mM $MgCl_2$, 0.2 mM each deoxynucleoside triphosphate (dNTP), 0.5 U Taq DNA polymerase, and 2 μL buffer as recommended by the manufacturer (Biotools), in a final volume of 20 μL. The PCR conditions were 2 min at 95 °C; followed by 30 cycles of 20 s at 95 °C, 10 s at 65.6 °C (*edd*) or 64 °C (*ara*C) and 1 min at 72 °C, followed by a 5 min extension at 72 °C.

3. Results

P. pseudoalcaligenes CECT5344 R1 was able to utilize furfural (up to 40 mM), furoic acid (up to 20 mM), and furfuryl alcohol (up to 20 mM) as the sole carbon and energy source, although after a very long lag phase (Figure 2A). The longest lag phase was observed with furfural as a C-source (Figure 2(A2)). In fact, furfural concentrations higher than 5 mM increased the lag phase, thus suggesting that this compound is toxic at a high concentration. Nevertheless the tolerance of *P. pseudoalcaligenes* CECT 5344 to furfural is relatively high (up to 40 mM) if compared to *C. basilensis* HMF14 (up to 12 mM) [16]. This toxic effect was not observed for furoic acid (FA), and furfuryl alcohol (FFA) was not toxic up to a concentration of 20 mM (Figure 2A). The maximum cell growth increased with the concentration of the furanic compound up to a concentration of 20 mM, indicating that below this concentration the growth was carbon limited. It is remarkable that at the same concentration of furanic compound, the maximum growth was highest with FFA, followed by F and finally FA. This is in agreement with the chemical compositions of these compounds, the alcohol being the most reduced, followed by the aldehyde and then the carboxylic acid. Both F and FFA were, in a first instance, almost stoichiometrically transformed into furoic acid (Figure 2B), thus indicating that the pathway for the assimilation of this compound is not active in the wild type strain of *Pseudomonas pseudoalcaligenes* CECT 5344 R1. Only after several days of incubation, a clear increase in cell growth was observed, which was concomitant with the assimilation of FA (Figure 2B).

P. pseudoalcaligenes CECT 5344 R1 failed to grow on 5-(hydroxymethyl)furfural (HMF), even after a prolonged incubation period of more than 10 days (Figure 3A). The analysis of the culture media by HPLC revealed that HMF was completely exhausted after 25 h, indicating that although this bacterial strain was unable to assimilate HMF, it had the capability of transforming it. Two new compounds were detected in the couture media after consumption of HMF, 2,5-furandicarboxylic acid (FDCA) in a proportion below 5% of the HMF consumed and an unknown aromatic compound (Figure 3A). In *C. basilensis* HMF14, HmfH catalyzes the oxidation of 5-hydroxymethyl-2-furoic acid (HMFA) to FDCA, whereas HmfFG catalyzes the decarboxylation of FDCA to FA [24]. Therefore, in this strain, the metabolic pathways for the assimilation of HMF, FFA and F converge in FA. In the genome of *P. pseudoalcaligenes* CECT 5344 R1, no homologous genes to *hmfH* and *hmfFG* of *C. basilensis* were observed. This genotype agrees with the fact that this bacterium does not assimilate HMF. Since HMFA is the substrate of HMFH, the unknown compound accumulated in the culture media from HMF (Figure 3A) could be HMFA. To test if the incapacity of *P. pseudoalcaligenes* was due to a problem of induction, the bacterium was inoculated in media containing both F and HMF (Figure 3B). The result was that the bacterium exclusively used F and that the presence of HMF resulted in being toxic (Figure 3B). Therefore, the absence of catalytic enzymes for the assimilation of FA was not the reason for the inability of *P. psedoalcaligens* to assimilate HMF, but the absence of reactions connecting HMFA and FDCA to FA. The transformation of FFA into FA takes place in two consecutive oxidative steps. In *Pseudomonas putida* Fu1, two different and inducible enzymes catalyze these reactions [23], but in *C. basilensis*, although no concrete genes have been assigned, it could be possible that the same enzyme catalyzes both oxidations [24]. In *P. pseudoalcaligenes* CECT5344 R1, both dehydrogenase activities (FFADH and FDH in the scheme of Figure 1D) co-eluted after anion exchange chromatography (not shown), thus suggesting that it is the same enzyme. Again, both activities had the same optimum temperature at 65 °C. Furfural dehydrogenase activity was clearly induced by furfuryl alcohol, if compared with acetate, FA or LB medium (not shown). Although these results suggest that the same enzyme could catalyze the oxidation of FFA to F and of F to FA, the only clear conclusion is that transformation of FFA to FA and conversion of FA to 2-oxoglutaric acid takes place through different pathways.

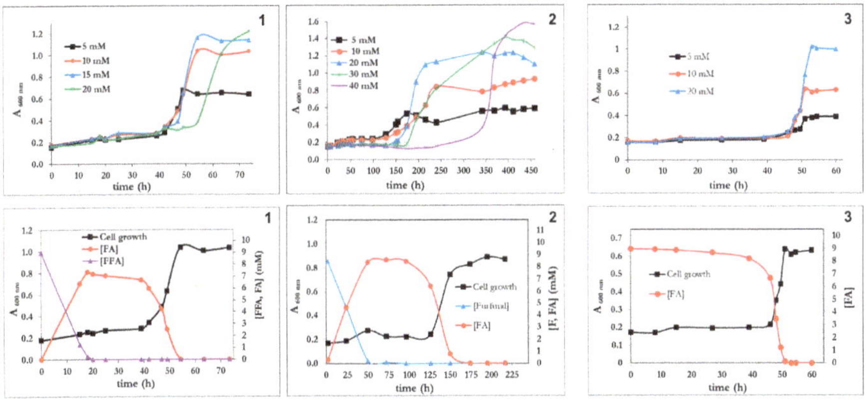

Figure 2. (**A**) Growth curves of *Pseudomonas pseudoalcaligenes* CECT 5344 R1 with different concentrations of furfuryl alcohol (panel A1), furfural (panel A2) or furoic acid (panel A3). (**B**) Growth curves and concentrations of furanic intermediates in the culture media of *P. pseudoalcaligenes* using as a carbon source either 5 mM furfuryl alcohol (panel B2), furfural (panel B2) or furoic acid (panel B3). Three independent experiments gave similar results.

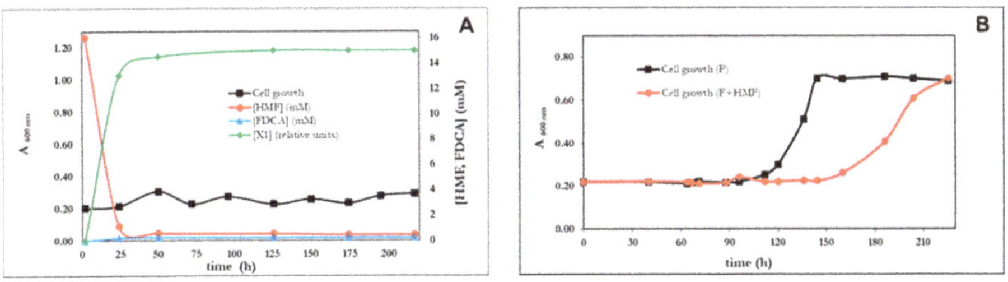

Figure 3. Biotransformation of HMF by *P. pseudoalcaligenes* CECT 5344 R1. (**A**) Cell growth (black line), concentration of HMF and FDCA (mM), and an unknown metabolite in cell cultures of *P. pseudoalcaligenes*. (**B**) Effect of HMF (5 mM) on the cell growth of *P. pseudoalcaligenes* at the expense of 10 mM furfural (F+HMF) in comparison with the cell growth with 10 mM furfural (F, black line). Similar results were obtained in three different experiments.

Once grown on furfural as a C-source, the serial dilutions of a culture of *P. pseudoalcaligenes* CECT 5344 R1 spread on solid minimal medium with furfural or FA giving colonies with two morphologies, big or small colonies. The small colonies had the same phenotype as the original stain (R1), but the big colonies grew faster on furfural even after successive generations on non-selective medium (LB). In fact, after four serial re-inoculations of the bacteria in in fresh media with furfural (10 mM) as the sole carbon source, allowed the selection of a mutant (big colonies in FA plates) in which the growth lag phase was drastically reduced and had a reproducible growth rate of 0.29 h^{-1} on furfural (5 mM). Figure 4A illustrates the isolation of the mutant, thereafter called R1D, and its phenotype in comparison to the *wt*. The selection of mutants with improved capacities has been widely used in biotechnological processes for the selection of evolved phenotypes, although having poor knowledge of their underlying genotype [31], also in the context of furfural assimilation [32]. For example, *P. putida* S12 expressing the *hmfABCDE* genes from *C. basilensis* HMF14 strain, was only able to efficiently assimilate furfural when adapted by repeated inoculation in media with furfural. The same has been recently described for *Pseudomonas putida* KT2440 expressing a 12 kb DNA fragment containing the *hmf* gene cluster from *Burkholderia phytofirmans* [33]. The next generation sequence (ngs) techniques open the possibility to analyze the genotypic variation associated with the evolved phenotypes. Adaptive laboratory evolution (ALE) takes profit of this advantage and may have a tremendous application in the rational design of genetically manipulated microorganisms, as well as in understanding some basic evolving mechanisms of living beings [31,34,35]. In our laboratory, the transcriptomic analysis of the R1D mutant in response to furfuryl alcohol was analyzed and it revealed to be more complex than expected (not shown). Interestingly, the transcriptomic reads sequences (RNA-seq) were obtained from the mutant strain (R1D), whereas the genome sequences available are from the wt strain [36]. The comparison of both sequences in the *hmf* locus revealed that the R1D mutant had a point mutation in a possible regulatory gene of the *araC* family. The presence of the mutation was confirmed by re-sequencing the *araC* gene (BN5_2307) (Figure 4B). The observed point mutation was a transversion (782T>G), leading to the non-conservative change of the triplet CTT (Leu) to CGT (Arg) (Figure 4B). The alignment of several AraC proteins revealed that the Arg in position 261 is a conserved residue in most members of this family of regulators (not shown). This position is located in the HTH domain of the protein and it is also conserved in *E. coli* [37], thus suggesting its essentiality. In conclusion, it seems that the pathway for the assimilation of FA is not active in the wt strain of *P. pseudoalcaligens* because AraC is not functional, and that the L262R mutation generates the active and functional regulator (AraC*) in the evolved R1D strain. In order to check this hypothesis, a mutant of the *araC** gene of *P. pseudoalcaligenes* CECCT 5344 R1D was generated by double recombination. As expected, the mutant strain was unable to use furfural as a C-source (Figure 4C).

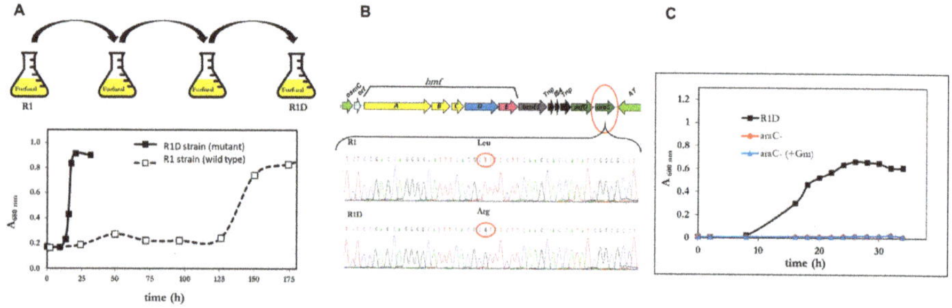

Figure 4. Role of *araC* in the assimilation of furfural by *P. pseudoalcaligenes*. (**A**) Scheme of the selection process (upper panel), and growth curve (lower panel) of the R1D mutant in comparison with the wt strain (R1), in media with furfural (10 mM) as the sole carbon source. (**B**) Scheme of the point mutation detected in the *araC* gene of the R1D mutant. (**C**) Growth curve of the R1D mutant and its derived mutant generated by the inactivation of the *araC** gene by insertion of the gentamicin resistance gene (*aacC1*) by double recombination. Similar results were obtained in three different experiments (panels A and C).

In addition to furfural, the mutant R1D was also capable to assimilate furfuryl alcohol and furoic acid (Figure 5). However, this strain remained unable to use HMF as a carbon source.

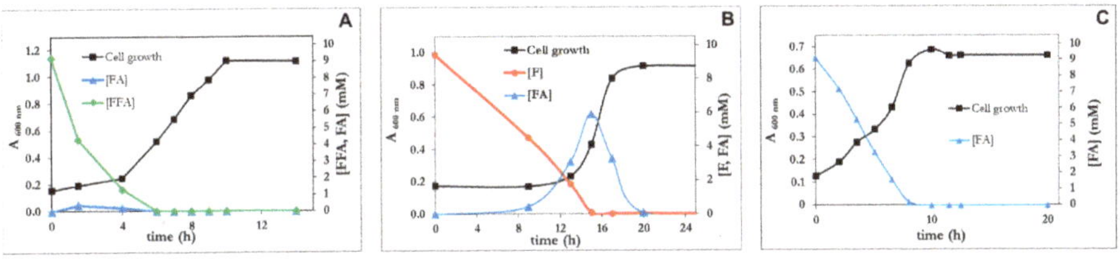

Figure 5. Growth curves of *P. pseudoalcaligenes* CECT 5344 R1D with furfuryl alcohol (**A**), furfural (**B**) or furoic acid (**C**) as the sole carbon sources (10 mM). The concentration of furanic intermediates in the culture media was measured by HPLC at the indicated times. Each experiment was done in triplicate giving similar results.

It is evident that R1D grew faster than R1 and with similar rates to other furfural-degrading strains reported in the literature. Table 2 summarizes the growth parameters of R1 and R1D strains of *P. pseudoalcaligenes* CECT 5344 in comparison to reference strains. In addition to the shorter lag phases and higher growth rates, a notable difference between the wild strain (R1) and the R1D mutant was the accumulation of FA during the lag phase. As shown in Figure 2, the wt strain stoichiometrically transformed FFA and F to FA. By contrast, in the RD1 mutant, FA was the only transiently accumulated form F (Figure 5B), whereas it was hardly detectable in media with FFA (Figure 5A). No lag phase was observed in the R1D strain growing with FA. Therefore, the presence of FA immediately induces its assimilation in R1D, thus suggesting that FA could be the inducer of the process. It is evident that the difference between R1 and R1D relies on the process of assimilation of furoic acid, not in the oxidation of alcohol and aldehyde to furoic acid. In analogy with the assimilation of toluene and xylenes in *P. putida* mt-2 [38], we can divide the pathway in two segments, the upper pathway and the lower pathway. We can consider the upper pathway as the oxidation of the alcohol group (-CH_2OH) and aldehyde (-CHO) to acid (-COOH). In the case of *C. basilensis*, the decarboxylase has to be included in the upper pathway, making the metabolism of HMF converge to FA. The lower pathway consists in the mineralization of furoic acid.

Table 2. Growth parameters of *P. pseudoalcaligenes* CECT 5344 R1 (wt) and the R1D mutant in comparison to bibliographic data.

Bacterial Strain	Growth Rate (h^{-1})			Lag Time (h)			Maximal $A_{600\ nm}$		
	FFA [1]	F [2]	FA [3]	FFA	F	FA	FFA	F	FA
P. pseudoalcaligenes R1	0.14	0.02	0.19	40	125	48	0.68	0.59	0.39
P. pseudoalcaligenes R1D	0.34	0.29	0.34	2	13	<1	0.72	0.55	0.41
C. basilensis HMF14 [4]	0.22	0.22	0.29	-	-	<1	1.1	1.09	0.99
P. putida S12 [5]	-	0.3	-	-	-	-	-	-	-

[1] Furfuryl alcohol, [2] Furfural or [3] furoic acid (5 mM). [4,5] Extracted form [16] and [24], respectively.

It is worth noting that these experiments confirmed the fact that the maximal cell growth of *P. pseudoalcaligenes* with FFA was higher than with F followed by FA (Table 2).

It is evident that R1D is an evolved strain much more efficient than the wt in the assimilation of FFA, F and FA. Nevertheless, the utilization of these capabilities for the elimination of the inhibitory compounds from lignocellulosic hydrolysates requires that the bacterium leave the sugars intact for their further fermentation process. *P. pseudoalcaligenes* CECT 5334 was unable to use as a C-source neither xylose, sucrose, arabinose, mannose, nor galactose. The capability of *P. pseudoalcaligenes* to use sugars was restricted to glucose. The chemical composition of the lignocellulosic hydrolysates depends on the raw material utilized and the treatment employed [39]. Therefore, the capacity of *P. pseudoalcaligenes* CECT 5344 R1D to use glucose and furfural simultaneously was studied. As expected, R1D assimilated both compounds simultaneously in blended media (Figure 6). From Figure 6, it became evident that F is toxic, since maintaining the concentration of glucose, the lag phase increased when furfural concentration increased (Figure 6A,B). In any case, since both glucose and F were assimilated simultaneously by *P. Pseudoalcaligenes*, the construction of a mutant impaired in glucose assimilation was designed. Most *Pseudomonas*, which have a relatively limited ability to assimilate sugars, usually assimilate glucose through the Entner-Doudoroff pathway [40], and the inactivation of the *edd* gene usually causes the inability to assimilate glucose [41]. As shown in Figure 6C, the *edd*$^-$ strain was as efficient as R1D assimilating furfural but glucose remains unaltered in the culture media. The *edd*$^-$ mutant was also able to use furfuryl alcohol and furoic acid as a C-source.

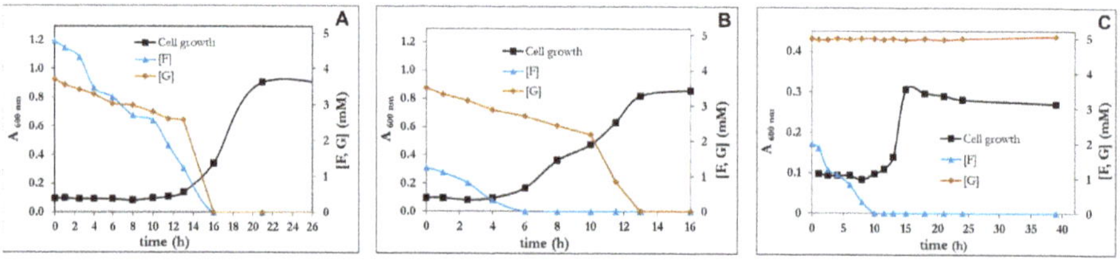

Figure 6. Cell growth of *P. pseudoalcaligenes* CECT 5344 R1D with glucose 5 mM, supplemented with furfural 5 mM (**A**) or 2 mM (**B**) as carbon sources. (**C**) Cell growth of *P. pseudoalcaligenes* CECT 5344 R1D *edd*$^-$ with furfural (2.5 mM) and glucose (5 mM) as carbon sources. Glucose and furfural concentration were determined at the indicated times. Similar results were obtained in three different experiments.

4. Discussion

The biodegradation of contaminants and the production of renewable energies are among the most important challenges of modern society. Bioethanol is a sustainable fuel that can be obtained by the yeast fermentation of sugars [42]. When using sugars from lignocellulosic by-products instead of sugars from food crops, the product is called second-generation bioethanol. However, the hydrolysis of these polymers generates, in addition to fermentable sugars, compounds that act as inhibitors of the yeast catalyzing further fermentation. The formation of furfural from pentoses was described many years ago [43]. Furfurals, and specially HMF and its derivatives, are feedstock for the synthesis of numerous valuable products. Therefore, their production in a sustainable manner is receiving an increasing interest [44]. On the other hand, from the bioethanol-production point of view, it is a challenge eliminating these inhibitors leaving sugars intact, with the purpose of increasing efficiency in second-generation ethanol production. Furfural and 5-hydroxymethyl furfural (HMF) are not the only inhibitory compounds present in the hydrolysates, but many others are not so well characterized, such as methanol or acetate [39]. The selection of yeast resistant to aldehydes may partially circumvent the problem. In fact, yeasts resistant up to 40 mM furfural and 80 mM HMF have been described recently [45,46]. Nevertheless, and taking into account the diversity of hydrolysates, the direct elimination of inhibitory compounds has to be taken into account as an alternative (bio)technology.

In this manuscript, we describe the selection of an adaptive mutant of *Pseudomonas pseudoalcaligenes* CECT 5344 able to assimilate furfural. The mutation was found to be located in a regulatory gene of the AraC/XylS family of activators [47]. In *E. coli*, *araC* is necessary for the assimilation of L-arabinose, a five carbon sugar [37]. Although speculative, perhaps it is not casual that the homologous gene in *P. pseudoalcaligenes* CECT 5344 could recognize as an activator the five-carbon dehydrated and oxidized pentose FA. This hypothesis has to be tested experimentally in the future. AraC homologous are proteins approximately 300 aa long, whose C-terminal segment is the HTH domain interacting with DNA. The mutation in the adapted strain was detected just in this domain. The substitution of a hydrophobic amino acid by a basic one (L262R) introduces a positive charge in the protein that seems to be essential for interacting with the negatively-charged DNA molecule. The less-conserved N-terminal domain of the AraC family is presumed to contain binding sites for specific activator molecules that confer specificity to each member [47].

Adaptive mutations frequently target regulatory genes [34], although there are interesting examples in which the genetic adaptation lies in catabolic genes. For example, in *P. putida* KT2440, a single point mutation was detected causing the suppression of a frameshift mutation in the transporter (*galT*), thus allowing the evolved strain to grow in gallic acid [48]. The *hmf* locus in *P. pseudoalcaligenes* contains two different modules, a catabolic one homologous to *C. baisilensis* HMF14 (plus *benE*) and another one homologous to *P. putida* Fu1 containing the regulatory gene *araC* (Figure 1B). As far as we know, no gene homologous to *benE* has been described in the context of FA assimilation. *benE* is a member of the MSF-family of transporters, homologous to the benzoate transporters. Since both compounds, benzoate and furoate, are aromatic monocarboxylic acids, it can be speculated that BenE is a furoate transporter, although this hypothesis needs further experimental evidence. The *hmf* locus (Figure 1) is flanked by mobile genetic elements, thus suggesting that it has been horizontally transferred. Furthermore, this hypothesis is in agreement with the sequence composition of the locus. The average GC content of the genes flanking the *hmf* locus in *P. pseudoalcaligenes* (green genes in the Figure 4B) is around 62%, which is also the %GC content of the genome of *P. mendocina* ymp (Figure 4A, 62.8%) and *P. pseudoalcaligenes* CECT 5344 itself (62.34%). By contrast, the average composition (%GC) of the *hmf* operon (BN5_2298 to BN5_2305) is 66.5%, close to the composition of *Cupriavidus basislensis* (65.3%) and other betaproteobacteria. The other two genes present in the island are BN5_2306 (*psfD*) and BN5_2307 (*araC*), whose %GC are 67.69% and 57.16%, respectively. It is also remarkable that all genes present in the *hmf* locus (Figure 1B) are singletons in relation to most Pseudomonaceae, except *psfD* and *benE*. In the evolution of prokaryotic metabolic networks and their regulation, the number of transcriptional regulators grows faster than the metabolic genes [49]. The horizontally transferred *hmf* pathway homologous to that described in *C. basilensis* HMF14 does not include its dedicated transcriptional regulator, but it seems that it has been acquired in a separate module from a second donor strain. Curiously, the regulator was originally in an inactive form, but evolved to the active form under selective pressure (Figure 4).

Even though *P. pseudoalcalignes* CEC T 5344 R1D assimilates furfural very efficiently, it could not be useful for the pre-treatment of lignocellulosic residues because it simultaneously assimilates glucose. By contrast, the *edd*$^-$ mutant assimilated furfural leaving the glucose intact (Figure 6). This is an important feature in comparison to the equivalent mutant in *P. putida* KT2440, that accumulates 6-phosphoogluconate from glucose [40]. On the other hand, *P. pseudoalcaligens* could not assimilate HMF (Figure 3), the other main inhibitory component of hydrolysates. In fact, the genome analyses anticipated this result due to the absence of homologous genes to *hmfH* and *hmfFG* genes. These genes in *C. basilensis* code for the enzymes catalyzing the oxidation of HMFA to FDCA and the decarboxylation of the latter to generate furoic acid, respectively. FA is the metabolite in which converges the assimilation of HMF and furfural in *C. basilensis* HMF14 [24]. To circumvent the problem, it would be convenient to have bacteria capable of eliminating HMF. We have isolated, by selective enrichment from ashes, some bacterial strains belonging to the genus *Pseudomonas*, which are able to assimilate HMF in addition to furfural, furoic acid or furfuryl alcohol [50]. These abilities could be used in mixed cultures, provided that they do not assimilate sugars, or by constructing the required traits using the available genetic modules. Even though *P. pseudomonas* CECT 5344 R1D fully transformed HMF into two new compounds, 5-Hydroxymethylfuranoic acid (HMFA) and 2,5-furandicarboxylic acid (FDCA), both of which are versatile chemical intermediates of high industrial potential [51,52]. Therefore, the *edd*$^-$ mutant of the evolved R1D strain of *P. pseudoalcaligenes* described in this manuscript may increase the productivity of second generation bioethanol both by eliminating yeasts' inhibitory chemicals and by producing value-added chemicals from biomass.

Supplementary Materials: The following are available online at http://www.mdpi.com/2073-4425/10/7/499/s1, Figure S1: Schematic construction of PGEM-TE *edd aacC1* plasmid. Figure S2. Schematic construction of PGEM-TE *araC aacC1* plasmid.

Author Contributions: Conceptualization, R.B. and M.I.I.; methodology, R.B. and M.I.I.; investigation, D.M. and M.I.I., writing, R.B.; visualization, R.B., D.M. and M.I.I.; supervision, R.B. and M.I.I.; project administration, R.B.; funding acquisition, R.B.

Funding: This research was funded by IB16062, Junta de Extremadura (Consejería de Economía e Infraestructuras), GR18031, Fondo Europeo de Desarrollo Regional (FEDER), European Union. The work of Daniel Macias was supported by a fellowship from Universidad de Extremadura (Accion III-41, UEX 2010-2014).

Acknowledgments: The technical support of Gloria Gutiérrez and Gracia Becerra are gratefully acknowledged.

Conflicts of Interest: The authors declare no conflict of interest. The funders had no role in the design of the study; in the collection, analyses, or interpretation of data; in the writing of the manuscript, or in the decision to publish the results.

References

1. Tavoni, M.; Kriegler, E.; Riahi, K.; van Vuuren, D.P.; Aboumahboub, T.; Bowen, A.; Calvin, K.; Campiglio, E.; Kober, T.; Jewell, J.; et al. Post-2020 climate agreements in the major economies assessed in the light of global models. *Nat. Clim. Chang.* **2014**, *5*, 119. [CrossRef]
2. Gray, K.A.; Zhao, L.; Emptage, M. Bioethanol. *Curr. Opin. Chem. Biol.* **2006**, *10*, 141–146. [CrossRef]
3. Blasco, R.; Castillo, F. Acerca de la biotecnología ambiental. *Arbor* **2014**, *190*, a157. [CrossRef]
4. Luque-Almagro, V.M.; Huertas, M.J.; Martinez-Luque, M.; Moreno-Vivian, C.; Roldan, M.D.; Garcia-Gil, L.J.; Castillo, F.; Blasco, R. Bacterial degradation of cyanide and its metal complexes under alkaline conditions. *Appl. Environ. Microbiol.* **2005**, *71*, 940–947. [CrossRef] [PubMed]
5. Akcil, A. Destruction of cyanide in gold mill effluents: Biological versus chemical treatments. *Biotechnol. Adv.* **2003**, *21*, 501–511. [CrossRef]
6. Akcil, A.; Mudder, T. Microbial destruction of cyanide wastes in gold mining: Process review. *Biotechnol. Lett.* **2003**, *25*, 445–450. [CrossRef] [PubMed]
7. Baxter, J.; Cummings, S.P. The current and future applications of microorganism in the bioremediation of cyanide contamination. *Antonie Leeuwenhoek* **2006**, *90*, 1–17. [CrossRef]
8. Raybuck, S.A. Microbes and microbial enzymes for cyanide degradation. *Biodegradation* **1992**, *3*, 3–18. [CrossRef]
9. Luque-Almagro, V.M.; Acera, F.; Igeño, M.I.; Wibberg, D.; Roldán, M.D.; Sáez, L.P.; Hennig, M.; Quesada, A.; Huertas, M.J.; Blom, J.; et al. Draft whole genome sequence of the cyanide-degrading bacterium *Pseudomonas pseudoalcaligenes* CECT5344. *Environ. Microbiol.* **2013**, *15*, 253–270. [CrossRef]
10. Barakat, A.; Monlau, F.; Steyer, J.-P.; Carrere, H. Effect of lignin-derived and furan compounds found in lignocellulosic hydrolysates on biomethane production. *Bioresour. Technol.* **2012**, *104*, 90–99. [CrossRef]
11. Palmqvist, E.; Hahn-Hägerdal, B. Fermentation of lignocellulosic hydrolysates. II: Inhibitors and mechanisms of inhibition. *Bioresour. Technol.* **2000**, *74*, 25–33. [CrossRef]
12. Alvira, P.; Tomas-Pejo, E.; Ballesteros, M.; Negro, M.J. Pretreatment technologies for an efficient bioethanol production process based on enzymatic hydrolysis: A review. *Bioresour. Technol.* **2010**, *101*, 4851–4861. [CrossRef] [PubMed]
13. Ortu, E.; Caboni, P. Levels of 5-hydroxymethylfurfural, furfural, 2-furoic acid in sapa syrup, Marsala wine and bakery products. *Int. J. Food Prop.* **2017**, *20*, S2543–S2551. [CrossRef]
14. Zhang, J.; Zhu, Z.; Wang, X.; Wang, N.; Wang, W.; Bao, J. Biodetoxification of toxins generated from lignocellulose pretreatment using a newly isolated fungus, Amorphotheca resinae ZN1, and the consequent ethanol fermentation. *Biotechnol. Biofuels* **2010**, *3*, 26. [CrossRef]
15. Ran, H.; Zhang, J.; Gao, Q.; Lin, Z.; Bao, J. Analysis of biodegradation performance of furfural and 5-hydroxymethylfurfural by *Amorphotheca resinae* ZN1. *Biotechnol. Biofuels* **2014**, *7*, 51. [CrossRef]
16. Wierckx, N.; Koopman, F.; Bandounas, L.; Winde, J.H.d.; Ruijssenaars, H.J. Isolation and characterization of *Cupriavidus basilensis* HMF14 for biological removal of inhibitors from lignocellulosic hydrolysate. *Microb. Biotechnol.* **2010**, *3*, 336–343. [CrossRef]
17. Wierckx, N.; Koopman, F.; Ruijssenaars, H.; de Winde, J. Microbial degradation of furanic compounds: Biochemistry, genetics, and impact. *Appl. Microbiol. Biotechnol.* **2011**, *92*, 1095–1105. [CrossRef] [PubMed]

18. Zhang, Y.; Han, B.; Ezeji, T.C. Biotransformation of furfural and 5-hydroxymethyl furfural (HMF) by Clostridium acetobutylicum ATCC 824 during butanol fermentation. *N. Biotechnol.* **2012**, *29*, 345–351. [CrossRef] [PubMed]
19. Koopman, F.; Wierckx, N.; de Winde, J.H.; Ruijssenaars, H.J. Efficient whole-cell biotransformation of 5-(hydroxymethyl)furfural into FDCA, 2,5-furandicarboxylic acid. *Bioresour. Technol.* **2010**, *101*, 6291–6296. [CrossRef]
20. Liu, Z.L.; Slininger, P.J.; Dien, B.S.; Berhow, M.A.; Kurtzman, C.P.; Gorsich, S.W. Adaptive response of yeasts to furfural and 5-hydroxymethylfurfural and new chemical evidence for HMF conversion to 2,5-bis-hydroxymethylfuran. *J. Ind. Microbiol. Biotechnol.* **2004**, *31*, 345–352. [CrossRef]
21. Taherzadeh, M.J.; Gustafsson, L.; Niklasson, C.; Liden, G. Physiological effects of 5-hydroxymethylfurfural on Saccharomyces cerevisiae. *Appl. Microbiol. Biotechnol.* **2000**, *53*, 701–708. [CrossRef] [PubMed]
22. Zheng, D.; Bao, J.; Lu, J.; Lv, Q. Biodegradation of furfural by Bacillus subtilis strain DS3. *J. Environ. Biol.* **2015**, *36*, 727–732. [PubMed]
23. Koenig, K.; Andreesen, J.R. Xanthine dehydrogenase and 2-furoyl-coenzyme A dehydrogenase from *Pseudomonas putida* Fu1: Two molybdenum-containing dehydrogenases of novel structural composition. *J. Bacteriol.* **1990**, *172*, 5999–6009. [CrossRef] [PubMed]
24. Koopman, F.; Wierckx, N.; de Winde, J.H.; Ruijssenaars, H.J. Identification and characterization of the furfural and 5-(hydroxymethyl)furfural degradation pathways of *Cupriavidus basilensis* HMF14. *Proc. Natl. Acad. Sci. USA* **2010**, *107*, 4919–4924. [CrossRef] [PubMed]
25. Nichols, N.N.; Mertens, J.A. Identification and transcriptional profiling of *Pseudomonas putida* genes involved in furoic acid metabolism. *FEMS Microbiol. Lett.* **2008**, *284*, 52–57. [CrossRef] [PubMed]
26. Koenig, K.; Andreesen, J.R. Molybdenum Involvement in Aerobic Degradation of 2-Furoic Acid by *Pseudomonas putida* Fu1. *Appl. Environ. Microbiol.* **1989**, *55*, 1829–1834. [PubMed]
27. Quesada, A.; Guijo, M.I.; Merchan, F.; Blazquez, B.; Igeño, M.I.; Blasco, R. Essential role of cytochrome *bd*-related oxidase in cyanide resistance of *Pseudomonas pseudoalcaligenes* CECT5344. *Appl. Environ. Microbiol.* **2007**, *73*, 5118–5124. [CrossRef]
28. Sambrook, J.; Russel, D.W. *Molecular Cloning: A Laboratory Manual*, 3rd ed.; Cold Spring Harbor Laboratory: New York, NY, USA, 2001.
29. Bradford, M.M. A rapid and sensitive method for the quantitation of microgram quantities of protein utilizing the principle of protein-dye binding. *Anal. Biochem.* **1976**, *72*, 248–254. [CrossRef]
30. Becker, A.; Schmidt, M.; Jäger, W.; Pühler, A. New gentamicin-resistance and *lacZ* promoter-probe cassettes suitable for insertion mutagenesis and generation of transcriptional fusions. *Gene* **1995**, *162*, 37–39. [CrossRef]
31. Winkler, J.; Reyes, L.H.; Kao, K.C. Adaptive Laboratory Evolution for Strain Engineering. *Syst. Metab. Eng.* **2013**, *985*, 211–222.
32. Abdulrashid, N.; Clark, D.P. Isolation and genetic analysis of mutations allowing the degradation of furans and thiophenes by *Escherichia coli*. *J. Bacteriol.* **1987**, *169*, 1267–1271. [CrossRef] [PubMed]
33. Guarnieri, M.T.; Ann Franden, M.; Johnson, C.W.; Beckham, G.T. Conversion and assimilation of furfural and 5-(hydroxymethyl)furfural by *Pseudomonas putida* KT2440. *Metab. Eng. Commun.* **2017**, *4*, 22–28. [CrossRef] [PubMed]
34. Conrad, T.M.; Lewis, N.E.; Palsson, B.O. Microbial laboratory evolution in the era of genome-scale science. *Mol. Syst. Biol.* **2011**, *7*, 509. [CrossRef] [PubMed]
35. Dragosits, M.; Mattanovich, D. Adaptive laboratory evolution—Principles and applications for biotechnology. *Microb. Cell Factories* **2013**, *12*, 64. [CrossRef] [PubMed]
36. Wibberg, D.; Luque-Almagro, V.M.; Igeño, M.I.; Bremges, A.; Roldán, M.D.; Merchán, F.; Sáez, L.P.; Guijo, M.I.; Manso, M.I.; Macías, D.; et al. Complete genome sequence of the cyanide-degrading bacterium *Pseudomonas pseudoalcaligenes* CECT5344. *J. Biotechnol.* **2014**, *175*, 67–68. [CrossRef] [PubMed]
37. Schleif, R. AraC protein, regulation of the l-arabinose operon in *Escherichia coli*, and the light switch mechanism of AraC action. *FEMS Microbiol. Rev.* **2010**, *34*, 779–796. [CrossRef] [PubMed]
38. Worsey, M.J.; Williams, P.A. Metabolism of toluene and xylenes by *Pseudomonas* (*putida* (*arvilla*) mt-2: Evidence for a new function of the TOL plasmid. *J. Bacteriol.* **1975**, *124*, 7–13. [PubMed]
39. Almeida, J.R.; Bertilsson, M.; Gorwa-Grauslund, M.F.; Gorsich, S.; Liden, G. Metabolic effects of furaldehydes and impacts on biotechnological processes. *Appl. Microbiol. Biotechnol.* **2009**, *82*, 625–638. [CrossRef] [PubMed]

40. Daddaoua, A.; Krell, T.; Ramos, J.-L. Regulation of Glucose Metabolism in Pseudomonas: The phosphorylative branch and entner-doudoroff enzymes are regulated by a repressor containing a sugar isomerase domain. *J. Biol. Chem.* **2009**, *284*, 21360–21368. [CrossRef] [PubMed]
41. Blevins, W.T.; Feary, T.W.; Phibbs, P.V., Jr. 6-Phosphogluconate dehydratase deficiency in pleiotropic carbohydrate-negative mutant strains of Pseudomonas aeruginosa. *J. Bacteriol.* **1975**, *121*, 942–949.
42. Mohd Azhar, S.H.; Abdulla, R.; Jambo, S.A.; Marbawi, H.; Gansau, J.A.; Mohd Faik, A.A.; Rodrigues, K.F. Yeasts in sustainable bioethanol production: A review. *Biochem. Biophys. Rep.* **2017**, *10*, 52–61. [CrossRef] [PubMed]
43. Hurd, C.D.; Isenhour, L.L. Pentose reacions. I. Furfural formation. *J. Am. Chem. Soc.* **1932**, *54*, 317–330. [CrossRef]
44. Hu, L.; Lin, L.; Wu, Z.; Zhou, S.; Liu, S. Recent advances in catalytic transformation of biomass-derived 5-hydroxymethylfurfural into the innovative fuels and chemicals. *Renew. Sustain. Energy Rev.* **2017**, *74*, 230–257. [CrossRef]
45. De Mello, F.d.S.B.; Coradini, A.L.V.; Tizei, P.A.G.; Carazzolle, M.F.; Pereira, G.A.G.; Teixeira, G.S. Static microplate fermentation and automated growth analysis approaches identified a highly-aldehyde resistant *Saccharomyces cerevisiae* strain. *Biomass Bioenergy* **2019**, *120*, 49–58. [CrossRef]
46. Nagamatsu, S.T.; Teixeira, G.S.; de Mello, F.d.S.B.; Tizei, P.A.G.; Nakagawa, B.T.G.; de Carvalho, L.M.; Pereira, G.A.G.; Carazzolle, M.F. Genome Assembly of a Highly Aldehyde-Resistant *Saccharomyces cerevisiae* SA1-Derived Industrial Strain. *Microbiol. Resour. Announc.* **2019**, *8*, e00071. [CrossRef] [PubMed]
47. Gallegos, M.T.; Schleif, R.; Bairoch, A.; Hofmann, K.; Ramos, J.L. Arac/XylS family of transcriptional regulators. *Microbiol. Mol. Biol. Rev.* **1997**, *61*, 393–410. [PubMed]
48. Nogales, J.; Canales, A.; Jimenez-Barbero, J.; Serra, B.; Pingarron, J.M.; Garcia, J.L.; Diaz, E. Unravelling the gallic acid degradation pathway in bacteria: The *gal* cluster from *Pseudomonas putida*. *Mol. Microbiol.* **2011**, *79*, 359–374. [CrossRef] [PubMed]
49. Maslov, S.; Krishna, S.; Pang, T.Y.; Sneppen, K. Toolbox model of evolution of prokaryotic metabolic networks and their regulation. *Proc. Natl. Acad. Sci. USA* **2009**, *106*, 9743–9748. [CrossRef] [PubMed]
50. Igeño, M.I.; Sánchez-Clemente, R.; Población, A.G.; Guijo, M.I.; Merchán, F.; Blasco, R. Biodegradation of 5-(Hydroxymethyl)-furfural and Furan Derivatives. *Proceedings* **2018**, *2*, 1283. [CrossRef]
51. Kang, E.-S.; Chae, D.W.; Kim, B.; Kim, Y.G. Efficient preparation of DHMF and HMFA from biomass-derived HMF via a Cannizzaro reaction in ionic liquids. *J. Ind. Eng. Chem.* **2012**, *18*, 174–177. [CrossRef]
52. Deng, J.; Liu, X.; Li, C.; Jiang, Y.; Zhu, J. Synthesis and properties of a bio-based epoxy resin from 2,5-furandicarboxylic acid (FDCA). *RSC Adv.* **2015**, *5*, 15930–15939. [CrossRef]

© 2019 by the authors. Licensee MDPI, Basel, Switzerland. This article is an open access article distributed under the terms and conditions of the Creative Commons Attribution (CC BY) license (http://creativecommons.org/licenses/by/4.0/).

genes

MDPI

Article

Further Insights into the Architecture of the P_N Promoter That Controls the Expression of the *bzd* Genes in *Azoarcus*

Gonzalo Durante-Rodríguez [1], Paloma Gutiérrez-del-Arroyo [2], Marisela Vélez [2], Eduardo Díaz [1] and Manuel Carmona [1,*]

1 Microbial and Plant Biotechnology Department, Centro de Investigaciones Biológicas-CSIC, Ramiro de Maeztu, 9, 28040 Madrid, Spain

2 Biocatalysis Department, Institute of Catalysis and Petrochemistry-CSIC, Marie Curie, 2, Cantoblanco, 28049 Madrid, Spain

* Correspondence: mcarmona@cib.csic.es; Tel.: +34-91-8373112

Received: 24 April 2019; Accepted: 26 June 2019; Published: 27 June 2019

Abstract: The anaerobic degradation of benzoate in bacteria involves the benzoyl-CoA central pathway. *Azoarcus/Aromatoleum* strains are a major group of anaerobic benzoate degraders, and the transcriptional regulation of the *bzd* genes was extensively studied in *Azoarcus* sp. CIB. In this work, we show that the *bzdR* regulatory gene and the P_N promoter can also be identified upstream of the catabolic *bzd* operon in all benzoate-degrader *Azoarcus/Aromatoleum* strains whose genome sequences are currently available. All the P_N promoters from *Azoarcus/Aromatoleum* strains described here show a conserved architecture including three operator regions (ORs), i.e., OR1 to OR3, for binding to the BzdR transcriptional repressor. Here, we demonstrate that, whereas OR1 is sufficient for the BzdR-mediated repression of the P_N promoter, the presence of OR2 and OR3 is required for de-repression promoted by the benzoyl-CoA inducer molecule. Our results reveal that BzdR binds to the P_N promoter in the form of four dimers, two of them binding to OR1. The BzdR/P_N complex formed induces a DNA loop that wraps around the BzdR dimers and generates a superstructure that was observed by atomic force microscopy. This work provides further insights into the existence of a conserved BzdR-dependent mechanism to control the expression of the *bzd* genes in *Azoarcus* strains.

Keywords: regulation; benzoate; anaerobic; *Azoarcus*; promoter architecture

1. Introduction

Aromatic compounds are the most widespread organic compounds in nature after carbohydrates. Moreover, the release of aromatic compounds into the biosphere increased considerably over the last century as a consequence of industrial activity [1]. Many of these compounds are toxic and/or carcinogenic, thus representing major persistent environmental pollutants. Some specialized microorganisms, mainly bacteria and fungi, adapted to degrade a wide variety of aromatic compounds aerobically and/or anaerobically [2]. Whereas the aerobic degradation of aromatics is extensively studied [3–5], the anaerobic degradation pathways are much less well studied, especially regarding the genes involved in these anaerobic processes [6,7]. However, many environments are anoxic and, thus, the anaerobic degradation of aromatic compounds has great importance at the ecological level [6–8]. The anaerobic degradation of a wide variety of aromatic compounds converges into a few central pathways (catabolic funnel) that carry out the reductive de-aromatization and further conversion of the intermediates to the central metabolism. Most monocyclic aromatic compounds are channeled and activated to the central intermediate benzoyl-CoA [2,8,9]. Benzoate was used over the two last decades as a model compound to study the main pathway for anaerobic degradation of aromatic compounds, i.e., the central route of benzoyl-CoA [7,9]. The anaerobic degradation of benzoate takes place through a peripheral route consisting of a single reaction that activates benzoate to benzoyl-CoA. Benzoyl-CoA is then de-aromatized by the action of a reductase, the only oxygen-sensitive enzyme within the benzoyl-CoA pathway, which generates cyclohexa-1,5-diene-1-carbonyl-CoA, which is further converted into 3-hydroxypimelyl-CoA through a modified β-oxidation mechanism (Figure 1A) [2,7–10]. The genes encoding the enzymes involved in the anaerobic degradation of benzoate were reported in some phototrophic bacteria, e.g., *Rhodopseudomonas palustris* strains (*bad* genes), facultative anaerobes, e.g., *Thauera*, *Azoarcus*, *Aromatoleum*, *Magnetospirillum*, and *Sedimenticola* strains (*bcr*/*bzd* genes), and in strict anaerobes, e.g., *Geobacter*, *Syntrophus*, and *Desulfobacula* strains (*bam* genes) [8–15]. The expression of the catabolic genes is controlled by specific transcriptional regulators, but only a few such regulators are described so far, mainly the BadM protein in *R. palustris* [16,17], the BamVW and BgeR proteins in *Geobacter* strains [18,19], and the BzdR protein in *Azoarcus* sp. CIB [20,21].

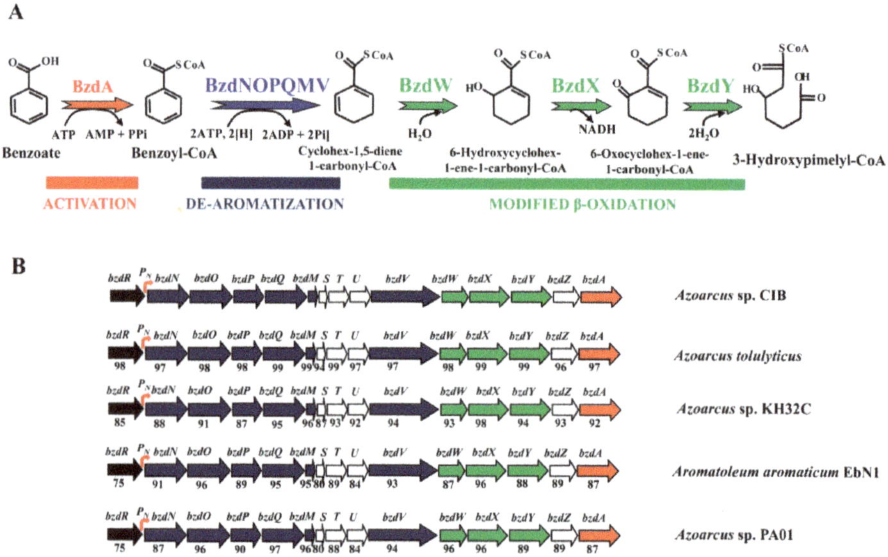

Figure 1. Scheme of anaerobic degradation of benzoate and gene organization of the *bzd* cluster in different *Azoarcus* and *Aromatoleum* strains. (**A**) Scheme of the anaerobic degradation pathway of benzoate into benzoyl-CoA (red), the de-aromatization of benzoyl-CoA (blue), and the modified β-oxidation that produces 3-hydroxypimelyl-CoA. Enzyme abbreviations: BzdA, benzoate-CoA ligase; BzdNOPQMV, benzoyl-CoA reductase, ferredoxin, and putative reduced nicotinamide adenine dinucleotide phosphate (NADPH)::ferredoxin oxidoreductase; BzdW, cyclohex-1,5-diene-1-carbonyl-CoA hydratase; BzdX, 6-hydroxycyclohex-1-ene-1-carbonyl-CoA dehydrogenase; BzdY, 6-oxocyclohexene-1-ene-carbonyl-CoA hydrolase. (**B**) Scheme of the *bzd* cluster in different *Azoarcus*/*Aromatoleum* strains. Genes are indicated in the same color code as the corresponding enzymes in panel (**A**), i.e., genes encoding the activation, de-aromatization, and modified β-oxidation are indicated in red, blue, and green color, respectively. The *bzdR* regulatory gene is shown in black, and the catabolic P_N promoter is shown in red as a curved arrow. Genes of unknown function are colored in white. Below each gene, the percentage of the amino-acid sequence identity to the corresponding *Azoarcus* sp. CIB ortholog is indicated. The accession numbers of the corresponding genome sequences are as follows: *Azoarcus* sp. CIB (CP011072), *Azoarcus* sp. KH32C (AP012304), *Aromatoleum aromaticum* EbN1 (CR555306), *Azoarcus tolulyticus* strain ATCC51758 (NZ_FTMD00000000), and *Azoarcus* sp. PA01 (NZ_LARU00000000).

Among facultative anaerobes, bacteria of the *Azoarcus* genus, recently re-classified within the new *Aromatoleum* genus [22], represent one of the major groups of aromatic degraders. *Azoarcus* sp. CIB is a denitrifying β-Proteobacterium able to aerobically and/or anaerobically degrade a wide variety of aromatic compounds, most of which are channeled to the benzoyl-CoA central pathway encoded by the *bzd* genes. The *bzd* genes from strain CIB are clustered together in a large catabolic operon (*bzdNOPQMSTUVWXYZA*) driven by the P_N promoter (Figure 1B). The specific transcriptional regulation of the *bzd* operon is carried out by the BzdR repressor encoded immediately upstream of the catabolic operon (Figure 1B) [20,23]. Benzoyl-CoA, the first intermediate during the anaerobic degradation of benzoate, is specifically recognized by BzdR and acts as the inducer molecule allowing the expression of the *bzd* catabolic genes [23]. BzdR binds to the P_N promoter at three different regions that contain a direct repetition of the sequence TGCA [20]. In addition to BzdR, two additional regulatory proteins account for an overimposed regulation of the P_N promoter. AcpR, a regulator belonging to the well-known fumarate and nitrate reductase (FNR)/cyclic AMP receptor protein (CRP) superfamily of transcriptional regulators, is essential for the activation of the P_N promoter under oxygen deprivation conditions [24]. Furthermore, the AccR response regulator interacts with P_N and inhibits its activity in response to several organic acids such as succinate, malate, or acetate, accounting for the carbon catabolite control of the *bzd* genes [23,25].

In this work, we expand the current knowledge on the organization of the *bzd* clusters in all *Azoarcus*/*Aromatoleum* strains sequenced so far that are able to degrade benzoate anaerobically. The evolutionary conservation of a common P_N promoter architecture with three BzdR operator regions can be highlighted. The role of the three BzdR operator regions was studied, and the formation of a BzdR/P_N superstructure was confirmed and visualized.

2. Material and Methods

2.1. Bacterial Strains, Plasmids, and Growth Conditions

Bacterial strains and plasmids used in this study are listed in Table 1.

Table 1. Bacterial strains and plasmids used in this study.

Strain or Plasmid	Relevant Genotype and Characteristic(s)	Reference
***Escherichia coli* strain**		
DH5α	*endA1 hsdR17 supE44 thi-1 recA1 gyrA*(Nal^r) *relA1* Δ(*argF-lac*) *U169 depR Φ80dlacd*(*lacZ*) *M15*	[26]
S17-1λ*pir*	Tp^r Sm^r *recA thi hsdRM*$^+$ RP4::2-Tc::Mu::Km λ*pir* phage lysogen	[27]
M15	Strain for regulated high-level expression with pQE vector	Qiagen *
MC4100	F$^-$, *araD319*, Δ(*argF-lac*)*U169 rpsL150* (Sm^r) *relA1 flbB5301 deoC1 ptsF25 rbs*	[28]
***Azoarcus* strain**		
CIB	Wild-type strain	[23]
Plasmids		
pQE32	Ap^r, *ori*ColE1 T5 promoter *lac* operator, λt_0/*E. coli rrnB T1* terminators, N-terminal His$_6$	Qiagen *
pQE32-His$_6$BzdR	Ap^r, pQE32 derivative harboring the His$_6$-*bzdR* gene	[20]
pQE60-His$_6$Fnr*	Ap^r, pQE60 derivative harboring the His$_6$- FNR* gene under the control of *T5* promoter *lac* operator	[29]
pQE32-His$_6$NBzdRL	Ap^r, pQE32 derivative harboring the His$_6$- NBzdRL fragment	[30]
pREP4	Km^r, plasmid that expresses the *lac I* repressor	Qiagen *

Table 1. *Cont.*

Strain or Plasmid	Relevant Genotype and Characteristic(s)	Reference
pECOR7	Ap^r, pUC19 harboring a 7.1-kb *Eco*RI DNA fragment containing the *bzdRNO* genes	[23]
pBBR1MCS-5	Gm^r, oripBBR1MCS Mob^+ *lacZα*, broad-host- range cloning and expression vector	[31]
pBBR5P_N	Gm^r, pBBR1MCS-5 derivative harboring a P_N::*lacZ* translational fusion	[20]
pBBR5P_NI	Gm^r, pBBR1MCS-5 derivative harboring a P_NI::*lacZ* translational fusion	This work
pBBR5P_NII	Gm^r, pBBR1MCS-5 derivative harboring a P_NII::*lacZ* translational fusion	This work
pJCD01	Ap^r, *ori*ColE1, polylinker of pUC19 flanked by *rpoC* and *rrnBT1T2* terminators	[32]
pJCD-P_N	Ap^r, pJCD01 derivative harboring a 585-bp *Eco*RI fragment that includes the P_N promoter	[24]
pJCD-P_NI	Ap^r, pJCD01 derivative harboring a 139-bp *Eco*RI fragment that includes the P_NI promoter	This work
pJCD-P_NII	Ap^r, pJCD01 derivative harboring a 209-bp *Eco*RI fragment that includes the P_NII promoter	This work
pSJ3-P_N	Ap^r, pSJ3 derivative harboring a P_N::*lacZ* translational fusion	[20]
pSJ3-P_NI	Ap^r, pSJ3 derivative harboring a P_NI::*lacZ* translational fusion	This work
pSJ3-P_NII	Ap^r, pSJ3 derivative harboring a P_NII::*lacZ* translational fusion	This work
pCK01	Cm^r, *ori*pSC101, low copy number cloning vector, polylinker flanked by *Not*I sites	[33]
pCK01-BzdR	Cm^r, pCK01 derivative harboring a DNA fragment containing the *bzdR* gene	[20]

Ap^r: ampicillin resistant; Km^r: kanamycin resistant; Cm^r: chloramphenicol resistant; Gm^r: gentamicin resistant.
*: Qiagen, Hilden, Germany.

Escherichia coli cells were grown at 37 °C in lysogeny broth (LB) medium [34]. When required, *E. coli* cells were grown anaerobically at 30 °C in LB medium supplemented with 0.1% casamino acids [34] and 10 mM KNO_3 as a terminal electron acceptor. *Azoarcus* strains were anaerobically grown at 30 °C in MC medium as described previously [23]. Where appropriate, antibiotics were added at the following concentrations: ampicillin, 100 $\mu g \cdot mL^{-1}$; chloramphenicol, 30 $\mu g \cdot mL^{-1}$; gentamicin, 7.5 $\mu g \cdot mL^{-1}$; and kanamycin, 50 $\mu g \cdot mL^{-1}$.

2.2. Molecular Biology Techniques

Recombinant DNA techniques were carried out according to published methods [26]. Plasmid DNA was prepared with a High Pure Plasmid Isolation Kit (Roche Applied Science, Penzberg, Germany). The DNA fragments were purified with Gene Clean Turbo (Q-BIOgene, Carlsbad, CA, USA). Oligonucleotides were supplied by Sigma (St. Louis, MO, USA). The sequence of inserts/DNA fragments was confirmed by DNA sequencing with an ABI Prism 377 automated DNA sequencer (Applied Biosystems, Foster City, CA, USA). Transformation of *E. coli* was carried out by using competent cells prepared by the RbCl method [34] or by electroporation (Gene Pulser, Bio-Rad, Cambridge, MA, USA) [34]. Plasmids were transferred from *E. coli* S17-1λ*pir* (donor strain) to *Azoarcus* sp. CIB recipient strains via biparental filter mating as previously reported [23]. Proteins were analyzed by SDS-PAGE as described previously [26].

2.3. Sequence Data Analyses

For bioinformatic inspection of genes and regulatory regions of interest, we employed the BioEdit Sequence Alignment Editor [35]. The Basic Local Alignment Search Tool (BLAST) platform [36] was used for studying the similarity/identity of proteins encoded in the *bzd* cluster. The amino-acid sequences of the open reading frames of proteins encoded in the *bzd* operon were compared with those present in databases using the TBLASTN algorithm at the National Center for Biotechnology Information (NCBI) server [37] (http://blast.ncbi.nlm.nih.gov/Blast.cgi). Nucleotide and protein alignments were done with ALIGN [38] and CLUSTALW [39], respectively, in the BioEdit editor.

2.4. Overproduction and Purification of His_6-BzdR, His_6-NBzdRL, and His_6-FNR* Proteins

Plasmid pQE32-His_6BzdR and pQE32-His_6NBzdRL produce N-terminally His_6-tagged BzdR and NBzdRL proteins (N-terminal domain plus the BzdR linker), respectively (Table 1). Plasmid pQE60-His_6FNR* produces a C-terminally His_6-tagged Fnr* (Table 1). All these plasmids express the cloned genes under control of the P_{T5} promoter and two *lac* operator boxes. The His_6-tagged proteins were overproduced in *E. coli* M15 strain also harboring plasmid pREP4, and they were purified following protocols previously established [20].

2.5. Analytical Ultracentrifugation Assays

To perform experiments of sedimentation equilibrium of His_6-NBzdRL protein bound to DNA, different concentrations of this protein (0.5 to 10 μM) were incubated with the DNA fragment (0.05 to 0.2 μM). The DNA fragments used were the P_NI promoter (145 bp) and the complete P_N promoter (P_N, 253 bp) that were obtained by PCR using the oligonucleotide pairs BLINE (5′–CGTGCCTGACATTTGACTTAGATC–3′) and 3IVTPN (5′–CGGGAATTCCATCGAACTATCTCCTCTGATG–3′, *Eco*RI site underlined), and RET1 (5′–CCGAGCCTCGCGTTTTACTGC–3′) and 3IVTPN, respectively. The ultracentrifugation experiments were conducted in a buffer containing 50 mM NaH_2PO_4, 300 mM KCl, 100 mM imidazole, pH 8, as previously described [21]. Since the BzdR protein aggregates in experiments of sedimentation equilibrium at concentrations higher than 2 μM, we used the NBzdRL protein that is totally soluble at 10 μM. The presence of DNA was checked by measuring the absorbance at 260 nm. The distribution of sediment coefficients obtained in the sedimentation velocity experiments was determined using the SEDFIT program as reported previously [21].

2.6. Construction of P_NI::lacZ and P_NII::lacZ Translational Fusions and β-Galactosidase Assays

The P_N::*lacZ* translational fusion was described previously [23]. To generate two truncated versions of the P_N promoter, P_NI and P_NII, fused to the *lacZ* reporter gene, we firstly PCR-amplified the truncated promoters. The oligonucleotides used for amplification of the P_NI promoter were BKIN (5′–GGGGTACCCGTGCCTGACATTTGACTTAGATC–3′; *Kpn*I site underlined) and 5BZN (5′–GCTCTAGACCCATCGAACTATCTCCTCTGATG–3′; *Xba*I site underlined). For the amplification of the P_NII promoter, the oligonucleotides used were NIU PN-II (5′–GGGGTACCCAAGAAAGATTGCAGTTTTCCATG; *Kpn*I site underlined) and 5BZN. The template used for amplification of the promoter regions was the pECOR7 plasmid (Table 1). After PCR-amplification, both DNA fragments were digested with *Xba*I and *Kpn*I restriction enzymes, and later cloned in plasmid pSJ3 in-frame with *lacZ* gene, thus generating the corresponding P_NI::*lacZ* and P_NII::*lacZ* translational fusions in plasmids pSJ3-P_NI and pSJ3-P_NII, respectively. These translational fusions were subsequently subcloned by *Eco*RI/*Hin*dIII double digestion into plasmid pBBR1MCS-5 generating plasmids, pBBR5-P_NI and pBBR5-P_NII, respectively. For determination of promoter activity, pSJ3-P_NI and pSJ3-P_NII were transformed in *E. coli* MC4100, while pBBR5-P_NI and pBBR5-P_NII were transferred to *Azoarcus* sp. CIB [23]. *E. coli* MC4100 carrying the P_N::*lacZ*, P_NI::*lacZ*, or P_NII::*lacZ* fusions was grown anaerobically in LB-rich medium, whereas *Azoarcus* sp. CIB carrying the P_N::*lacZ*, P_NI::*lacZ*, or P_N*II::lacZ* fusions was grown anaerobically in MC medium in the presence of 3 mM benzoate or 0.2% (*w/v*) succinate. Cultures were incubated at 30 °C until the mid-exponential phase. The β-galactosidase activity (in Miller Units) was determined in permeabilized cells (using 0.1% SDS and chloroform) according to the method described by Miller [34].

2.7. *In Vitro Transcription Assays*

In vitro transcription assays were performed as previously published [40,41] using plasmids pJCD-P_N, pJCD-P_NI, and pJCD-P_NII (0.5 nM) as supercoiled templates. To construct these plasmids, firstly, 585-bp, 139-bp, and 209-bp DNA fragments containing the P_N, P_N*I*, and P_N*II* promoters, respectively, were PCR-amplified from the *Azoarcus* sp. strain CIB chromosome. The forward and reverse oligonucleotides 5IVTPN (5′–CGGAATTCCGTGCATCAATGATCCGGCAAG–3′; *Eco*RI site underlined) and 3IVTPN (5′–CGGAATTCCATCGAACTATCTCCTCTGATG–3′; *Eco*RI site underlined), BIN (5′–CGGAATTCCGTGCCTGACATTTGACTTAGATC–3′; *Eco*RI site underlined) and 3IVTPN, and BIIN (5′–CGGAATTCCAAGAAAGATTGCAGTTTTCCATG–3′; *Eco*RI site underlined) and 3IVTPN, respectively, were used for PCR-amplification. Then, the DNA fragments were *Eco*RI restricted, and cloned into the *Eco*RI-restricted pJCD01 cloning vector, giving rise to plasmids pJCD-P_N, pJCD-P_NI, and pJCD-P_NII, respectively (Table 1). Reactions (50-μL mixtures) were performed in a buffer containing 50 mM Tris-HCl (pH 7.5), 50 mM KCl, 10 mM $MgCl_2$, 0.1 mM bovine serum albumin, 10 mM dithiothreitol, and 1 mM ethylenediaminetetraacetic acid (EDTA). Unless otherwise indicated, each DNA template was premixed with 100 nM σ^{70}-containing *E. coli* RNA polymerase (RNAP) holoenzyme (Amersham plc, Amersham, UK) and different amounts of purified His_6-Fnr*. For multiple-round assays, transcription was then initiated by adding a mixture of 500 μM (each) adenine, cytosine, and guanine triphosphate (ATP, CTP, and GTP); 50 μM uridine triphosphate (UTP); and 2.5 μCi of [α-^{32}P]-UTP (3000 Ci·mmol^{-1}). After incubation for 15 min at 37 °C, the reactions were stopped with an equal volume of a solution containing 50 mM EDTA, 350 mM NaCl, and 0.5 mg of carrier transfer RNA (tRNA; yeast tRNA from Thermo Fisher Scientific, Whatham, MA, USA) per ml. The messenger RNA (mRNA) produced was then precipitated with ethanol, separated on a denaturing 7 M urea/4% polyacrylamide gel, and visualized by autoradiography. Transcript levels were quantified with a Bio-Rad Molecular Imager FX system and using the ImageJ software [42].

2.8. *Gel Retardation Assays*

The P_N*I* DNA probe (145 bp) was obtained as described previously [21], digested with *Eco*RI, and labeled by filling in the overhanging *Eco*RI-digested end with [α-^{32}P]dATP (6000 Ci·mmol^{-1}; Amersham Biosciences, Amersham, UK) and the Klenow fragment of *E. coli* DNA polymerase as described previously [20]. The DNA probe was mixed with the purified proteins at the concentration indicated in each assay. The retardation reaction mixtures contained 20 mM Tris-HCl pH 7.5, 10% glycerol, 2 mM β-mercaptoethanol, 50 mM KCl, 0.05 nM DNA probe, 250 μg·mL^{-1} bovine serum albumin, and purified His_6-BzdR protein in a 9-μL final volume. The samples were fractionated by electrophoresis in 5% polyacrylamide gels buffered with 0.5× TBE (45 mM Tris borate, 1 mM EDTA). The gels were dried onto Whatman 3MM paper and exposed to Hyperfilm MP (Amersham Biosciences, Little Chalfont, UK).

2.9. *Analysis of Protein–DNA Interaction by Atomic Force Microscopy*

To perform atomic force microscopy (AFM) experiments, a DNA fragment of 1225 bp, termed P_NL, from position −505 to +720 and covering the intergenic region between *bzdR* and *bzdN* genes, as well as the 5′-end of the *bzdN* gene, was PCR-amplified using as template the plasmid pECOR7, and the oligonucleotides 3REG (5′–GCGGTACCCGTGCATCAATGATCCGGCAAG–3′; *Kpn*I site underlined) and N3 (5′–TTCAGCATCTCGTTGTGCTC–3′). The purified His_6-BzdR protein was obtained as detailed above. The binding reactions were performed for 15 min at room temperature using retardation buffer (20 mM Tris·HCl, pH 7.5, 10% glycerol, 2 mM β-mercaptoethanol, 50 mM

KCl). The protein and DNA concentrations were optimized to ensure that the unbound protein (probably adsorbed to the surface) did not disturb the image. The incubation of the complexes was performed at a final concentration of 2 nM P_NL fragment with [P_NL]:[BzdR] ratios of 1:28 and 1:70. The samples of complexes were diluted 10 times in retardation buffer, in the presence of $MgCl_2$ or $NiCl_2$ at concentrations between 3 and 10 mM. Each reaction mixture was then deposited on a freshly exfoliated mica surface (which provides a clean and atomically flat substrate). After 1 min of incubation on the mica, the non-adsorbed material was removed by successive washes with the same adsorption buffer. Finally, the samples, imaged while immersed in buffer, were visualized with an atomic force microscope from Nanotec Electronica S.L (Tres Cantos, Spain) operated in jump mode [43]; the images were processed with the WSxM program [44].

3. Results and Discussion

3.1. *The Genetic Organization of the* bzd *Cluster and the Architecture of the* P_N *Promoter Are Conserved in the Azoarcus/Aromatoleum Genus*

The anaerobic degradation of benzoate is a common feature among many strains of the *Azoarcus/Aromatoleum* genus [22]. By mining in the available genome sequences of *Azoarcus* aromatic degrader strains, i.e., those of *A. tolulyticus*, *A. toluclasticus*, *Azoarcus* sp. KH32C, and *Azoarcus* sp. PA01, we found the presence of genes homologous to the *bzd* genes responsible of the anaerobic degradation of benzoate in *A. evansii*, *Azoarcus* sp. CIB, and *A. aromaticum* EbN1 [8,45]. Interestingly, in all these *Azoarcus* strains, the *bzd* genes are arranged in a cluster that contains a *bzdR* regulatory gene located upstream of a putative *bzd* catabolic operon. The *bzd* operon is organized into at least three different functional modules: (i) the activation module (*bzdA*) that codes for the enzyme that generates benzoyl-CoA; (ii) the de-aromatization module (*bzdNOPQMV*) that encodes the benzoyl-CoA reductase and auxiliary enzymes; and (iii) the modified β-oxidation module (*bzdWXY*) that encodes the enzymes that generate 3-hydroxypimelyl-CoA (Figure 1) [8,10]. A close inspection of the promoter region of the predicted *bzd* operons from all *Azoarcus* strains revealed the conservation of the P_N promoter described in *Azoarcus* sp. CIB [20] (Figure 2). Remarkably, it should be noted that this organization found in *Azoarcus/Aromatoleum* was not observed in closely related aromatic-compound-degrading denitrifying bacteria such as those of the genus *Thauera*.

As described before, the P_N promoter of strain CIB shows the transcription start site located minus 75 nucleotides with respect to the ATG start codon of the *bzdN* gene, and the −10 and −35 boxes of interaction with the σ^{70} subunit of the RNAP were also identified [20] (Figure 2). Three BzdR operator regions were detected in the P_N promoter by DNase I footprinting, i.e., operator region I (OR1), located between positions −32 and +31 (with respect to the transcription start site), operator region II (OR2) (−83 to −63), and operator region III (OR3) (−146 to −126) [20]. The operator region recognized by the AcpR activator protein overlaps the −35 box of RNAP [24] (green box, Figure 2). Interestingly, the P_N promoter sequences of all *bzd* operons from *Azoarcus* strains exhibited the three BzdR operator regions (OR1, OR2, and OR3) and the operator region that binds to the AcpR transcriptional activator (Figure 2). The OR2 and OR3 regions of P_N contain the conserved TGCA(N_6)TGCA palindromic sequence where N_6 is a six-nucleotide A-rich region in OR3. OR1 is longer and it contains two conserved palindromic sequences, TGCA(C)T(G/C)(C/G)A and TGCA(N_{15})TGCA, located between the −35 and −10 boxes and downstream of the transcription start site, respectively (Figure 2).

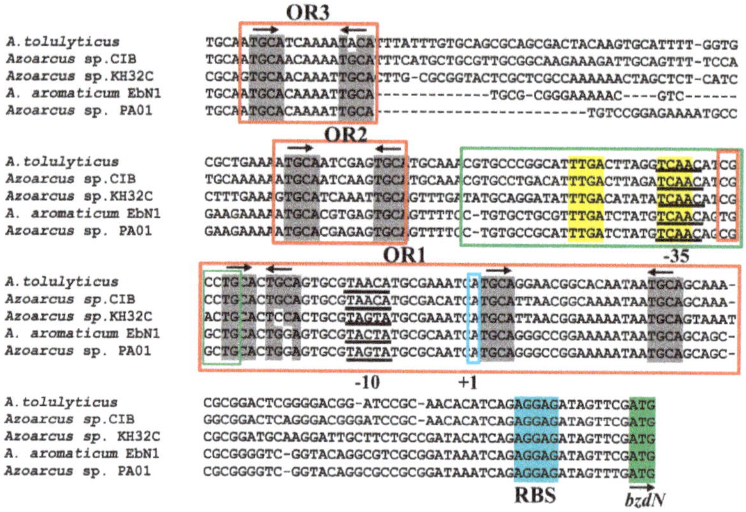

Figure 2. Sequence comparison analysis between the *Azoarcus* sp. CIB P_N promoter [20] and the P_N promoters from other *Azoarcus*/*Aromatoleum* strains. The sequence of the P_N promoter from strain CIB spanning from position −174 to +79, with respect to the transcription start site (+1, indicated) is shown and compared with that of P_N promoters from other *Azoarcus* strains. The σ^{70}-RNA polymerase (RNAP) recognition sequences −10 and −35 are underlined, the ribosome-binding sequence (RBS) is marked in cyan, and the ATG initiation codon of the *bzdN* gene is shown in green. The operator regions I, II, and III (OR1, OR2, and OR3), involved in the interaction with the BzdR protein, are boxed in red. The arrows indicate the palindromic TGCA sequences (in gray) present in each of the operator regions. The green box shows the AcpR-binding region (palindromic sequences are indicated in yellow).

The TGCA palindromic structures were also described in other operator regions such as those recognized by the CopG repressor from *Streptococcus* strains, a 45 amino-acid homodimer encoded in the pMV158 plasmid family, at the *Pcr* promoter that drives the expression of the *copG* and *repB* genes involved in plasmid replication [46,47]. Another example of a regulatory protein that recognizes operator boxes that include the TGCA sequence is the XylS regulator, which controls the expression of the operon for *m*-xylene and toluene catabolism in *Pseudomonas putida*. XylS acts as an activator binding to a region of the *Pm* promoter that includes two imperfect palindromes of the TGCA(N_6)GGNTA type [48]. The regulatory protein BenR, involved in the regulation of the aerobic catabolism of benzoate and other analogues in *P. putida*, also seems to recognize this type of imperfect palindrome [49].

The conservation of the three BzdR operator regions in all P_N promoters identified so far suggests that they are required for the accurate regulation of this promoter. To further study the role of the three BzdR operator regions of P_N, we performed in vivo and in vitro approaches with a set of truncated versions of this promoter.

3.2. In Vivo Studies on the Activity of Truncated P_N Promoters

As described previously, the activity of the P_N promoter requires the RNAP and the interaction with the AcpR, a transcriptional activator from the well-known FNR/CRP superfamily [50] that binds at a consensus sequence centered at position −41.5 from the transcription start site and overlapping the RNAP −35 box [24] (Figure 2). We demonstrated previously that the FNR protein from *E. coli* is able to bind and activate the P_N promoter [24]. On the other hand, despite the P_N promoter containing three BzdR operators, the OR1 region itself overlaps the RNAP −10 box and +1 sites (Figure 2), thus having the requested sequence for the interaction with RNAP. Therefore, it could be expected that a truncated promoter just containing OR1 and the AcpR operator could behave as a functional

BzdR-controlled minimal promoter. To check this assumption, we engineered a set of truncated P_N promoters (Figure 3A) and compared their behavior with that of the complete P_N promoter by in vivo assays. To this end, the truncated P_NI promoter (spanning from position −61 to +79; carries OR1) and truncated P_NII promoter (spanning from position −112 to +79; carries OR1 and OR2) (Figure 3A) were cloned in plasmid pSJ3, rendering plasmids pSJ3-P_NI (contains the P_NI::*lacZ* translational fusion) and pSJ3-P_NII (contains the P_NII::*lacZ* translational fusion) (Table 1). These plasmid constructions were independently transformed into *E. coli* MC4100 (pCK01) and *E. coli* MC4100 (pCK01-BzdR), respectively (Table 1). The β-galactosidase activity assays revealed that promoters P_NI and P_NII showed activity levels similar to those obtained with the complete P_N promoter cloned in plasmid pSJ3P_N (Table 1) when present in the *E. coli* MC4100 (pCK01) strain (Figure 3B). Moreover, the BzdR protein was able to inhibit the activity of P_NI and P_NII as efficiently as in the case of the wild-type P_N promoter in *E. coli* MC4100 (pCK01-BzdR) cells (Figure 3B). Therefore, these results show that P_NI behaves as a functional minimal promoter that becomes fully active in the absence of BzdR and fully repressed in the presence of the BzdR protein.

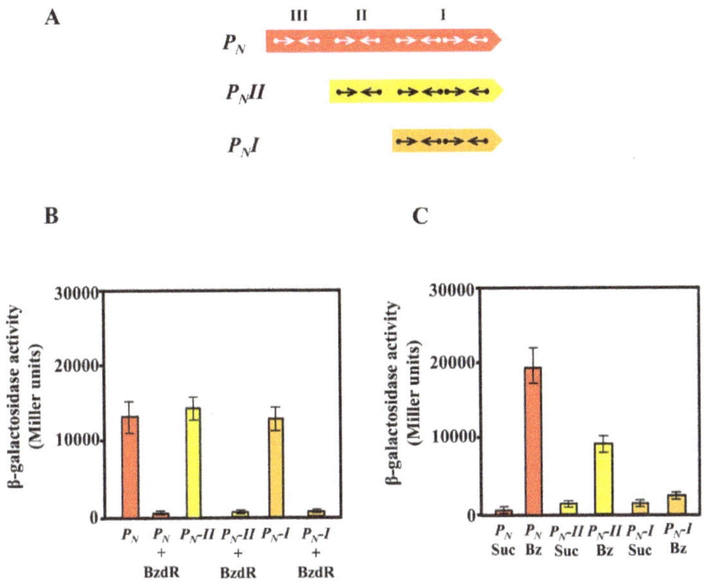

Figure 3. Activity of the P_N, P_NI, and P_NII promoters in *Escherichia coli* and *Azoarcus* sp. CIB cells. (**A**) Schematic representation of the P_N, P_NII, and P_NI promoters. Arrows represent the inverted repeat TGCA sequences for BzdR binding at operator regions I, II, and III. (**B**) The activity of the promoters was analyzed in *E. coli* MC4100 cells harboring the control plasmid pCK01 or plasmid pCK01-BzdR that expresses the *bzdR* gene (BzdR), and the plasmids pSJ3-P_N, pSJ3-P_NI, or pSJ3-P_NII that harbor the P_N::*lacZ* (P_N), P_NI::*lacZ* (P_N-*I*), or P_NII::*lacZ* (P_N-*II*) translational fusions, respectively. Growth was performed in lysogeny broth (LB) in anaerobic conditions for 16 h as described in Section 2.6. The β-galactosidase activity was measured as detailed in Section 2.6 [34]. Graphed values are the averages from three independent experiments ± SD (error bars). (**C**) Activity of the P_N, P_NII, and P_NI promoters in *Azoarcus* sp. CIB cells. *Azoarcus* sp. CIB cells harboring plasmids pBBR5-P_N, pBBR5-P_NI, or pBBR5-P_NII, which express the P_N::*lacZ* (P_N), P_NI::*lacZ* (P_N-*I*), or P_NII::*lacZ* (P_N-*II*) translational fusions, respectively, were anaerobically grown for 72 h in MC medium supplemented with 0.2% succinate (Suc) or 3 mM benzoate (Bz), and the β-galactosidase activity was measured as detailed in Section 2.6. Graphed values are the averages from three independent experiments ± SD (error bars).

To check the de-repression (induction) of the wild-type and truncated P_N promoters by benzoyl-CoA, we expressed the P_N::*lacZ*, P_NI::*lacZ*, and P_NII::*lacZ* fusions in *Azoarcus* sp. CIB. *Azoarcus* sp. CIB is able to produce benzoyl-CoA when benzoate is present in the culture medium. Benzoyl-CoA acts as inducer molecule allowing the de-repression of the system by inducing conformational changes in BzdR that lead to the release of the repressor from the promoter [21]. To perform the analysis in *Azoarcus*, the *lacZ* fusions were independently subcloned into the broad-host-range pBBR1MCS-5 vector generating plasmids pBBR5-P_N, pBBR5-P_NI, and pBBR5-P_NII, respectively (Table 1). The plasmids were introduced by conjugation into *Azoarcus* sp. CIB, and the level of activity of each promoter was determined by β-galactosidase assays. As expected, the activity of the P_N-derivative promoters expressed in *Azoarcus* cells grown in succinate (non-induction conditions) led to very low activity of all translational fusions (Figure 3C), revealing an efficient repression of the three P_N promoters by the chromosomally encoded BzdR protein. When *Azoarcus* sp. CIB cells were grown in 3 mM benzoate (induction conditions), the complete P_N promoter showed the expected activation (Figure 3C), thereby revealing the benzoyl-CoA-induced release of the BzdR repressor from the target promoter [20]. However, in contrast to the clear de-repression of the wild-type P_N promoter in *Azoarcus* sp. CIB cells grown in benzoate, the P_NII promoter showed a limited de-repression (Figure 3C), and almost no de-repression was observed with the truncated P_NI promoter (Figure 3C).

In summary, the in vivo assays revealed that P_NI is a fully active promoter in the absence of BzdR but a constitutively repressed promoter in the presence of BzdR. The OR2 and OR3 regions appear to be essential for the full de-repression (induction) of the P_N promoter when benzoyl-CoA is produced in *Azoarcus* sp. CIB cells grown in benzoate. To further confirm the role of OR2 and OR3 in the de-repression of P_N by benzoyl-CoA, several in vitro assays were performed.

3.3. Operator Regions II and III Are Needed for the Benzoyl-CoA-Dependent De-Repression of the P_N *Promoter*

To further study the truncated P_N promoters, we compared their activity levels to that of the wild-type P_N promoter by performing in vitro transcription assays using the supercoiled plasmids pJCD-P_N, pJCD-P_NI, and pJCD-P_NII (Table 1) as DNA templates, and the *E. coli* FNR* protein (a constitutively active AcpR ortholog [24]) as an activator. As shown in Figure 4, all three promoters generated the expected 184-nucleotide transcript in the absence of BzdR, but this transcript was lacking in the presence of the BzdR repressor. However, whereas a clear de-repression at the P_N promoter was observed (about five times) when benzoyl-CoA was added to the BzdR-containing reaction, a slight de-repression (about two times) was observed with P_NII, and benzoyl-CoA did not produce any de-repression effect at P_NI (Figure 4). Thus, these results suggest that benzoyl-CoA was not able to alleviate the BzdR-mediated repression at P_N when both OR2 and OR3 were missing. To check that the lack of de-repression at promoter P_NI was a consequence of the promoter occupancy by BzdR even in the presence of the benzoyl-CoA inducer, gel shift assays were performed using the P_NI probe in the presence of BzdR and increasing concentrations of benzoyl-CoA (Figure 5B). As expected, BzdR produced retardation of the P_NI promoter probe, but increasing concentrations of benzoyl-CoA up to 2 mM, which released the BzdR repressor from the wild-type P_N promoter (Figure 5A), were unable to induce the dissociation of the BzdR/P_NI complex (Figure 5B). Since benzoyl-CoA promotes the release of BzdR from the P_N promoter (Figure 5A) [20], this result suggests that the constitutive repression of P_NI by BzdR is due to the irreversible binding of the repressor to the target promoter.

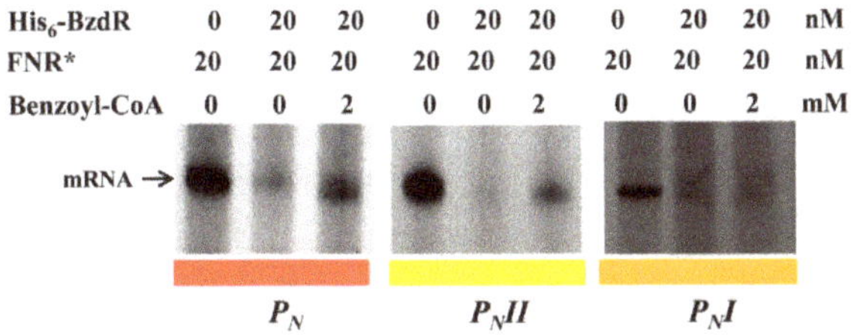

Figure 4. In vitro activity of the P_N, P_NII, and P_NI promoters. Multiple-round transcription reactions were carried out as detailed in Section 2.7 by using pJCD-P_N, pJCD-P_NI, and pJCD-P_NII plasmids harboring P_N, P_NI, and P_NII promoter templates, respectively, which produced the corresponding messenger RNA (mRNA). All the in vitro transcription reactions were performed with 100 nM *E. coli* σ^{70}-RNAP holoenzyme. Purified FNR* was used at 20 nM. When required, His$_6$-BzdR protein was used at 20 nM, and benzoyl-CoA was added at 2 mM.

We demonstrated previously that the binding of BzdR to its cognate promoter is cooperative [21]. Sedimentation velocity analyses showed that 8–10 molecules of BzdR bind to the P_N promoter, and it was postulated that each of the four TGCA palindromic regions bind a BzdR dimer, i.e., four BzdR molecules bound at OR1, two at OR2, and another two at OR3 [21]. Thus, it could be expected that the lack of de-repression at the P_NI promoter in the presence of benzoyl-CoA inducer could be due to an excess of BzdR molecules bound to OR1 when both OR2 and OR3 are missing in the target promoter.

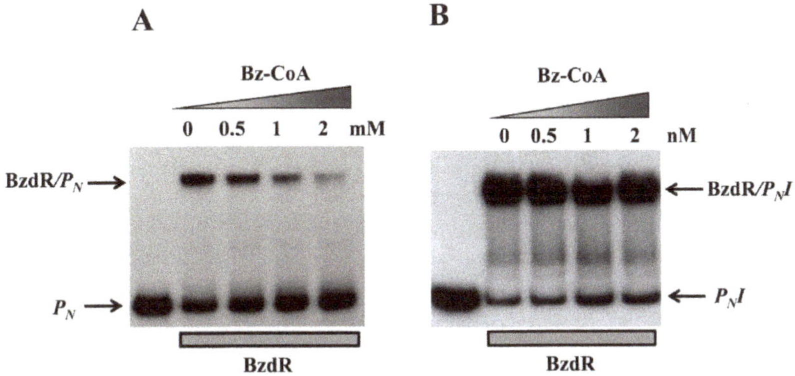

Figure 5. Gel retardation analysis of BzdR binding to the P_N (**A**) or the P_NI promoter (**B**). Gel retardation analysis was performed as described in Section 2.8 by using 10 nM purified His$_6$-BzdR in the presence of increasing concentrations (from 0 to 2 mM) of benzoyl-CoA (Bz-CoA). The free P_N probe and the BzdR/P_N complex (**A**) or P_NI probe and the BzdR/P_NI complex (**B**) are indicated by arrows.

To check this hypothesis, we performed analytical ultracentrifugation experiments with the N-terminal domain of the BzdR protein (NBzdRL protein; residues 1–90) that is able to bind to P_N [30]. Previous sedimentation velocity experiments revealed that NBzdRL was a dimer with S-values of 2.3 S, demonstrating that BzdR dimerization is an intrinsic property of the N-terminal domain [30]. The results presented here revealed that eight molecules of NBzdRL were bound to the wild-type P_N promoter (Figure 6A), confirming the previous results obtained with the complete BzdR protein [30]. More relevant, the ultracentrifugation analyses showed that four molecules of NBzdRL were bound to the P_NI fragment (Figure 6B). Hence, these results provide the first experimental evidence that two BzdR dimers bind to OR1 and support the previous assumption of the distribution of the eight BzdR molecules along the P_N promoter [21].

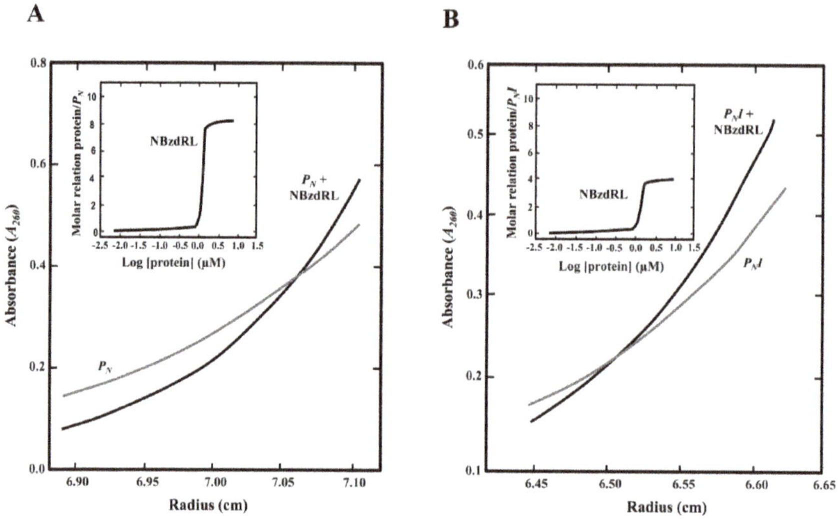

Figure 6. Sedimentation equilibrium analysis to study the interaction of purified NBzdRL protein with P_N and P_NI DNA fragments. (**A**) Distribution of the gradient in sedimentation equilibrium of the P_N fragment (P_N, gray line) or P_N fragment with His$_6$-NBzdRL protein (black line). The inset graph shows the protein/DNA molar ratio as the concentration of His$_6$-NBzdRL increases. (**B**) Distribution of the gradient in sedimentation equilibrium of the P_NI fragment (gray line) and P_NI + His$_6$-NBzdRL complex (black line). The inset graph shows the protein/DNA molar ratio as the concentration of NBzdRL increases.

3.4. Visualization of the BzdR/P$_N$ Complex

The results presented above indicate that three BzdR operator regions are conserved in all P_N promoters identified so far, and they suggest that binding of four BzdR dimers to the wild-type P_N promoter should lead to the formation of a big protein–DNA complex. Atomic force microscopy (AFM) permits direct visualization of the tridimensional structure of protein/DNA complexes arranged on an atomically flat surface and supplies data about the topology and stoichiometry of the complexes [51–54]. A 1225-bp DNA sequence (414 nm in length) spanning from position −505 to +720 with respect to the transcription start site was used as template (P_NL fragment) to visualize the structure of P_N interacting with BzdR when using a 50:1 [BzdR]:[DNA] ratio (Figure 7A). Images obtained when the three operator boxes were occupied by BzdR revealed a compact structure (Figure 7B). The lengths of the free (unoccupied) DNA that emerged at both sides of the compact structure were 108.9 ± 20 nm and 199.9 ± 44 nm (n = 16), respectively (Figure 7C, shown for one of the 16 samples). These sizes are slightly lower than those theoretically expected, i.e., 121 nm and 233 nm, respectively (Figure 7A). The total extension of the BzdR–DNA complex was reduced by 14 ± 4% with respect to the length of the naked P_NL DNA fragment (Figure 7C).

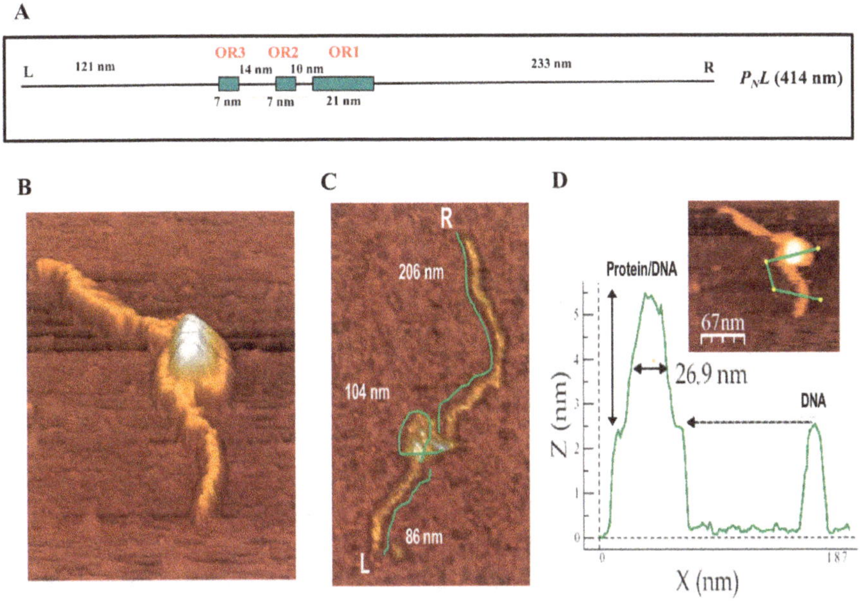

Figure 7. Analysis of the binding of BzdR to the P_N promoter by atomic force microscopy (AFM). (**A**) Scheme of the P_NL DNA fragment used as a template. The DNA fragment extends from position −505 to +720 with respect to the transcription start site of the P_N promoter. The operator regions recognized by BzdR are represented by green boxes with their corresponding numbers (OR1, OR2, and OR3). The distances separating each of the indicated regions, the extension of each operator region in nm, and the left (L) and right (R) ends of the P_NL fragment are also indicated. (**B**) Image of a superstructure of the BzdR/P_NL complex. (**C**) Analysis of distances in a BzdR/P_NL superstructure (one image, n = 16). The distances of both arms and the loop of the superstructure are detailed. L and R indicate the left and right ends of the P_NL DNA fragment, respectively. (**D**) Image of a BzdR/P_NL superstructure and its corresponding height profile. The green line of the image corresponds to the profile outlined in the lower graph. In the graph, the length (in nm) of the line is represented on the x-axis and the height on the y-axis. The first peak corresponds to the BzdR/P_NL superstructure and the second peak corresponds to the naked DNA.

These results are compatible with the formation of one loop of DNA around the BzdR proteins in the BzdR–DNA complex. The length of the DNA, including the DNA that is forming the loop, was estimated to be 398.6 ± 28.9 nm, which is compatible with the total length of the P_NL DNA fragment (414 nm) and with the previous length calculated for the P_NL naked template (400 ± 13 nm; [55]). Furthermore, the width of the BzdR–DNA "superstructure" had a value of 26.9 ± 6 nm (n = 10) (Figure 7D). Since a BzdR dimer resolved by electron microscopy is like a cylinder of 9 nm × 7 nm [21], the width of the superstructure observed by AFM is compatible with 4–5 dimers (8–10 molecules) of BzdR protein interacting with OR regions I, II, and III. These results are in good agreement with those obtained by analytical ultracentrifugation experiments (Figure 6A) [21]. In summary, the AFM results strongly suggest that the interaction of BzdR with the three operator regions of P_N leads to the formation of a superstructure compatible with a DNA loop embracing 4–5 BzdR dimers by turning on itself (Figure 7). This type of superstructure was also described for other DNA–protein complexes such as that formed by the DNA/MutS protein [56], although other transcriptional regulators, such as the TodT protein that controls the aerobic toluene degradation pathway in *P. putida*, interact with the target DNA following a fork-type model [57]. The existence of a DNA loop wrapping the BzdR dimers as observed by AFM is also consistent with the DNase I footprinting studies that showed hyper-reactive nucleotides between OR1 and OR2 and between OR2 and OR3 [18], which might reflect BzdR-induced DNA bends leading to the loop formation.

4. Conclusions

In this work, we showed that the genes responsible for the anaerobic degradation of benzoate in all aromatic-degrader *Azoarcus/Aromatoleum* strains whose genome sequences are currently available are organized in a well-conserved *bzd* cluster. Thus, a *bzdR* regulatory gene and a P_N promoter could be identified upstream of the catabolic *bzd* operon in all *bzd* clusters analyzed. Interestingly, all P_N promoters show a conserved architecture that includes three BzdR operator regions (OR1–3). Whereas OR1 is sufficient for the BzdR-mediated repression of the P_N promoter, the presence of OR2 and OR3 is required for the de-repression promoted by the benzoyl-CoA inducer molecule. Our results suggest that the dimeric protein BzdR binds to the P_N promoter in the form of four dimers; two of the dimers are initially bound to OR1, and the complex formed may favor the interaction of two additional dimers at OR2 and OR3. The BzdR/P_N complex formed induces a DNA loop that wraps around the BzdR dimers and generates a superstructure that was observed by AFM. This three-dimensional (3D) configuration may keep the −10/−35 sequences of P_N inaccessible to the RNAP for achieving an efficient repression and, at the same time, could facilitate the P_N de-repression when benzoyl-CoA is generated. However, the elucidation of the molecular mechanism underlying the participation of OR2 and OR3 in the de-repression of P_N requires further research. In this sense, it should be taken into account that regulatory proteins other than BzdR, e.g., AcpR and AccR, are also involved in the control of the P_N promoter [24,25] and, hence, more complex structure–function relationships at the P_N promoter should be expected. In any case, the work presented here strongly suggests the existence of a common BzdR-dependent mechanism to control the expression of the *bzd* genes in *Azoarcus/Aromatoleum* strains.

Author Contributions: G.D.-R. carried out the genetic manipulations, physiology experiments, and in vitro enzyme assays. P.G.-d.-A. and M.V. designed and performed the AFM experiments and interpreted all the data obtained from this technique. G.D.-R., E.D., and M.C. conceived the whole study, designed the experiments, and contributed to the discussion of the research and interpretation of the data. G.D.-R., E.D., and M.C. wrote the article.

Funding: This work was supported by grant BIO2016-79736-R from the Ministry of Economy and Competitiveness of Spain, by grant CSIC 2016 20E093 from the CSIC, and by European Union H2020 grant 760994.

Acknowledgments: The author thanks A. Valencia for the technical assistance and S.L. Secugen for DNA sequencing. We are indebted to C. Alfonso and G. Rivas for their help in the analytical ultracentrifugation experiments.

Conflicts of Interest: The authors declare that there are no competing interests.

References

1. Pieper, D.H.; Reineke, W. Engineering bacteria for bioremediation. *Curr. Opin. Biotechnol.* **2000**, *11*, 262–270. [CrossRef]
2. Fuchs, G.; Boll, M.; Heider, J. Microbial degradation of aromatic compounds—From one strategy to four. *Nat. Rev. Microbiol.* **2011**, *9*, 803–816. [CrossRef] [PubMed]
3. Brzeszcz, J.; Kaszycki, P. Aerobic bacteria degrading both n-alkanes and aromatic hydrocarbons: And undervalued strategy for metabolic diversity and flexibility. *Biodegradation* **2018**, *29*, 359–407. [CrossRef] [PubMed]
4. Díaz, E.; Jiménez, J.I.; Nogales, J. Aerobic degradation of aromatic compounds. *Curr. Opin. Biotechnol.* **2013**, *24*, 431–442. [CrossRef] [PubMed]
5. Duarte, M.; Jauregui, R.; Vilchez-Vargas, R.; Junca, H.; Pieper, D.H. AromaDeg, a novel database for phylogenomics of aerobic bacterial degradation of aromatics. *Database* **2014**, *2014*, bau118. [CrossRef] [PubMed]
6. Rabus, R.; Boll, M.; Heider, J.; Meckenstock, R.U.; Buckel, W.; Einsle, O.; Ermler, U.; Golding, B.T.; Gunsalus, R.P.; Kroneck, P.M.H.; et al. Anaerobic microbial degradation of hydrocarbons: From enzymatic reactions to the environment. *J. Mol. Microbiol. Biotechnol.* **2016**, *26*, 5–28. [CrossRef]
7. Lovley, D.R. Biorremediation. Anaerobes to the rescue. *Science* **2001**, *293*, 1444–1446. [CrossRef]

8. Carmona, M.; Zamarro, M.T.; Blázquez, B.; Durante-Rodríguez, G.; Juárez, J.F.; Valderrama, J.A.; López-Barragán, M.J.; García, J.L.; Díaz, E. Anaerobic catabolism of aromatic compounds: A genetic and genomic view. *Microbiol. Mol. Biol. Rev.* **2009**, *73*, 71–133. [CrossRef]
9. Boll, M.; Löffler, C.; Morris, B.E.; Kung, J.W. Anaerobic degradation of homocyclic aromatic compounds via arylcarboxyl-coenzyme A esters: Organisms, strategies and key enzymes. *Environ. Microbiol.* **2014**, *16*, 612–627. [CrossRef]
10. Durante-Rodríguez, G.; Gómez-Álvarez, H.; Blázquez, B.; Fernández-Llamosas, H.; Martín-Moldes, Z.; Sanz, D.; Nogales, J.; Carmona, M.; Díaz, E. Anaerobic pathways for the catabolism of aromatic compounds. In *Lignin Valorization: Emerging Approaches*; Beckham, T., Ed.; The Royal Society of Chemistry Publishing: Cambridge, UK, 2018; pp. 333–390.
11. Egland, P.G.; Pelletier, D.A.; Dispensa, M.; Gibson, J.; Harwood, C.S. A cluster of bacterial genes for anaerobic benzene ring biodegradation. *Proc. Natl. Acad. Sci. USA* **1997**, *94*, 6484–6489. [CrossRef]
12. Wischgoll, S.; Heintz, D.; Peters, F.; Erxleben, A.; Sarnighausen, E.; Reski, R.; van Dorsselaer, A.; Boll, M. Gene clusters involved in anaerobic benzoate degradation in *Geobacter metallireducens*. *Mol. Microbiol.* **2005**, *58*, 1238–1252. [CrossRef] [PubMed]
13. Butler, J.E.; He, Q.; Nevin, K.P.; He, Z.; Zhou, J.; Lovley, D.R. Genomic and microarray analysis of aromatics degradation in *Geobacter metallireducens* and comparison to a *Geobacter* isolate from a contaminated field site. *BMC Genom.* **2007**, *8*, 180. [CrossRef] [PubMed]
14. Mclnerney, M.J.; Rohlin, L.; Mouttaki, H.; Kim, U.; Krupp, R.S.; Rios-Hernandez, L.; Sieber, J.; Struchtemeyer, C.G.; Bhattacharyya, A.; Campbell, J.W.; et al. The genome of *Syntrophus aciditrophicus*: Life at the thermodynamic limit of microbial growth. *Proc. Natl. Acad. Sci. USA* **2007**, *104*, 7600–7605. [CrossRef] [PubMed]
15. Wöhlbrand, L.; Jacob, J.H.; Kube, M.; Mussmann, M.; Jarling, R.; Beck, A.; Amann, R.; Wilkes, H.; Reinhardt, R.; Rabus, R. Complete genome, catabolic sub-proteomes and key-metabolites of *Desulfobacula toluolica* Tol2, a marine, aromatic compound-degrading, sulfate-reducing bacterium. *Environ. Microbiol.* **2013**, *15*, 1334–1355. [CrossRef] [PubMed]
16. Hirakawa, H.; Hirakawa, Y.; Greenberg, E.P.; Harwood, C.S. BadR and BadM proteins transcriptionally regulate two operons needed for anaerobic benzoate degradation by *Rhodopseudomonas palustris*. *Appl. Environ. Microbiol.* **2015**, *81*, 4253–4262. [CrossRef] [PubMed]
17. Van Drisse, C.M.; Escalante-Semerena, J.C. Small-molecule acetylation controls the degradation of benzoate and photosynthesis in *Rhodopseudomonas palustris*. *MBio* **2018**, *16*, e01895-18.
18. Ueki, T. Identification of a transcriptional repressor involved in benzoate metabolism in *Geobacter bemidjiensis*. *Appl. Environ. Microbiol.* **2011**, *77*, 7058–7062. [CrossRef] [PubMed]
19. Juárez, J.F.; Zamarro, M.T.; Barragán, M.J.; Blázquez, B.; Boll, M.; Kuntze, K.; García, J.L.; Díaz, E.; Carmona, M. Identification of the *Geobacter metallireducens bamVW* two-component system, involved in transcriptional regulation of aromatic degradation. *Appl. Environ. Microbiol.* **2010**, *76*, 383–385. [CrossRef]
20. Barragán, M.J.; Blázquez, B.; Zamarro, M.T.; Mancheño, J.M.; García, J.L.; Díaz, E.; Carmona, M. BzdR, a repressor that controls the anaerobic catabolism of benzoate in *Azoarcus* sp. CIB, is the first member of a new subfamily of transcriptional regulators. *J. Biol. Chem.* **2005**, *280*, 10683–10694. [CrossRef]
21. Durante-Rodríguez, G.; Valderrama, J.A.; Mancheño, J.M.; Rivas, G.; Alfonso, C.; Arias-Palomo, E.; Llorca, O.; García, J.L.; Díaz, E.; Carmona, M. Biochemical characterization of the transcriptional regulator BzdR from *Azoarcus* sp. CIB. *J. Biol. Chem.* **2010**, *285*, 35694–35705. [CrossRef]
22. Rabus, R.; Wöhlbrand, L.; Thies, D.; Meyer, M.; Reinhold-Hurek, B.; Kämpfer, P. *Aromatoleum* gen. nov., a novel genus accommodating the phylogenetic lineage including *Azoarcus evansii* and related species, and proposal of *Aromatoleum* aromaticum sp. nov., *Aromatoleum petrolei* sp. nov., *Aromatoleum bremense* sp. nov., *Aromatoleum toluolicum* sp. nov. and *Aromatoleum diolicum* sp. nov. *Int. J. Syst. Evol. Microbiol.* **2019**, *69*, 982–997. [PubMed]
23. López-Barragán, M.J.; Carmona, M.; Zamarro, M.T.; Thiele, B.; Boll, M.; Fuchs, G.; García, J.L.; Díaz, E. The *bzd* gene cluster, coding for anaerobic benzoate catabolism, in *Azoarcus* sp. strain CIB. *J. Bacteriol.* **2004**, *186*, 5762–5774. [CrossRef] [PubMed]
24. Durante-Rodríguez, G.; Zamarro, M.T.; García, J.L.; Díaz, E.; Carmona, M. Oxygen-dependent regulation of the central pathway for the anaerobic catabolism of aromatic compounds in *Azoarcus* sp. strain CIB. *J. Bacteriol.* **2006**, *188*, 2343–2354. [CrossRef] [PubMed]

25. Valderrama, J.A.; Shingler, V.; Carmona, M.; Díaz, E. AccR is a master regulator involved in carbon catabolite repression of the anaerobic catabolism of aromatic compounds in *Azoarcus* sp. CIB. *J. Biol. Chem.* **2014**, *289*, 1892–1904. [CrossRef] [PubMed]
26. Sambrook, J.; Rusell, D. *Molecular Cloning: A Laboratory Manual*; Cold Spring Harbor Laboratory Press: New York, NY, USA, 2001.
27. De Lorenzo, V.; Timmis, K.N. Analysis and construction of stable phenotypes in gram-negative bacteria with Tn*5*- and Tn*10*-derived minitransposons. *Methods Enzymol.* **1994**, *235*, 386–405. [PubMed]
28. Casadaban, M.J. Transposition and fusion of the lac genes to selected promoters in *Escherichia coli* using bacteriophage lambda and Mu. *J. Mol. Biol.* **1976**, *104*, 541–555. [CrossRef]
29. Wing, H.J.; Green, J.; Guest, J.R.; Busby, S.J. Role of activating region 1 of *Escherichia coli* FNR protein in transcription activation at class II promoter. *J. Biol. Chem.* **2000**, *275*, 29061–29065. [CrossRef]
30. Durante-Rodríguez, G.; Mancheño, J.M.; Rivas, G.; Alfonso, C.; García, J.L.; Díaz, E.; Carmona, M. Identification of a missing link in the evolution of an enzyme into a transcriptional regulator. *PLoS ONE* **2013**, *8*, e57518. [CrossRef]
31. Kovach, M.; Elzer, P.H.; Hill, D.S.; Robertson, G.T.; Farris, M.A.; Roop, R.M.I.; Peterson, K.M. Four new derivatives of the broad-host-range cloning vector pBBR1MCS, carrying different antibiotic-resistance cassettes. *Gene* **1995**, *166*, 175–176. [CrossRef]
32. Marshall, C.; Labrousse, V.; Kreimer, M.; Weichart, D.; Kolb, A.; Hengge-Aronis, R. Molecular analysis of the regulation of *csiD*, a carbon starvation-inducible gene in *Escherichia coli* that is exclusively dependent on σ^s and requires activation by cAMP-CRP. *J. Mol. Biol.* **1998**, *276*, 339–353. [CrossRef]
33. Ferrández, A.; Miñambres, B.; García, B.; Olivera, E.R.; Luengo, J.M.; García, J.L.; Díaz, E. Catabolism of phenylacetic acid in *Escherichia coli*. Characterization of a new aerobic hybrid pathway. *J. Biol. Chem.* **1998**, *273*, 25974–26986. [CrossRef] [PubMed]
34. Miller, J.H. *Experiments in Molecular Genetics*; Cold Spring Harbor Press: New York, NY, USA, 1972; pp. 352–355.
35. Hall, T.A. BioEdit: A user-friendly biological sequence alignment editor and analysis program for Windows 95/98/NT. *Nucleic Acids Symp. Ser.* **1999**, *41*, 95–98.
36. Altschul, S.F.; Gish, W.; Miller, W.; Myers, E.W.; Lipman, D.J. Basic local alignment search tool. *J. Mol. Biol.* **1990**, *215*, 403–410. [CrossRef]
37. National Center for Biotechnology Information (NCBI) Server. Available online: http://blast.ncbi.nlm.nih.gov/Blast.cgi (accessed on 26 June 2019).
38. Wilbur, W.J.; Lipman, D.J. Rapid similarity searches of nucleic acid and protein data banks. *Proc. Natl. Acad. Sci. USA* **1983**, *80*, 726–730. [CrossRef] [PubMed]
39. Thompson, J.D.; Higgins, D.G.; Gibson, T.J. CLUSTAL W: Improving the sensitivity of progressive multiple sequence alignment through sequence weighting, position-specific gap penalties and weight matrix choice. *Nucleic Acids Res.* **1994**, *22*, 4673–4680. [CrossRef] [PubMed]
40. Claverie, F.; Magasanik, B. Positive and negative effects of DNA bending on activation of transcription from a distant site. *J. Mol. Biol.* **1992**, *227*, 996–1008. [CrossRef]
41. Carmona, M.; Magasanik, B. Activation of transcription at σ^{54}-dependent promoters on linear templates requires intrinsic or induced bending of the DNA. *J. Mol. Biol.* **1996**, *261*, 348–356. [CrossRef]
42. Collins, T.J. ImageJ for microscopy. *Biotechniques* **2007**, *43*, 25–30. [CrossRef]
43. Moreno-Herrero, F.; De Pablo, P.J.; Fernández-Sánchez, R.; Colchero, J.; Gómez-Herrero, J.; Baró, A.M. Scanning force microscopy jumping and tapping modes in liquids. *Appl. Phys. Lett.* **2002**, *81*, 2620–2622. [CrossRef]
44. Horcas, I.; Fernández, R.; Gómez-Rodríguez, J.M.; Colchero, J.; Gómez-Herrero, J.; Baro, A.M. WSXM: A software for scanning probe microscopy and a tool for nanotechnology. *Rev. Sci. Instrum.* **2007**, *78*, 013705. [CrossRef]
45. Rabus, R.; Trautwein, K.; Wöhlbrand, L. Towards habitat-oriented systems biology of "*Aromatoleum aromaticum*" EbN1: Chemical sensing, catabolic network modulation and growth control in anaerobic aromatic compound degradation. *Appl. Microbiol. Biotechnol.* **2014**, *98*, 3371–3388. [CrossRef] [PubMed]
46. Del Solar, G.H.; Pérez-Martín, J.; Espinosa, M. Plasmid pLS1-enconded RepA protein regulates transcription from *repAB* promoter by binding to a DNA sequence containing a 13-base pair symmetric element. *J. Biol. Chem.* **1990**, *265*, 12569–12575. [PubMed]

47. Gomis-Ruth, F.X.; Sola, M.; Acebo, P.; Parraga, A.; Guasch, A.; Eritja, R.; González, A.; Espinosa, M.; del Solar, G.; Coll, M. The structure of plasmid-encoded transcriptional repressor CopG unligated and bound to its operator. *EMBO J.* **1998**, *17*, 7404–7415. [CrossRef] [PubMed]
48. González-Pérez, M.M.; Ramos, J.L.; Gallegos, M.T.; Marqués, S. Critical nucleotides in the upstream region of the XylS-dependent TOL meta-cleavage pathway operon promoter as deduced from analysis of mutants. *J. Biol. Chem.* **1999**, *274*, 2286–2290. [CrossRef] [PubMed]
49. Cowles, C.E.; Nichols, N.N.; Harwood, C.S. BenR, a XylS homologue regulates three different pathways of aromatic acid degradation in *Pseudomonas putida*. *J. Bacteriol.* **2000**, *182*, 6339–6346. [CrossRef] [PubMed]
50. Körner, H.; Sofia, H.J.; Zumft, W.G. Phylogeny of the bacterial superfamily of Crp-Fnr transcriptional regulators: Exploiting the metabolic spectrum by controlling alterative gene programas. *FEMS Microbiol. Rev.* **2003**, *27*, 559–592. [CrossRef]
51. Hansma, H. Surface biology of DNA by atomic force microscopy. *Annu. Rev. Phys. Chem.* **2001**, *52*, 71–92. [CrossRef] [PubMed]
52. Rivetti, C.; Guthold, M. Single DNA molecule analysis of transcription complexes. *Methods Enzymol.* **2003**, *371*, 34–50.
53. Yang, Y.; Sass, L.; Du, C.; Hsieh, P.; Erie, D. Determination of protein-DNA binding constants and specificities from statistical analyses of single molecule: MutS-DNA interactions. *Nucleic Acids Res.* **2005**, *33*, 4322–4334. [CrossRef]
54. Zhang, J.; Zeuner, Y.; Kleefeld, A.; Unden, G.; Janshoff, A. Multiple site-specific binding of Fis protein to *Escherichia coli nuoA-N* promoter DNA and its impact on DNA topology visualized by means of scanning force microscopy. *ChemBioChem* **2004**, *5*, 1286–1289. [CrossRef]
55. Gutiérrez-del-Arroyo, P. Caracterización Mediante Microscopía de Fuerzas Atómicas de Complejos de Regulación Transcripcional. Ph.D. Thesis, Universidad Autónoma de Madrid, Madrid, Spain, 2009.
56. Jia, Y.; Bi, L.; Li, F.; Chen, Y.; Zhang, C.; Zhang, X. α-shaped DNA loops induced by MutS. *Biochem. Biophys. Res. Commun.* **2008**, *372*, 618–622. [CrossRef] [PubMed]
57. Lacal, J.; Guazzaroni, M.E.; Gutiérrez del Arroyo, P.; Busch, A.; Vélez, M.; Krell, T.; Ramos, J.L. Two level of comparativeness in the binding of TodT to the *tod* operon promoter. *J. Mol. Biol.* **2008**, *384*, 1037–1047. [CrossRef] [PubMed]

© 2019 by the authors. Licensee MDPI, Basel, Switzerland. This article is an open access article distributed under the terms and conditions of the Creative Commons Attribution (CC BY) license (http://creativecommons.org/licenses/by/4.0/).

genes

MDPI

Article

Engineering CatM, a LysR-Type Transcriptional Regulator, to Respond Synergistically to Two Effectors

Melissa P. Tumen-Velasquez [1], Nicole S. Laniohan [1,2], Cory Momany [1,2] and Ellen L. Neidle [1,*]

1 Departments of Microbiology, University of Georgia, Athens, GA 30602, USA; dnavgais@uga.edu (M.P.T.-V.); nlanio@uga.edu (N.S.L.); cmomany@uga.edu (C.M.)
2 Pharmaceutical and Biomedical Sciences, University of Georgia, Athens, GA 30602, USA
* Correspondence: eneidle@uga.edu; Tel.: +1-706-542-2852

Received: 29 April 2019; Accepted: 30 May 2019; Published: 31 May 2019

Abstract: The simultaneous response of one transcriptional regulator to different effectors remains largely unexplored. Nevertheless, such interactions can substantially impact gene expression by rapidly integrating cellular signals and by expanding the range of transcriptional responses. In this study, similarities between paralogs were exploited to engineer novel responses in CatM, a regulator that controls benzoate degradation in *Acinetobacter baylyi* ADP1. One goal was to improve understanding of how its paralog, BenM, activates transcription in response to two compounds (*cis,cis*-muconate and benzoate) at levels significantly greater than with either alone. Despite the overlapping functions of BenM and CatM, which regulate many of the same *ben* and *cat* genes, CatM normally responds only to *cis,cis*-muconate. Using domain swapping and site-directed amino acid replacements, CatM variants were generated and assessed for the ability to activate transcription. To create a variant that responds synergistically to both effectors required alteration of both the effector-binding region and the DNA-binding domain. These studies help define the interconnected roles of protein domains and extend understanding of LysR-type proteins, the largest family of transcriptional regulators in bacteria. Additionally, renewed interest in the modular functionality of transcription factors stems from their potential use as biosensors.

Keywords: LysR; transcription factor; *Acinetobacter*; LTTR; benzoate; muconate; synergism; biosensor

1. Introduction

Two LysR-type transcriptional regulators (LTTRs) control benzoate degradation by a soil bacterium, *Acinetobacter baylyi* ADP1 [1]. These paralogs, BenM and CatM, have overlapping but distinct functions (Figure 1). BenM controls the initial steps in benzoate consumption by activating transcription of the *benABCDE* operon in response to benzoate and one of its metabolites, *cis,cis*-muconate, hereafter designated muconate [2,3]. CatM, which responds solely to muconate, activates low-level expression from this promoter, P_{benA} [4]. We sought to create a benzoate-responsive CatM that mimics an unusual characteristic of BenM, namely the ability to activate transcription synergistically in response to two effectors. These studies should improve understanding of the molecular basis of this type of transcriptional activation and, in general, facilitate the engineering of LTTRs to respond to novel effectors when designed for varied biotechnology applications.

BenM and CatM have similar N-terminal DNA-binding domains (DBDs, Figure 2), and they bind to the same regions of P_{benA} [2,4]. Whereas LTTR-DNA interactions repress basal transcription, conformational changes occur in response to effectors to activate transcription (depicted in Figure S1). Although CatM alone does not activate sufficient transcription for growth on benzoate as the carbon source, mutations can increase CatM-regulated P_{benA} transcription by augmenting its response to muconate or by enabling transcription without an effector [3,4]. Two amino acid replacements in the effector-binding domain (EBD) of CatM each enable BenM-independent growth on benzoate, yet neither of these, nor any other known CatM variant, responds to benzoate.

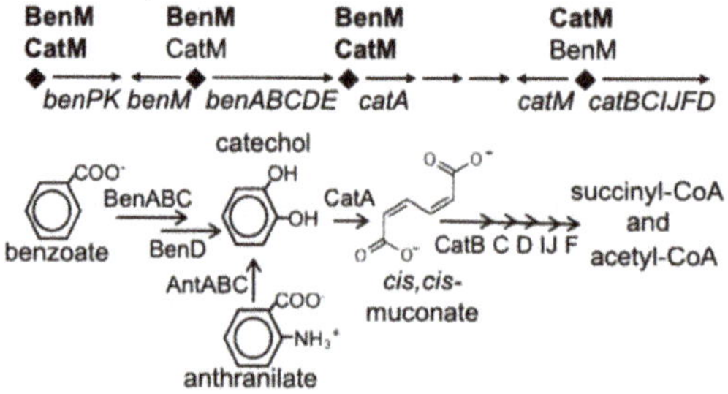

Figure 1. BenM and CatM regulate the *ben* and *cat* genes from four promoter regions (diamonds). The *benABCDE* operon is primarily regulated by BenM from P_{benA}, and the *catBCIJFD* operon is primarily regulated by CatM from P_{catB}. The encoded enzymes are used for the degradation of benzoate.

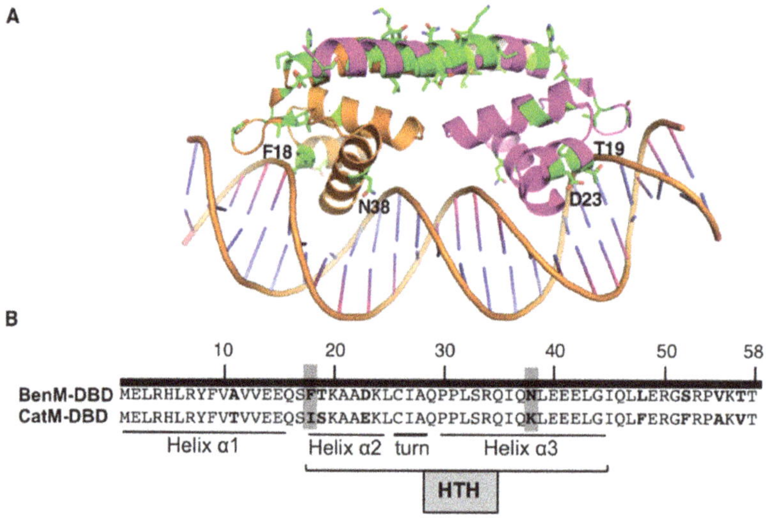

Figure 2. (**A**) Ribbon representation of BenM-DBD and the adjacent "linker" helix bound to DNA (PDB ID 4IHS). Green side chains show differences between BenM and CatM. Labeled residues are discussed in the text. (**B**) Nine amino acids differences in the DBDs (bold), including those at positions 18 and 38 (highlighted). The Helix (α2)-Turn-Helix (α3) (HTH) motif is involved in DNA recognition.

The structures of BenM-EBD and CatM-EBD are highly conserved (Figure S2) [5,6]. Although benzoate binds in a hydrophobic pocket of BenM, this compound has not been detected in structures of the corresponding region of CatM [5,6]. This pocket is distinct from an inter-domain cleft in BenM and CatM that binds muconate and serves as the typical effector-binding site in LTTRs [6]. The EBDs of LTTRs usually assume the conformation of a periplasmic-binding protein [7–9]. However, BenM is the only LTTR known to have a secondary effector-binding site that enables synergistic activation of transcription with different metabolites [2]. In an effort to understand this synergism, we engineered amino acid replacements to make CatM more similar to BenM.

Two residues in the hydrophobic binding pocket of BenM are critical for benzoate-activated transcription, R160 and Y293 [5,6]. When these amino acids are replaced with those at the comparable positions of CatM, H160 and F293, BenM fails to activate transcription in response to benzoate as a sole effector or in combination with muconate [5]. Moreover, benzoate inhibits muconate-activated gene expression in these BenM variants [3,5]. As described here, the converse changes in CatM (H160R and F293Y) did not initially generate a benzoate-responsive CatM. However, efforts to alter CatM were expanded to increase understanding of these representative members of the LTTR family, the largest group of homologous transcriptional regulators in bacteria [10]. After multiple attempts, CatM variants were isolated that respond to benzoate and that activate increased transcription in response to both effectors. As discussed below, changes were required in both the DBD and EBD regions of CatM for this new functionality.

2. Materials and Methods

2.1. Bacterial Strains and Growth Conditions

A. baylyi strains, derived from the wild-type ADP1 [11,12] are listed in Table S1. *Escherichia coli* DH5α (Thermo Fisher Scientific, Waltham, MA, USA) and XL-1 blue (Agilent Technologies, Santa Clara, CA, USA) were used as plasmid hosts. Bacteria were grown on Luria-Bertani (LB) medium at 37 °C [13]. In some cases, *A. baylyi* strains were grown in minimal medium [14,15] with succinate (10 mM), pyruvate (20 mM), benzoate (2 mM), muconate, (2.5 mM), or anthranilate (1.5 mM) as the carbon source. Antibiotics were added when needed at the following final concentrations: ampicillin, 150 μg/mL, kanamycin, 25 μg/mL, spectinomycin, 13 μg/mL, and streptomycin, 13 μg/mL. Growth was monitored by turbidity (OD_{600}).

2.2. Site-Directed Mutagenesis and Strain Construction by Allelic Replacement

Site-directed mutagenesis of plasmid DNA was conducted with mutagenic primers and methods based on the QuickChange II protocol (Agilent Technologies, Santa Clara, CA, USA [5]). The primers, and the mutations they introduce, are listed in Tables S2 and S3. In some cases, plasmids were constructed using splicing by overlap extension PCR (SOEing) [16]. Linearized plasmid-borne alleles were used to replace chromosomal genes in *A. baylyi* recipient strains by homologous recombination [17,18]. Transformants were identified by phenotypic changes in antibiotic resistance, carbon source utilization or loss of the *sacB* marker (in the presence of 10% sucrose and no NaCl in the medium) [5,18]. The genotypes of mutant strains were confirmed by PCR analysis and DNA sequencing (Genewiz laboratories, South Plainfield, NJ, USA) of the chromosomal regions where changes were introduced.

2.3. Selection for BenM-Independent Growth on Benzoate

Strains that form colonies on plates with benzoate as the sole carbon source were defined as Ben^+. Spontaneous Ben^+ mutants arising from strains lacking BenM were isolated as described [3,4]. Chromosomal *catM* DNA was recovered from Ben^+ strains using the gap-repair method [18,19]. Briefly, cells were grown on benzoate medium to mid-log phase, mixed with linearized pBAC184 (Table S2) and plated on LB medium. Transformants with circularized plasmids, resulting from homologous recombination, were selected in medium with ampicillin. Drug-resistant cells were pooled, and plasmid DNA was extracted and used to transform *E. coli*. Recovered *A. baylyi* DNA was tested for the ability to confer a Ben^+ phenotype to recipient strains (without BenM) by allelic replacement. Mutations were identified by DNA sequencing (Genewiz laboratories, South Plainfield, NJ, USA).

2.4. β-Galactosidase (LacZ) Assays

Transcriptional *lacZ* fusions were constructed as described [3,4]. For cultures grown on LB, effectors were added at final concentrations of 500 μM of benzoate or muconate, or when added together, 250 μM of each. Some cultures were grown with pyruvate (20 mM) or muconate (3 mM) as the carbon source. Effectors added to pyruvate-grown cultures were added at the following concentrations: 65 μM benzoate or muconate, or 32.5 μM of each when added together. Growth was measured by optical density (OD_{600}), and assays were done when cultures reached late-exponential phase as described [4,5]. Directions from the FlourAce β-galactosidase reporter kit (BioRad, Hercules, CA, USA) were followed. The hydrolysis of the substrate, 4 methylumbelliferyl-galactopyranoside (MUG) to the product 4-methyllumbelliferone (4MU) was detected with a TD-360 miniflourometer (Turner Designs, San Jose, CA, USA). A standard curve was used to quantify 4MU.

2.5. Purification of BenM and CatM and Variant Proteins

Plasmids, pBAC433 and pBAC430, were used to express full-length regulators with C-terminal histidine tags, BenM-His and CatM-His, respectively [2]. Plasmids were made to encode variants, pBAC1027 ($_{\text{BenM-DBD}}$CatM-His), pBAC1045 (CatM(I18F,K38N)-His), and pBAC1086 ($_{\text{Ben-DBD}}$CatM(H160R,F293Y)-His). BenM-His was purified as described [20]. CatM-His and CatM variants were purified similarly but were eluted in a different buffer (30 mM Tris, 500 mM NaCl, 30% glycerol (*v*/*v*), 500 mM imidazole, and 10 mM β-mercaptoethanol (pH 7.9)). Fractions with pure CatM were pooled and dialyzed against a buffer (20 mM Tris-HCl, 250 imidazole (pH 7.9), 500 NaCl, 10% (*v*/*v*) glycerol) to increase solubility, and then were concentrated to 2–10 mg·mL^{-1}. Protein concentrations were determined by the method of Bradford with bovine serum albumin as the standard [21]. Proteins fractions were frozen with liquid nitrogen and stored at −70 °C until use.

2.6. Electrophoretic Mobility Shift Assays (EMSAs)

Operator-promoter DNA (P_{benA} and P_{catB}) was PCR amplified with 5′-6-carboxyfluorescein (6-FAM) labeled primers (Table S3). PCR products, approximately 150–250 bp, were extracted with a gel DNA recovery kit (Zymo Research, Irvine, CA, USA). For EMSAs, 1 nM DNA was incubated with different concentrations of protein (0, 2.5 nM, 5 nM, 10 nM, 20 nM, 40 nM, 80 nM, 160 nM, 320 nM, 640 nM and 1.28 μM) for 1 h at 37 °C with or without muconate, benzoate or both. Effectors in the reaction were present at a total concentration of 1.6 mM individually or at a concentration of 800 μM each when combined. DNA-protein samples were resolved in 6% polyacrylamide gels. Electrophoresis was performed in buffer (40 mM Tris (pH 7.6), 20 mM acetic acid, 1 mM EDTA) for 1 h at 185 volts at 4 °C. When indicated, effectors were added to this buffer at the same concentrations described above. Fluorescently labeled bands were detected using the Amersham Typhoon PhosphorImager system (GE Healthcare Bio-Sciences, Pittsburgh, PA, USA) at 526 nm using the short-pass emission filter. The bound DNA relative to the unbound DNA was quantified by Gel-Pro analyzer (Media Cybernetics, Rockville, MD, USA). Values were fitted into a saturation curve using the equation for one site binding with total accounting for ligand depletion to determine the equilibrium constant (K_d) (Prism 8 software, GraphPad, San Diego, CA, USA) [22].

2.7. Modeling the Transcription Activation Complex at P_{benA} and P_{catB}

An atomic model of P_{benA} was created by merging the structures of BenM-DBD bound to DNA [23] and a transcription initiation complex of *E. coli* RNA polymerase (RNAP) [24] using the X3DNA suite [25] with the DNA atoms as alignment guides. Since the Site 1 DNA of P_{catB} has the most sequence identity with *benA* Site 2, the structure of BenM-DBD complexed with P_{catB} Site 1 (PDB ID 4IHS, chains A, B, E, and F) was used to model atoms locally at *benA* Site 2. The structure of BenM-DBD complexed with P_{benA} Site 1 (PDB ID 4IHT, chains A, B, E, and F) was used to model *benA* Site 1. The model for RNAP (PDB ID 4YLN) was positioned assuming that the TTGAAC downstream from P_{benA} Site 2 corresponds to the *E. coli* σ70 binding site with the sequence TTGACA in the 4IHT complex. DNA residues in the three structures were changed and renumbered to match the P_{benA} sequence with the program *mutate_bases*. A composite DNA backbone was generated by calculating the helical parameter sets (shear, stretch, stagger, buckle, prop-tw, opening, shift, slide, rise, tilt, roll, and twist) with program *find_pair* from the three individual DNA double helical segments and then merging the parameters into one file. The two helical parameter sets at the transition regions were averaged. A DNA model was generated from the combined helical parameters using the program *rebuild*. The three atomic structures containing both the DNA and protein residues were aligned on the composite

DNA backbone using *align* command of PyMOL [26]. Similar structures of the P_{catB} and $P_{benA5146}$ transcriptional activation complexes were generated by mutating the *benA* DNA sequences in the composite structure. The EBD domains were not fitted to the model because the close proximity of the two binding sites is not consistent with any full-length structure of an LTTR at this time.

3. Results

3.1. Engineered CatM Variants: Amino Acid Replacements at Positions 160 and 293

In an effort to make CatM respond to benzoate, *A. baylyi* strains were engineered to encode CatM (H160R), CatM (F293Y), or CatM (H160R, F293Y) (designated ACN662, ACN682, and ACN685, Table S1). These amino acid replacements were designed to match residues in BenM-EBD that interact with benzoate (Figure S2). To ensure that transcriptional regulation could be attributable to the CatM variants, *benM* was also disrupted. The engineered strains all grew on anthranilate, catechol, and muconate as sole carbon sources. Such growth requires transcription from the *cat*-operon promoter, P_{catB} (Figure 1), indicating that the variants are functional as *cat*-gene regulators. To assess P_{benA} regulation, growth on benzoate was evaluated. Without BenM, neither wild-type CatM, in strain ISA36 [3] nor CatM (F293Y) in ACN682 supported growth on benzoate as the carbon source. In contrast, CatM (H160R) in ACN662, and CatM (H160R, F293Y), in ACN685, conferred growth on benzoate, suggesting that H160R enables higher than normal levels of P_{benA} transcription.

3.2. Transcriptional Regulation of P_{benA} by CatM Variants

A *benA*::*lacZ* fusion was used to replace *benA*, thereby preventing growth on benzoate as the carbon source and enabling benzoate to be tested as a non-metabolized effector [3]. β-Galactosidase (LacZ) activity reflects P_{benA} regulation by CatM(F293Y), CatM(H160R), CatM(H160R,F293Y), or wild-type CatM in strains ACN717, ACN673, ACN694, and ACN1307, respectively (Figure 3). BenM is the major regulator of P_{benA}. Without effectors, a tetramer of BenM or CatM, can bind Site 1 and Site 3 to repress transcription of *benA* or *catB* (Figure 3C). Effectors cause a shift in which protein binding to Sites 1 and 2 improves RNAP access to Site 3 [2]. Despite possible cross-regulation, the maximum level of CatM-activated transcription was 18-fold lower than for BenM (ACN1307 versus ACN1232). However, the H160R replacement increased this low-level CatM-mediated response to muconate (ACN673 and ACN694). The elevated transcription levels remained significantly below that for BenM (ACN1232), yet they are comparable to those of other CatM variants that enable BenM-independent growth on benzoate [4]. Regulation by CatM(F293Y), which did not permit growth on benzoate, was comparable to wild-type CatM, which also fails to allow such growth.

While the *catM* changes failed to recapitulate the effects of benzoate with BenM, this failure might be due to problems interacting with P_{benA} DNA as opposed to problems with binding benzoate. In previous studies, CatM activated higher levels of *benA* transcription when there was a transversion (T-to-A) at position −40 relative to the transcriptional start site [3,4]. Here we tested regulation of this promoter ($P_{benA5146}$) by CatM and its variants CatM(F293Y), CatM(H160R), and CatM(H160R,F293Y) using the *lacZ* fusion (in ACN157, ACN827, ACN832, and ACN839, respectively). CatM yielded higher levels of transcription from $P_{benA5146}$ than P_{benA} under all conditions (ACN157 versus ACN1307, Figure 3). At this promoter, the variants all increased transcription even further than wild-type CatM. Benzoate led to expression levels 170% or 155% of the non-induced levels for CatM(H160R) or CatM(H160R,F293Y), respectively (Figure 3). Despite this suggestion of a minor response to benzoate, muconate-mediated transcription was inhibited by benzoate. In contrast, for BenM, benzoate works synergistically with muconate (ACN1232) [2].

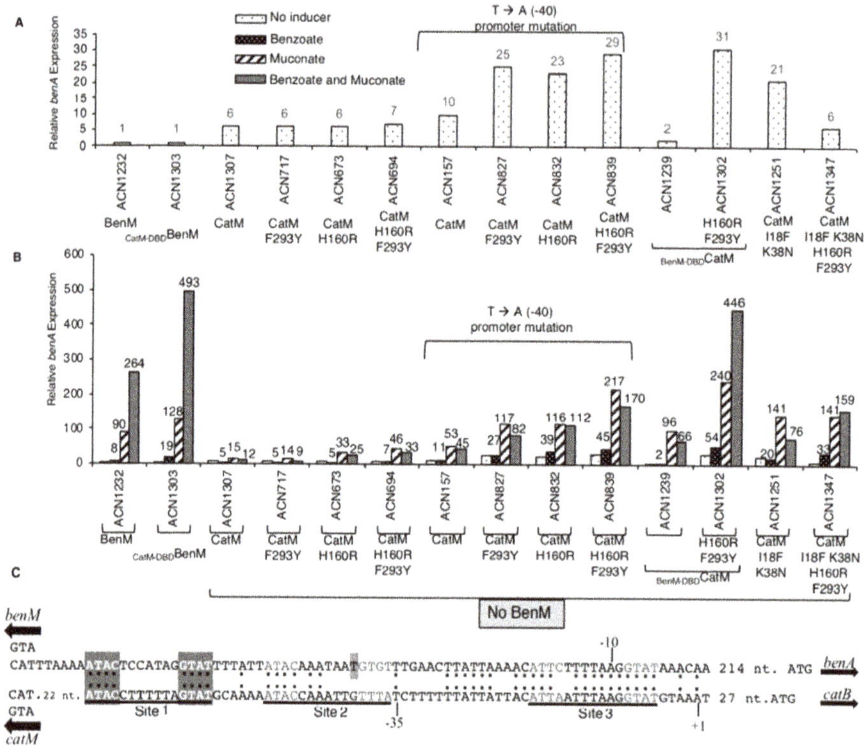

Figure 3. Relative expression of LacZ. Transcription was controlled by P_{benA}, or, $P_{benA5146}$, which differs by a mutation at −40 relative to the *benA* transcriptional start site. (**A**) Enlarged scale displays basal activity (no added effectors). (**B**) Effectors were added (or not) as indicated. ACN1232 encodes BenM but not CatM, and ACN1307 encodes CatM but not BenM. Other strains encode a CatM variant. Cultures were grown in LB. LacZ activity is reported relative to uninduced ACN1232 (2.6 ± 0.51 nmol/min/mL/OD600). Activities are averages of at least four repetitions; standard deviations were <20% of the average value. (**C**) Identical nucleotides in aligned P_{benA} and P_{catB} regions are indicated (:). The transcriptional start site (+1) and promoter (−10 and −35 regions) are shown for *catB*. For both regions, Site 1 matches the consensus LTTR-binding motif (T-N_{11}-A, within dyad symmetry, ATAC-N_7-GTAT). Site 2 and Site 3 differ slightly from this consensus. The $P_{benA5146}$ mutation (T) is marked.

3.3. Spontaneous Mutant with Increased ben-Gene Expression: Changesne in the CatM-DBD

A different approach was used in another attempt to isolate a benzoate-responsive CatM. While ACN682, encoding CatM(F293Y), does not grow on benzoate, its *catM* mutation might facilitate the ability of additional mutations to confer this trait. Based on this rationale, Ben^+ colonies were directly selected from ACN682, as was done previously for wild-type CatM [3,27]. One ACN682-derived mutant encoded two changes, F293Y and I18F. To test the role of I18F, a strain was made to encode only this change which matches residue 18 of BenM-DBD (Figure 2). CatM(I18F) conferred Ben^+ growth (ACN1095), indicating increased transcription of P_{benA}. When tested with the *benA::lacZ* fusion in ACN1111, there was no response to benzoate as a sole effector, and benzoate decreased the muconate-inducible expression (data not shown). It appears that the Ben^+ phenotype arises from an increase in muconate-induced P_{benA} expression (ACN1111 compared to ACN1307; Figure 4).

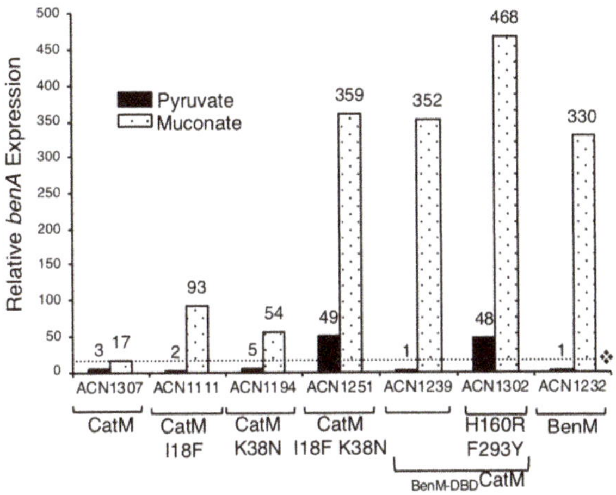

Figure 4. Relative LacZ activity of a chromosomal *benA::lacZ* fusion. The only strain encoding BenM is ACN1232, which has no CatM. The dotted line shows the level of the CatM response to muconate (ACN1307). Cultures were grown on pyruvate or muconate as the sole carbon source. Expression is reported relative to the basal level of ACN1232 (6 ± 2 nmol/min/mL/OD600). Values represent the average of at least three independent replicates. Standard deviations were <20% of the average value.

Since I18F makes the variant more like BenM, all nine DBD differences were considered (Figure 2). Residue 38, in the recognition helix of the helix-turn-helix (HTH) DNA-binding motif, is implicated in BenM-DNA interactions [23]. Therefore, strains were made to encode CatM(K38N) and CatM(I18F, K38N) (ACN1193 and ACN1249, respectively). Both strains grew on anthranilate, muconate, and benzoate, indicating that the CatM variants are functional. Muconate enabled CatM(K38N) and CatM(I18F, K38N) to activate higher levels of P_{benA} transcription than CatM (ACN1194, ACN1251 versus ACN1307, Figure 4). For the double-replacement variant, gene expression was comparable to that mediated by BenM (in ACN1232). CatM(I18F, K38N) also resulted in high P_{benA} basal expression (Figures 3 and 4). However, there was no transcriptional activation in response to benzoate. Moreover, for both variants with K38N, benzoate decreased the ability of muconate to activate transcription (ACN1251 in Figure 3).

3.4. Further Investigation of DBD Residues in BenM and CatM

The ability of BenM to regulate P_{benA} might be weakened by having the residues of CatM at positions 18 and 38. To test this possibility, strains were made to encode BenM(F18I,N38K). When tested with a *benA::lacZ* reporter, regulation by this variant was significantly decreased under all conditions relative to BenM. Nevertheless, BenM(F18I,N38K) remained capable of activating transcription synergistically in response to benzoate and muconate (Supplementary Figure S3).

The entire DBD of CatM was replaced with that of BenM by changing nine amino acids. This variant ($_{\text{BenM-DBD}}$CatM) enabled growth on muconate, anthranilate, and benzoate as sole carbon sources (ACN1234). When regulating a *benA::lacZ* reporter, this variant activated transcription similarly to BenM with muconate (ACN1239 versus ACN1232). However, there was no response to benzoate as the sole effector, and benzoate decreased muconate-inducible expression as observed for the previously studied CatM variants (Figure 3).

3.5. Benzoate-Responsive CatM Variants Obtained with Combinations of DBD and EBD Changes

Starting with $_{\text{BenM-DBD}}$CatM, we altered the EBD in further attempts to create a response to benzoate. $_{\text{BenM-DBD}}$CatM (H160R, F293Y) enabled growth on benzoate (ACN1301, Table 1). This variant (in ACN1302) also enabled muconate or benzoate to activate high-level transcription (Figures 3 and 4). Notably, there was a BenM-like regulatory pattern when benzoate and muconate were added together. This pattern indicated that 11 amino acid changes are sufficient for a benzoate-responsive CatM capable of synergistic transcriptional activation.

Table 1. Growth on Benzoate as the Sole Carbon Source [a].

Strain	Relevant Characteristics	Generation Time (min) [b]	Lag Time (h) [c]
ADP1	Wild-type	70 ± 5	4.5 ± 1
ISA36	No BenM	No growth	No growth
ACN682	No BenM, CatM(F293Y)	No growth	No growth
ACN1095	No BenM, CatM(I18F)	175 ± 3	18 ± 3
ACN1193	No BenM, CatM(K38N)	145 ± 6	11 ± 1
ACN1249	No BenM, CatM(I18F,K38N)	85 ± 2	5.5 ± 2
ACN1234	No BenM, $_{\text{BenM-DBD}}$CatM	82 ± 3	5 ± 1.5
ACN1301	No BenM, $_{\text{BenM-DBD}}$CatM(F293Y,H160R)	81 ± 4	5 ± 1
ACN1345	No BenM, CatM(I18F,K38N,H160R,F293Y)	83 ± 3	5 ± 1.5
ACN1294	$_{\text{CatM-DBD}}$BenM, No CatM	186 ± 6	21 ± 3

[a] Strains had comparable growth rates with succinate as the sole carbon source (data not shown); [b] Averages of at least four determinations; [c] Time between inoculation and start of exponential growth.

Another combination of changes, CatM(I18F,K38N,H160R,F293Y), generated a Ben$^+$ strain (ACN1345, Table 1). While two DBD changes enabled muconate-inducible expression, the two additional EBD changes (in ACN1347) were required for induction by benzoate (a 5.5-fold increase in expression, Figure 3). Moreover, when benzoate and muconate were both provided, gene expression increased for CatM(I18F,K38N,H160R,F293Y). In contrast, for CatM(I18F,K38N), benzoate inhibited the response to muconate, Figure 3.

3.6. Promoter Specificity and Regulator-DNA Binding Affinities

We also studied P_{catB}. While CatM-DBD should recognize this region, CatM(H160R,F293Y) did not activate transcription of a *catB::lacZ* fusion in response to benzoate (ACN1389, Figure 5). With BenM, P_{catB} transcription was high, but the pattern of effector responses differed from what occurred at P_{benA}. Notably, benzoate failed to enhance transcriptional activation in response to muconate (ACN1375, Figure 5). When the BenM-EBD was combined with the CatM-DBD, the overall response to effectors was lessened and the response pattern was altered (ACN1367 compared to ACN1375, Figure 5). As observed at P_{benA}, a combination of DBD and EBD changes enabled a response to benzoate. Two CatM variants increased gene expression from P_{catB} in response to benzoate, $_{\text{BenM-DBD}}$CatM(F293Y,H160R) in ACN1369 and CatM(I18F,K38N,H160R,F293Y) in ACN1393, Figure 5. With these variants, benzoate enhanced transcriptional activation by muconate.

To evaluate binding to the operator–promoter regions, an electrophoretic mobility gel shift assay (EMSA) was used. However, because the DNA in the assay includes all three regulatory binding sites (Figure 3), the measured K_d coefficients do not distinguish between repression and activation. Thus, the affinity of BenM for P_{benA}, appears to be of the same order of magnitude regardless of the presence of effectors (Table 2). CatM appeared to have a slightly higher K_d, which corresponds to a lower affinity, for this DNA. Interestingly, the CatM variants that carry the DBD of BenM have a higher affinity for P_{benA} than either CatM or BenM. This increased affinity appears to require more than the two amino acid changes in the HTH region of the DBD (I18F and K38N). Moreover, increased affinity does not correlate with benzoate responsiveness. These results suggest the hybrid proteins have a conformation that differs from those of the wild-type CatM or BenM.

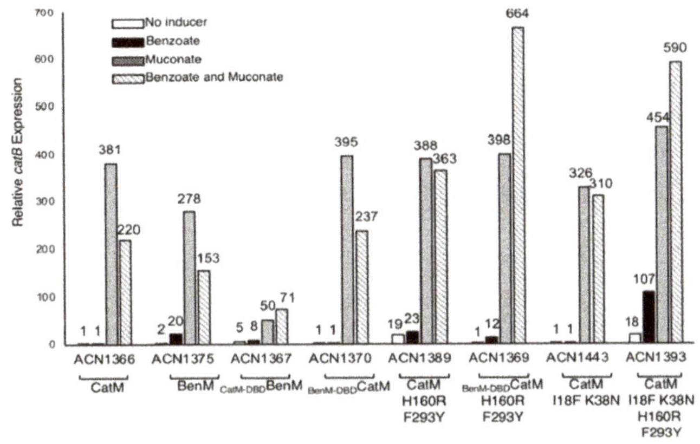

Figure 5. Expression from a chromosomal *catB::lacZ* transcriptional reporter. ACN1375 encodes BenM but not CatM. All other strains lack BenM. All have a *benA* disruption to prevent benzoate catabolism. Cultures were grown on 20 mM pyruvate with or without added effectors. β-Galactosidase (LacZ) activity is reported relative to uninduced ACN1366 (0.71 ± 0.5 nmol/min/mL/OD600). Activities are the average of at least three repetitions, and standard deviations were <10% of the average value.

Table 2. Estimated Binding Affinities of Regulators to DNA With and Without Effectors [a].

DNA	Transcriptional Regulator	Effector Added to Protein-DNA Complex			
		No Effector K_d (nM)	Benzoate K_d (nM)	Muconate K_d (nM)	Muconate and Benzoate K_d (nM)
P_{benA}	Wild-type BenM	32 ± 3	17 ± 2	41 ± 3	46 ± 3
	Wild-type CatM	70 ± 3	71 ± 4	67 ± 3	71 ± 4
	BenM-DBDCatM	2 ± 0.5	4 ± 1	3 ± 0.5	9 ± 3
	BenM-DBDCatM(F293Y,H160R)	1 ± 0.2	4 ± 1	2 ± 0.5	4 ± 2
	CatM(18F,K38N)	42 ± 2	32 ± 3	34 ± 1	32 ± 4
P_{catB}	Wild-type BenM	26 ± 3	30 ± 2	16 ± 2	19 ± 1
	Wild-type CatM	11 ± 2	19 ± 3	10 ± 1	19 ± 2
	BenM-DBDCatM	1 ± 0.2	3 ± 0.5	3 ± 0.2	4 ± 1
	BenM-DBDCatM(F293Y,H160R)	3 ± 0.5	3 ± 0.5	3 ± 0.5	3 ± 0.5
	CatM(18F,K38N)	28 ± 2	13 ± 2	8 ± 2	14 ± 0.5

[a] Values represent averages of four replicates with $p > 0.001$.

In experiments with P_{catB}, CatM had a higher affinity than BenM for this DNA (Table 2). As for the P_{benA} region, the CatM variants with the entire BenM-DBD region had a higher affinity for the promoter DNA than did the variant with two replacements in the DBD. While it is difficult to infer the significance of the variations in K_d values, there appears to be no correlation between the ability of a CatM variant to respond to benzoate and its binding affinity (high or low) for P_{benA} or P_{catB}.

3.7. Regulator-DNA Interactions at P_{benA}: A Structural Model with RNAP

To understand protein-DNA interactions better, a model was built with available structures of BenM-DBD bound to P_{benA} and P_{catB} and *E. coli* RNAP (Figure 6) [23,24]. In this model, Site 2 overlaps the −35 region of the promoter. A run of adenosine nucleotides within Site 2, with a T nucleotide in the middle, appears to create a promoter feature that may increase transcription by interacting with the RNAP α-subunits, an UP element. The relative position of BenM-DBD to the promoter is typical of a class II σ70 promoter in which a regulator directly contacts domain 4 of the sigma factor of bound RNAP to stabilize the initiation complex [28]. However, the model suggests that the BenM-DBDs would not directly contact σ70. The EBD (not in the model) could make contacts with the sigma factor, but the bulk of the EBD units must sit on the opposite face of the RNAP if BenM is a tetramer. In contrast, for class I σ70 promoters, regulators do contact the α-subunit C-terminal domains (α-CTD), but typically the regulatory proteins bind further upstream of bound RNAP. Protein surface residues of the DBD-dimers that flank the UP-element, such as F31 of GcvA (C26 of BenM), have been implicated in α-CTD interactions [29]. The conserved −35 σ70 recognition-sequence in P_{benA} (TTGAAC vs. consensus TTGACA) suggests that de-repression by conformational change from Site 1 and Site 3 to Site 1 and Site 2 may be a significant aspect of transcriptional activation along with UP-element interactions.

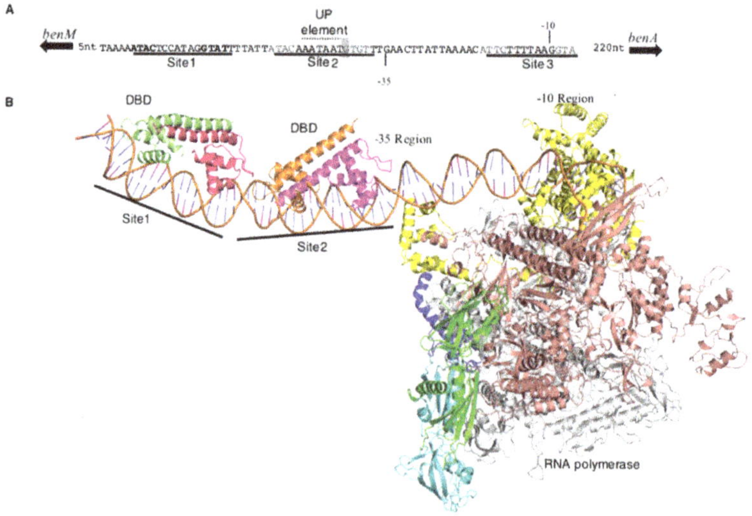

Figure 6. Structural model of BenM-P_{benA} interactions. (**A**). P_{benA} with binding sites for BenM (underlined). A potential UP element [30] is indicated and discussed in the text. The boxed nucleotide (T) corresponds to the mutation in $P_{benA5146}$. (**B**). Model of the initiation complex at P_{benA} rendered as a ribbon representation. BenM-DBD subunits are colored pale green, dark magenta, gold, and magenta going from the 5′ (left) to 3′ of the *benA* promoter; RNA polymerase (RNAP) subunits are green and cyan (α), salmon (β), grey (β′), blue (ω) and yellow (σ).

4. Discussion

4.1. Comparisons of BenM and CatM Provide a Framework to Engineer Effector-Binding Changes in CatM

Residues R160 and Y293 in BenM-EBD, which are critical for benzoate-induced transcriptional activation of P_{benA}, were introduced into CatM. These replacements were designed to create a hydrophobic binding pocket resembling one in BenM that binds benzoate, termed the secondary effector-binding site (Figure S2) [5,6]. Since benzoate decreased the muconate-activated gene expression for these CatM variants (Figure 3), it is likely that benzoate competes with muconate for binding in the primary effector-binding site of this LTTR rather than binding to a secondary site. In CatM(H160R,F293Y) and CatM(H160R), R160 increased the muconate-activated transcription of P_{benA}. In BenM, responses to effectors in the primary and secondary binding sites appear to connect through charge-based interactions among three residues that separate muconate and benzoate in the protein structure (R146, E162 and R160) [1].Without benzoate in the secondary binding site of CatM variants, it is unclear how the replacement of H160 with R160 increases the response to muconate. Changes in the local environment may mimic what occurs in BenM upon binding benzoate. Regardless of the mechanism, the increased muconate-responsive transcription due to R160 is consistent with our previous conclusion that this residue in BenM can impact the nearby primary binding site.

The importance of the DBD in P_{benA} regulation was revealed by a spontaneous Ben$^+$ mutant. In this mutant, one amino acid change in the HTH region altered transcriptional activation in response to muconate. CatM(I18F), in strain ACN1111, increased muconate-dependent P_{benA} expression more than five-fold relative to CatM, in ACN1307 (Figure 4). A more BenM-like DBD may improve promoter recognition at P_{benA} and thus enable the CatM-EBD to display its typical effector specificity. Some data support the interpretation that DBDs and EBDs follow expected patterns, whereby promoter recognition is governed by the DBD and effector specificity, is controlled by the EBD. For example, replacing the entire DBD of CatM with that of BenM significantly improves regulation of P_{benA} (Figures 3 and 4, ACN1239 compared to ACN1307).

Most results suggest a more complicated relationship among the domains. For example, if the low level of CatM-mediated transcription from P_{benA} were due solely to poor interaction with CatM-DBD, then $_{\text{CatM-DBD}}$BenM would be expected to display a typical BenM pattern of effector response but yield low levels of transcription. Instead, the transcriptional levels mediated by this chimeric protein at P_{benA} were higher than for BenM (ACN1303 compared to ACN1232, Figure 3). At P_{catB} this hybrid variant ($_{\text{CatM-DBD}}$BenM) might be expected to increase transcriptional activation compared to BenM because CatM is the cognate regulator of *catB*. Instead, the overall expression levels were lowered, and the BenM-mediated pattern of response was altered (ACN1367 compared to ACN1375, Figure 5). Another example that is counter to simple predictions resulted from replacing CatM-DBD with BenM-DBD. This alteration to CatM had little effect at P_{catB} (ACN1366 versus ACN1370, Figure 5). Experimental data are lacking to show if the variant proteins are produced at comparable levels in vivo or whether these proteins are equally stable. Nevertheless, the regulatory patterns, in most cases, are complex. An additional layer of variability results from differences in the flexible helix (the linker helix) that connects the DBD and EBD. In the 30 amino acid residues corresponding to this helix, BenM and CatM are 50% identical and 83% similar in sequence.

4.2. Complex Patterns of Regulation

Even without effectors, DBD alterations impacted transcription. Since BenM and CatM repress transcription in the absence of effectors, a several-fold increase in P_{benA} basal expression, which results from the loss of BenM, is interpreted as de-repression [2,3]. Consistent with this model, basal expression in most strains encoding CatM or a CatM variant was approximately 6-fold higher than for the strain encoding BenM (e.g., ACN1307, ACN717, ACN673, or ACN694 compared to ACN1232, Figure 3A). In one notable exception, two replacements in the CatM-DBD, I18F and K38N, caused a 21-fold expression increase in basal expression relative to BenM. This high increase suggests that this variant activates transcription without effectors (ACN1251 compared to ACN1232, Figure 3A). This variant may cause a change in protein structure that relieves binding to Site 3 and helps recruit RNAP. When these two DBD changes were accompanied by changes in the EBD or the DBD, high-level basal expression was no longer observed (ACN1347 and ACN1239, respectively, Figure 3A). Yet when all these EBD and DBD changes were combined in a single variant, $_{\text{BenM-DBD}}$CatM(H160R,F293Y), transcriptional activation in the absence of effectors resulted again (ACN1302, Figure 3A). These transcriptional patterns did not correlate meaningfully with changes in the affinity of these proteins for the operator–promoter DNA (Table 2). Better interpretation awaits additional crystal structures and experiments with RNAP. Until then, the crystal structures of BenM-DBD with the Site 1 region of P_{benA} or P_{catB} can provide a framework for considering LTTR-DNA interactions.

4.3. Interactions Between DBDs and Operator–Promoter Regions of P_{benA} and $P_{benA5146}$

The I18F replacement in CatM-DBD alters transcription sufficiently to enable Ben$^+$ growth. Crystal structures of BenM-DBD-$P_{\text{benA-Site 1}}$ and a similar structure, CbnR-DBD-$P_{\text{cbnA-Site}}$ indicate that this residue is in helix $\alpha 2$, where it is involved in indirect readout [23,31]. Indirect readout refers to the effects of local nucleotides that cause sequence-dependent deformations of the phosphate backbone to control binding. In BenM, F18 is grouped with residues whose main chain amide N atoms form hydrogen bonds and van der Waals contacts with DNA. The equivalent residue in CbnR, M18, is involved in sugar–phosphate recognition and contributes to DNA-binding strength rather than promoter specificity [31]. CatM binding to P_{benA} is weaker than for BenM (Table 2), which may be due in part from I18 distorting the packing of the helix against the DNA. While binding affinity was not tested for CatM(I18F), the CatM(I18F,K38N) was evaluated (Table 2). Consistent with our interpretation, this variant had an affinity for P_{benA} that was intermediate between that of CatM and BenM. The entire BenM-DBD not only further increased the affinity of CatM variants for P_{benA}, but this affinity actually surpassed that of the native BenM (Table 2). These results indicate that binding affinity is not determined solely by the DBD and also suggest there are important variations in the oligomeric conformation of these regulators.

Residue 38 in BenM and CatM is in the recognition helix (α3) of the HTH motif. N38 in BenM can form a hydrogen bond with the DNA phosphate backbone, although this type of non-specific interaction is unlikely to confer specificity for P_{benA}. Instead, the importance of this residue may derive from a dynamic interaction with R34, which interacts directly with DNA and provides specificity in DNA recognition. The surrounding residues orient the side chain of R34. An interaction network between these residues and others, including E41 and Q37, might contribute to a conformational switch controlling promoter recognition or RNAP activation at P_{benA} [23]. K38 in CatM would form an electrostatic interaction with the phosphate backbone of P_{catB} and would project deeper into the major grove if a purine were positioned at the interaction point, e.g., nucleotide A36 in the BenM-DBD-*catB* DNA structure. However, the methyl-group of thymine, as found in this nucleotide position of P_{benA}, would impede this direct interaction. Consistent with this possibility, CatM(K38N) increased muconate-activated transcription of P_{benA} better than the wild-type CatM (Figure 4). Similar interactions are observed in other LTTRs, such as CbnR (A38, R34, and D42) [31], MetR (S38, S34, H35, and Q42) [32], DntR (R43, N39, T46 and A47) [33] and Tsar (Q38, D42, and S34) [34].

DNA interactions with residue 38 in the DBD may influence effects at $P_{benA5146}$. Binding at Site 2 of this promoter (which has a T-to-A transversion) by CatM will be enhanced compared to P_{benA} due to K38 interactions with the partial negative charge on the N7 atom of the adenine base. Interactions of BenM with $P_{benMA5146}$ may not be negatively impacted because a water molecule can bridge to the N7. Thus, the interaction of CatM with $P_{benMA5146}$ may be stabilized at Site 2 but not Site 3, thereby disrupting the equilibrium between the protein binding to Site 2 (stabilized by effectors) and Site 3 (stabilized by lack of effectors). This alteration could result in the de-repression of basal activity observed at this promoter for CatM and help explain transcriptional activation in the absence of effectors for the CatM variants with F293Y and/or H160R replacements in the CatM-EBD (Figure 3A).

4.4. Interactions Between DBDs and Operator–Promoter Regions of P_{catB}

Factors affecting K38 interactions with $P_{benA5146}$ may also affect recognition at P_{catB}. Between the two half sites of dyad symmetry in P_{catB} Site 2, a G corresponds to the position of the T-to-A mutation of $P_{benMA5146}$ (Figure 3C). In general, the sequence ATAC-pyrimidine-N_5-purine-GTAT will favor binding at P_{catB} by CatM. This pattern is observed for Sites 1 and 2. At P_{benA}, Site 2 swaps the arrangement of the purine/pyrimidine pair, a situation that may partially explain why wild-type CatM shows significantly reduced responses at P_{benA}.

BenM activated transcription relatively strongly at P_{catB} despite a different pattern of response to benzoate compared to its activity at P_{benA} (Figure 5). As measured by EMSAs, the affinity of BenM for P_{benA} and P_{catB} is comparable. While the native CatM has a higher affinity for its cognate P_{catB} promoter than for P_{benA}, all proteins with a native BenM-DBD have a comparable affinity for both promoters (Table 2). It is not yet possible to discern the features that control such variation in binding and transcriptional control, and a better understanding of LTTR interactions with RNAP is needed.

4.5. A Model of P_{benA} and P_{catB} Promoters with RNAP and the DBDs of BenM and CatM

Aspects of the model described above for protein interactions with P_{benA} also apply to P_{catB}. However, some important differences stem from DNA sequence variation. P_{catB} differs in the region of the 3′ half-site of symmetry in Site 2 from that of P_{benA}. The sequence of P_{catB} in this region (TCTTTT, Figure 3) does not match the −35 consensus sequence of σ70 promoters. Further, the 3′ half-site of Site 2 (TTTA) lacks similarity to the canonical binding 3′ half-site shared by BenM and CatM, GTAT (P_{benA}, GTGT). This sequence divergence suggests that the regulators at P_{catB} may play a more direct role in transcription initiation by replacing the -35 region DNA interactions with σ70 with compensating LTTR-RNAP contact. Analysis of the electrostatic surfaces of σ70 and BenM-DBD at the *benA* promoter show a strongly negative surface created by D23 in BenM (E23 in CatM) and a strongly positive surface defined by residues K593, R596, and K597 of σ70 which are part of the conserved regulatory region 4.2 of the sigma factor. Clearly, rotation of the RNAP holoenzyme toward BenM by moving forward a few base pairs along P_{catB} will position the RNAP against BenM or CatM in a complementary fashion. The amino acids in this region 4 of sigma have been mapped as binding residues in other sigma regulatory molecules [35]. Between the half-sites of symmetry of Site 2 in P_{catB}, there is a mix of A and T nucleotides that could act as an UP-element to stabilize αCTD interactions, as observed in P_{benA}. This combination of σ70 and αCTD interactions would make the P_{catB} a mixed class I–class II promoter. However, the UP-element sequence differs from that of P_{benA} because of the flanking C/G nucleotide pair at the 5′ half-site of Site 2. Alternatively, if CatM and BenM bind only to the 5′ half-site sequence, the 3′ area of Site 2 might be occupied by σ70 (perhaps at the intervening TTGTT) and allow some conformational flexibility for the previously mentioned electrostatic interactions to be favored.

4.6. Broader Implications and Conclusions

Although LTTRs have been studied for more than 30 years, many aspects of their structure and function remain unclear [8]. Because of their similarities and the overlap in their control of a complex regulon, BenM and CatM provide unique opportunities for comparative investigation [1]. Furthermore, their role in aromatic compound catabolism holds promise for biotechnology applications, including lignin valorization [18]. Because of such applications, these and related regulators are receiving renewed attention. For example, transcriptional regulators that respond to aromatic compounds are useful as biosensors [36]. In a recent study, high throughput methods were developed to combine protein domains for DNA binding and those for effector responses with the aim of creating benzoate-responsive biosensors [37]. This approach builds on the modular nature of bacterial transcriptional regulators.

Our studies demonstrate that regulatory alterations can be engineered to integrate multiple signals by enabling the simultaneous response to more than one effector molecule. Although it took repeated efforts, we obtained CatM variants that recognize and respond to benzoate. In some variants, benzoate increased the transcriptional response to muconate, thereby approximating the synergistic transcriptional regulation mediated by BenM. This type of rapid signal amplification has potential uses in metabolic engineering and synthetic biology. While it is not yet possible to interpret some characteristics of the variant regulators that were studied, our results lay the foundation for continued investigation. LTTRs are more than the sum of their parts; both the DBDs and the EBDs affected promoter specificity as well as transcriptional responses to effectors. The abundance of this family of regulators and the importance of LTTR-regulated processes underscore the value of continued research.

Supplementary Materials: The following are available online at http://www.mdpi.com/2073-4425/10/6/421/s1, Figure S1. Regulatory model for P_{benA}. Figure S2. Structural comparisons of BenM-EBD and CatM-EBD. Figure S3. Changes in BenM-DBD affect relative *benA* expression. Table S1. *Acinetobacter baylyi* strains. Table S2. Plasmids. Table S3. Primers.

Author Contributions: Conception and design: M.P.T.-V., C.M., and E.L.N.; Data collection and analysis: M.P.T.-V., C.M., E.L.N; Plasmid and strain construction: M.P.T.-V. and N.S.L.; Lab experiments: M.P.T.-V.; Manuscript writing: M.P.T.-V., E.L.N and C.M.

Funding: This research was supported by NSF grants MCB1024108 (to C.M. and E.L.N) and MCB1615365 (to E.L.N.).

Acknowledgments: We thank Ajchareeya Ruangprasert and Melesse Nune for assistance in protein purification and electromobility shift assays. We also thank Curtis Bacon for help with plasmid and strain construction. Jenifer Morgan, Chelsea Kline, James Valle and Walker Whitley assisted with the selection of *A. baylyi* strains.

Conflicts of Interest: The authors declare no conflict of interest.

References

1. Craven, S.H.; Ezezika, O.C.; Momany, C.; Neidle, E.L. LysR homologs in Acinetobacter: Insights into a diverse and prevalent family of transcriptional regulators. In *Acinetobacter Molecular Biology*; Gerischer, U., Ed.; Caister Academic Press: Norfolk, UK, 2008; pp. 163–202.
2. Bundy, B.M.; Collier, L.S.; Hoover, T.R.; Neidle, E.L. Synergistic transcriptional activation by one regulatory protein in response to two metabolites. *Proc. Natl. Acad. Sci. USA* **2002**, *99*, 7693–7698. [CrossRef]
3. Collier, L.S.; Gaines, G.L., 3rd; Neidle, E.L. Regulation of benzoate degradation in *Acinetobacter* sp. strain ADP1 by BenM, a LysR-type transcriptional activator. *J. Bacteriol.* **1998**, *180*, 2493–2501.
4. Ezezika, O.C.; Collier-Hyams, L.S.; Dale, H.A.; Burk, A.C.; Neidle, E.L. CatM regulation of the *benABCDE* operon: functional divergence of two LysR-type paralogs in *Acinetobacter baylyi* ADP1. *Appl. Environ. Microbiol.* **2006**, *72*, 1749–1758. [CrossRef]
5. Craven, S.H.; Ezezika, O.C.; Haddad, S.; Hall, R.A.; Momany, C.; Neidle, E.L. Inducer responses of BenM, a LysR-type transcriptional regulator from *Acinetobacter baylyi* ADP1. *Mol. Microbiol.* **2009**, *72*, 881–894. [CrossRef] [PubMed]
6. Ezezika, O.; Haddad, S.; Clark, T.; Neidle, E.; Momany, C. Distinct effector-binding sites enable synergistic transcriptional activation by BenM, a LysR-type regulator. *J. Mol. Biol.* **2007**, *367*, 616–629. [CrossRef] [PubMed]
7. Maddocks, S.E.; Oyston, P.C. Structure and function of the LysR-type transcriptional regulator (LTTR) family proteins. *Microbiology* **2008**, *154*, 3609–3623. [CrossRef]
8. Momany, C.; Neidle, E.L. Defying stereotypes: the elusive search for a universal model of LysR-type regulation. *Mol. Microbiol.* **2012**, *83*, 453–456. [CrossRef] [PubMed]
9. Quiocho, F.A.; Ledvina, P.S. Atomic structure and specificity of bacterial periplasmic receptors for active transport and chemotaxis: Variation of common themes. *Mol. Microbiol.* **1996**, *20*, 17–25. [CrossRef] [PubMed]
10. Schell, M.A. Molecular biology of the LysR family of transcriptional regulators. *Annu. Rev. Microbiol.* **1993**, *47*, 597–626. [CrossRef]
11. Juni, E.; Janik, A. Transformation of Acinetobacter calco-aceticus (Bacterium anitratum). *J. Bacteriol.* **1969**, *98*, 281–288.
12. Vaneechoutte, M.; Young, D.M.; Ornston, L.N.; De Baere, T.; Nemec, A.; Van Der Reijden, T.; Carr, E.; Tjernberg, I.; Dijkshoorn, L. Naturally transformable *Acinetobacter* sp. strain ADP1 belongs to the newly described species *Acinetobacter baylyi*. *Appl. Environ. Microbiol.* **2006**, *72*, 932–936. [CrossRef]
13. Sambrook, J.; Fritsch, F.; Maniatis, T. *Molecular cloning: A laboratory manual*, 2nd ed.; Cold Spring Harbor Laboratory Press: Cold Spring Harbor, NY, USA, 1989.
14. Shanley, M.S.; Neidle, E.L.; Parales, R.E.; Ornston, L.N. Cloning and expression of *Acinetobacter calcoaceticus catBCDE* genes in *Pseudomonas putida* and *Escherichia coli*. *J. Bacteriol.* **1986**, *165*, 557–563. [CrossRef] [PubMed]
15. Singh, A.; Bedore, S.R.; Sharma, N.K.; Lee, S.A.; Eiteman, M.A.; Neidle, E.L. Removal of aromatic inhibitors produced from lignocellulosic hydrolysates by *Acinetobacter baylyi* ADP1 with formation of ethanol by Kluyveromyces marxianus. *Biotechnol. Biofuels.* **2019**, *12*. [CrossRef] [PubMed]
16. Horton, R.M.; Cai, Z.L.; Ho, S.N.; Pease, L.R. Gene splicing by overlap extension: tailor-made genes using the polymerase chain reaction. *Biotechniques* **1990**, *8*, 528–535. [CrossRef] [PubMed]
17. Neidle, E.L.; Hartnett, C.; Ornston, L.N. Characterization of *Acinetobacter calcoaceticus catM*, a repressor gene homologous in sequence to transcriptional activator genes. *J. Bacteriol.* **1989**, *171*, 5410–5421. [CrossRef] [PubMed]
18. Tumen-Velasquez, M.; Johnson, C.W.; Ahmed, A.; Dominick, G.; Fulk, E.M.; Khanna, P.; Lee, S.A.; Schmidt, A.L.; Linger, J.G.; Eiteman, M.A.; et al. Accelerating pathway evolution by increasing the gene dosage of chromosomal segments. *Proc. Natl. Acad. Sci. USA* **2018**, *115*, 7105–7110. [CrossRef]

19. Gregg-Jolly, L.A.; Ornston, L.N. Recovery of DNA from the *Acinetobacter calcoaceticus* chromosome by gap repair. *J. Bacteriol.* **1990**, *172*, 6169–6172. [CrossRef] [PubMed]
20. Ruangprasert, A.; Craven, S.H.; Neidle, E.L.; Momany, C. Full-length structures of BenM and two variants reveal different oligomerization schemes for LysR-type transcriptional regulators. *J. Mol. Biol.* **2010**, *404*, 568–586. [CrossRef] [PubMed]
21. Bradford, M.M. A rapid and sensitive method for the quantitation of microgram quantities of protein utilizing the principle of protein-dye binding. *Anal. Biochem.* **1976**, *72*, 248–254. [CrossRef]
22. Swillens, S. Interpretation of binding curves obtained with high receptor concentrations: Practical aid for computer analysis. *Mol. Pharmacol.* **1995**, *47*, 1197–1203.
23. Alanazi, A.M.; Neidle, E.L.; Momany, C. The DNA-binding domain of BenM reveals the structural basis for the recognition of a T-N11-A sequence motif by LysR-type transcriptional regulators. *Acta Crystallogr. D. Biol. Crystallogr.* **2013**, *69*, 1995–2007. [CrossRef] [PubMed]
24. Zuo, Y.; Steitz, T.A. Crystal structures of the *E. coli* transcription initiation complexes with a complete bubble. *Mol. Cell.* **2015**, *58*, 534–540. [CrossRef]
25. Lu, X.J.; Olson, W.K. 3DNA: a software package for the analysis, rebuilding and visualization of three-dimensional nucleic acid structures. *Nucleic. Acids. Res.* **2003**, *31*, 5108–5121. [CrossRef] [PubMed]
26. DeLano, W.L. The Pymol User's Manual. Delano Scientific: San Carlos, CA, USA, 2002.
27. Cosper, N.J.; Collier, L.S.; Clark, T.J.; Scott, R.A.; Neidle, E.L. Mutations in *catB*, the gene encoding muconate cycloisomerase, activate transcription of the distal ben genes and contribute to a complex regulatory circuit in *Acinetobacter* sp. strain ADP1. *J. Bacteriol.* **2000**, *182*, 7044–7052. [CrossRef] [PubMed]
28. Busby, S.; Ebright, R.H. Transcription activation at class II CAP-dependent promoters. *Mol. Microbiol.* **1997**, *23*, 853–859. [CrossRef] [PubMed]
29. Stauffer, L.T.; Stauffer, G.V. GcvA interacts with both the alpha and sigma subunits of RNA polymerase to activate the *Escherichia coli gcvB* gene and the *gcvTHP* operon. *FEMS Microbiol. Lett.* **2005**, *242*, 333–338. [CrossRef]
30. Ross, W.; Gosink, K.K.; Salomon, J.; Igarashi, K.; Zou, C.; Ishihama, A.; Severinov, K.; Gourse, R.L. A third recognition element in bacterial promoters: DNA binding by the alpha subunit of RNA polymerase. *Science* **1993**, *262*, 1407–1413. [CrossRef] [PubMed]
31. Koentjoro, M.P.; Adachi, N.; Senda, M.; Ogawa, N.; Senda, T. Crystal structure of the DNA-binding domain of the LysR-type transcriptional regulator CbnR in complex with a DNA fragment of the recognition-binding site in the promoter region. *FEBS J.* **2018**, *285*, 977–989. [CrossRef]
32. Punekar, A.S.; Porter, J.; Carr, S.B.; Phillips, S.E. Structural basis for DNA recognition by the transcription regulator MetR. *Acta Crystallogr. F. Struct. Biol. Commun.* **2016**, *72*, 417–426. [CrossRef] [PubMed]
33. Lerche, M.; Dian, C.; Round, A.; Lonneborg, R.; Brzezinski, P.; Leonard, G.A. The solution configurations of inactive and activated DntR have implications for the sliding dimer mechanism of LysR transcription factors. *Sci. Rep.* **2016**, *6*, 19988. [CrossRef]
34. Monferrer, D.; Tralau, T.; Kertesz, M.A.; Dix, I.; Sola, M.; Uson, I. Structural studies on the full-length LysR-type regulator TsaR from *Comamonas testosteroni* T-2 reveal a novel open conformation of the tetrameric LTTR fold. *Mol. Microbiol.* **2010**, *75*, 1199–1214. [CrossRef]
35. Lonetto, M.A.; Rhodius, V.; Lamberg, K.; Kiley, P.; Busby, S.; Gross, C. Identification of a contact site for different transcription activators in region 4 of the *Escherichia coli* RNA polymerase sigma70 subunit. *J. Mol. Biol.* **1998**, *284*, 1353–1365. [CrossRef] [PubMed]
36. Jha, R.K.; Bingen, J.M.; Johnson, C.W.; Kern, T.L.; Khanna, P.; Trettel, D.S.; Strauss, C.E.; Beckham, G.T.; Dale, T. A protocatechuate biosensor for *Pseudomonas putida* KT2440 via promoter and protein evolution. *Metab. Eng. Commun.* **2018**, *6*, 33–38. [CrossRef] [PubMed]
37. Juárez, J.F.; Lecube-Azpeitia, B.; Brown, S.L.; Johnston, C.D.; Church, G.M. Biosensor libraries harness large classes of binding domains for construction of allosteric transcriptional regulators. *Nat. Commun.* **2018**, *9*. [CrossRef] [PubMed]

© 2019 by the authors. Licensee MDPI, Basel, Switzerland. This article is an open access article distributed under the terms and conditions of the Creative Commons Attribution (CC BY) license (http://creativecommons.org/licenses/by/4.0/).

genes

MDPI

Article

Genomic Diversity of Two Hydrocarbon-Degrading and Plant Growth-Promoting *Pseudomonas* Species Isolated from the Oil Field of Bóbrka (Poland)

Valeria Imperato [1], Miguel Portillo-Estrada [2], Breanne M. McAmmond [3], Yorben Douwen [1], Jonathan D. Van Hamme [3], Stanislaw W. Gawronski [4], Jaco Vangronsveld [1,5] and Sofie Thijs [1,*]

1 Environmental Biology, Centre for Environmental Sciences, Hasselt University, Agoralaan Building D, 3590 Diepenbeek, Belgium; valeria.imperato@uhasselt.be (V.I.); d_yorben@hotmail.com (Y.D.); jaco.vangronsveld@uhasselt.be (Y.V.)
2 Centre of Excellence PLECO, Department of Biology, University of Antwerp, Universiteitsplein 1, B-2610 Wilrijk, Belgium; Miguel.PortilloEstrada@uantwerpen.be
3 Department of Biology Sciences, Thompson Rivers University, 805 TRU Way, Kamloops, BC V2C 0C8, Canada; Breannemcammond95@gmail.com (B.M.M.); Jvanhamme@tru.ca (J.D.V.H.)
4 Warsaw University of Life Sciences, Faculty of Horticulture, Biotechnology and Landscape Architecture, Nowoursynowska Road 159, 02-776 Warsaw, Poland; stanislaw_gawronski@sggw.pl
5 Department of Plant Physiology, Faculty of Biology and Biotechnology, Maria Skłodowska-Curie University, 20-950 Lublin, Poland
* Correspondence: sofie.thijs@uhasselt.be; Tel.: +32-11-26-82-25

Received: 13 April 2019; Accepted: 4 June 2019; Published: 11 June 2019

Abstract: Hydrocarbon-degrading bacteria are important resources for use in phytoremediation applications. Yet, for many hydrocarbonoclastic strains the genetic information regarding pollutant degradation and detoxification has not been thoroughly revealed. In this study, hydrocarbon-degrading bacteria were isolated from a long-term oil-polluted soil in Bóbrka, Poland. *Pseudomonas* spp. was the most dominant species. Of all 69 isolated strains tested in the laboratory using qualitative biochemical assays, 61% showed the capability to use diesel as sole carbon source, 33% could produce indole, 19% produced siderophores, 36% produced organic acids, and 54% were capable of producing 1-aminocyclopropane-1-carboxylate (ACC)-deaminase. From all morphologically and genetically different strains, two representative *Pseudomonas* spp., strain VI4.1 and VI4T1, were selected for genome sequencing. Genomic analyses indicated the presence of the full naphthalene dioxygenase operon (plasmid and chromosomal), of genes involved in the degradation of BTEX compounds (Benzene, Toluene, Ethylbenzene, Xylene) and alkanes (*alkB* gene) as well as the anthranilate degradation pathway (strain VI4T1) and terephthalate dioxygenase protein (strain VI4.1). Proton transfer reaction time-of-flight mass spectrometry (PTR-TOF-MS) analyses confirmed naphthalene and BTEX degradation within seven days. Motility, resistance to abiotic stresses, high and low temperatures, low pH, and salinity were confirmed at the genetic level and experimentally verified. The presence of multiple degradative and plant growth promotion genes, together with the in vitro experimental evidence, indicates the high value of these two strains and their potential use for sustainable site clean-up.

Keywords: naphthalene; toluene; hydrocarbons; plant growth promotion; bioremediation; *Pseudomonas*; soil pollution; phytoremediation

---

1. Introduction

Environmental spills with recalcitrant pollutants such as mono- and polycyclic aromatic hydrocarbons (MAHs, PAHs) are a problem worldwide, threatening the environment and human health [1]. Many of the diverse group of aromatic hydrocarbons are classified as priority pollutants by the US Environmental Protection Agency [2]. Naphthalene has also been classified as a group C possible human carcinogen. When oil spill accidents occur on land, degradation of petroleum hydrocarbons by indigenous microorganisms is often a slower process to remediate contaminated areas in comparison to traditional physical and chemical remediation treatments, often due to low microbial numbers and activity [3]. Microbial bioremediation, the process of pollutant degradation by microorganisms, is a green, cheap, and safe approach to cleaning up polluted sites [4,5]. This process can be improved by promoting the growth of endogenous bacteria in the polluted soil itself (biostimulation) or by introducing hydrocarbon-degrading bacteria (bioaugmentation) [6,7]. Frequently, due to strict legislations that impede the introduction of non-indigenous microorganisms at the site of interest, bioaugmentation is implemented by the use of native strains, or strains which carry the degradative genes on plasmids which can be transferred to the indigenous population though means of natural gene transfer. Studies demonstrated how this approach is considered the best choice for bio-detoxification of soils with a low degradation potential by the indigenous bacterial communities [8,9].

Undoubtedly, it is a challenge to find an effective and efficient method to remediate polluted soils [10], especially because of the complexity and specificity of each site. Oil spillages cause profound and persistent changes in soil properties. pH may vary, as well as nutrient availability (e.g., carbon, nitrogen), with high concentrations of mono- and polycyclic aromatic hydrocarbons, often in combination with heavy metal pollution (nickel, lead, chromium, copper, zinc, cadmium). All these factors affect and re-shape the structure of indigenous microbial communities [11,12]. In this highly complex framework, the study and isolation of novel native ring hydroxylating bacterial strains constitutes an approach with high potential and a powerful alternative to the traditional physical and chemical remediation approaches.

In this study we combined traditional microbiological methods to isolate and characterize potential degraders from an ancient oilfield, combined with a bacterial genomic investigation aiming to reveal the genetic backgrounds of two interesting hydrocarbon-degrading bacteria belonging to the genus *Pseudomonas*. Our study aims to elucidate (1) the degradation pathways of linear and aromatic hydrocarbons, (2) plant growth promotion traits (hormones and stress enzyme reduction), and (3) the adaptation of the two identified *Pseudomonas* spp. to life in soil, tolerance to various stresses, and extraordinary degradation capacities of micropollutants.

2. Materials and Methods

2.1. Isolation of Hydrocarbon-Degrading Bacteria Using a Sandwich Diffusion System

In June 2016, we sampled the historical oilfield forest soil in Bóbrka, Poland (Latitude: 49.616449; Longitude: 21.710454; Altitude: 350.45 m). The Bóbrka sampling site is unique in the world since it has one of the oldest oil wells still in production since 1854, and still pumps one barrel a day. The black top soil layer near an oil pump (Figure 1) was sampled with a sterilized shovel and stored in self-sealed bags. These bags were kept in temperature-controlled boxes and shipped to Hasselt University.

In the lab, one gram of soil was added into 15-mL sterile falcon tubes containing 5 mL of 0.1 M phosphate-buffered saline (PBS, pH = 7.0). After vortexing, tubes were incubated for one hour on an orbital shaker at 120 rpm to allow the microorganisms to effectively being released from the soil particles. In the meantime, Teflon components, steel screws, and hydrophilic membranes of a custom-made sandwich diffusion system (Figure S1) were sterilized by autoclaving (120 psi, 30 min). Each sterilized "perforated plate" was dipped into a mixture of Bushnell Haas medium [13] with 0.7% w/v Gelzan (G1910; Sigma-Aldrich, St. Louis, Missouri, USA) supplied with 0.1% $CaCl_2$ and the soil dilution to obtain a 10^{-5} concentration of the soil suspension. This dilution was chosen to obtain the growth of one or few bacterial cells per well. The mixture was left to solidify under a sterile laminar flow after which the membranes were placed on both sides of the agar-filled plates (WHA111103; Nucleopore®polycarbonate track-etched membrane; 47-mm diameter, 0.05-µm pore size, Whatman, Maidstone, UK). The membranes allow the exchange of nutrients and molecules from the soil to the water-agar containing bacterial cells but not the passage of bacterial cells. Finally, the sandwich diffusion systems were closed with screws and buried into the wetted soil and incubated at 20 °C for one month.

Figure 1. Sampling site of Bóbrka with spilled black crude oil.

After one month of incubation, the sandwich systems were disassembled and each agar plug was pushed out with sterile toothpicks into a sterile deep-well masterblock filled with Bushnell Haas medium and incubated overnight at 30 °C on an orbital shaker. The next day, 100 µL of each bacterial suspension were plated onto 1/10 869 rich medium [14] supplemented with 2 mM indole according to Nagayama et al. 2015 [15]. After incubation for seven days at 30 °C, the colonies producing a dark-blue pigment were selected as positive for aromatic compound ring hydroxylase activity (conversion of the colourless indigo to dark-blue indole by dioxygenase enzymes) [15]. *Pseudomonas putida* f1 [16] and *Pseudomonas aeruginosa* WatG [17] were used as positive controls.

2.2. Growth Conditions and Genotypic Characterization of the Bacterial Strains

Bacteria stored in a 96 well masterblock (Greiner Bio-One, Kremsmünster, Austria) and those scoring positive on the indole test were grown in 5 mL of 869 medium [14] and incubated for 24 h at 30 °C while shaking at 200 rpm. Two mL of culture broth were used for total DNA extraction using the Qiagen Blood and Tissue kit (Qiagen, Venlo, Netherlands). The quality of the purified genomic DNA was checked using a NanoDrop 1000 UV-Vis spectrophotometer (Thermo Fisher Scientific, Wilmington, DE, USA). The 27F (5′ AGAGTTTGATCMTGGCTCAG 3′) and 1492R (5′ TACGGYTACCTTGTTACGACTT 3′) primers were used for the amplification of the 16S rRNA gene [18]. The PCR master mix consisted of: DNA template (10 ng μL^{-1}), 1 × High Fidelity PCR buffer (Invitrogen, Carlsbad, CA, USA), 0.2 mM dNTPs, 2 mM $MgCl_2$, 0.2 µM each of the forward and reverse primers, and 1 U High Fidelity Platinum Taq DNA polymerase (Invitrogen, Carlsbad, CA, USA) per 50 µL. PCR conditions were set as follow: denaturation at 94 °C for 5 min, 30 cycles of 94 °C for 1 min, 54 °C for 45 s, and 72 °C for 1.5 min, followed by a final extension of 10 min at 72 °C. Confirmation of PCR product amplification was tested by running a 1% agarose gel electrophoresis. The PCR products for each strain were sent to Macrogen (Amsterdam, The Netherlands) for 16S rRNA gene Sanger sequencing. Sequences were quality checked using Geneious v4.8.5 (Biomatters ApS, Aarhus, Denmark) and blasted against the nucleotide sequences present in the Ribosomal Database Project (RDP Release 11).

2.3. BOX-Fingerprinting of Potential Hydrocarbon Degraders and nahAc PCR Amplification

BOX-fingerprinting was performed on the dark blue pure colonies from the indole test. The BOX_A1R primer (5′-CTACGGCAAGGCGACGCTGACG-3′), as described by Rademaker (1997), was used in the PCR reaction mixture described above. The reaction started with denaturation for 10 min at 95 °C, followed by 30 cycles of denaturation for 1 min at 95 °C, annealing for 1 min at 50 °C, and elongation for 8 min at 68 °C, with the final elongation for 8 min at 68 °C. PCR products were visualized on a 1.5% agarose gel stained with gel-red [19].

Naphthalene dioxygenase Fe-S large subunit (NahAc) genes were amplified with the primers nah-F (5′ CAAAARCACCTGATTYATGG 3′) and nah-R (5′ AYRCGRGSGACTTCTTTCAA 3′) using PCR conditions as described by Baldwin et al. 2003 [20]. PCR products of the correct size were purified and sent to Macrogen (Amsterdam, The Netherlands) for Sanger sequencing. Multiple sequence alignment was performed with Clustal Omega.

2.4. Genome Sequencing and Assembly

RNA-free DNA was extracted from stationary phase cells of strains *Pseudomonas* sp. VI4.1 and *Pseudomonas veronii* VI4.1 grown in 869 rich broth prior to digesting and ligating sequencing adaptors and barcodes using an Ion Xpress Plus Fragment Library Kit (Life Technologies Inc., Burlington, ON, Canada). Adaptor-ligated DNA was size-selected to 480 bp on a 2% E-Gel SizeSelect agarose gel, and Agencourt MAPure XP beads (Beckman Coulter, Mississauga, ON, Canada) were used for purification. The library dilution factor was determined using an Ion Library Quantitation Kit prior to amplification and enrichment with an Ion PGM Template Hi-Q OT2 kit on an Ion OneTouch 2 system. The enriched Ion Sphere Particles were quantified using an Ion Sphere Quality Control Kit prior to sequencing on a 316v2 chip with an Ion PGM Hi-Q View Sequencing Kit on an Ion Torrent PGM (Life Technologies Inc., Carlsbad, CA, USA).

Sequencing of strain VI4T1 generated a total of 1.1 million reads (mean length 300 bases, or 331 Mb of data (>307 M Q20 bases) in Torrent Suite 5.0.5. For strain VI4.1, 1.24 million reads (mean length 299 bases) generated 371 Mb data (>344 M Q20 bases). Reads were assembled using SPAdes 3.11.1 (uniform coverage mode; kmers 21, 33, 55, 77, 99, 127) into contigs greater than 1000 bp [21]. The assembly of strain VI4T1 resulted in 211 contigs with consensus length of 7,150,343 bp (>1000 bp) at 27x coverage (60.54% G+C content, largest contig of 237,158 bp; and N50 = 68,401 bp). Assembly of VI4.1 resulted in 185 contigs with a consensus length of 7,346,306 bp at 26.0X coverage (60.02% G+C content; largest contig of 285,953 bp; N50=73,645 bp). Plasmids were predicted using plasmidSPAdes, with kmers 21, 33, 55, 77, 99, 127 [22].

2.5. Genome Annotation and Phylogenetic Tree of the 16S rRNA Gene Sequences

Open Reading Frame (ORF) prediction and gene annotation were performed using the RAST annotation system (Overbeek et al. 2014) [23], Prokaryotic Genome Automatic Annotation Pipeline (PGAP) of the National Center for Biotechnology Information (NCBI) [24] and the platform MicroScope using the tool Magnifying Genomes, MaGe (http://www.genoscope.cns.fr) [25]. Clusters of Orthologous Genes (COGs) [26] and metabolic pathway reconstruction was carried out using the databases of Kyoto Encyclopedia of Genes and Genomes (KEGG) [27] and MetaCyc [28].

Comparative genome analyses were performed in SimpleSynteny, which uses NCBI BLAST, BioRuby and RMagick to map genes onto genomes and generate figures [29]. Default settings were used with BLAST E-value of 0.001 and 50% minimum query coverage.

The in silico DNA–DNA hybridisation (DDH) values between VI4.1 and VI4T1 and closely related strains was calculated using the online Genome-to-Genome distance calculator version 2.1 (GGDC) (http://ggdc.dsmz.de/ggdc_background.php#) [30,31]. Circular maps of the chromosomes and plasmids were generated in MaGe and in Bandage version 0.8.0 [32]. Clustered regularly interspaced short palindromic repeats (CRISPRs) were identified using the web service (http://crispr.u-psud.fr/Server/CRISPRfinder.php).

The bioinformatics prediction of genes associated with the biodegradation of naphthalene, BTEX (Benzene, Toluene, Ethylbenzene, Xylene), and aliphatic hydrocarbons was based on Basic Local Alignment Search Tool (BLAST) searches, Protein family (PFAM) queries and conserved domain searches, using the databases MetaCyc, UniProt database (http://www.uniprot.org/) and NCBI. Visualisations of the plasmid and operon structures was performed with SnapGene version 3.2.1.

A phylogenetic tree with VI4.1, VI4T1, and closely related strains evaluated against the Ribosomal Database was built in MEGA7 [33]. The evolutionary distances were computed using the Maximum Composite Likelihood method [34].

2.6. Biochemical and Chemotaxonomic Identification

The biochemical metabolite profile of VI4T1 and VI4.1 was determined using Gen-III MicroPlates (Biolog, Hayward, USA). Bacteria were grown in 869 rich medium overnight, subsequently resuspended in sterile 10 mM sodium phosphate buffer (pH 7.2) to an Optical Density (OD) of one checked at 600 nm. Next day, 100 μL of each strain was inoculated into each well of the Gen-III microplate and incubated at 30 °C for five days. Each microplate allows to carry out 94 phenotypic tests: 71 carbon source utilization assays and 23 chemical sensitivity assays. Each well contains a redox tetrazolium dye that changes colour as a result of cellular respiration providing a reliable metabolic fingerprint, which can be evaluated by measuring absorbance at 595 nm.

In addition, a GC Analysis of total fatty acid methyl esters (FAMEs) was performed at EMSL Analytical (Cinnaminson, NJ, USA) to identify the strains based on the FAME fingerprint and using the Sherlock Microbial Identification system (MIS).

2.7. In Vitro Estimation of Diesel Oil Degradation Capabilities

The bacterial strains were tested in triplicates for their capability to use diesel as the only carbon source using the 2,6-dichlorophenol indophenol (DCPIP) assay [35]. DCPIP is a compound with high affinity for electrons which when oxidized by metabolic reactions turns from blue to colourless. In brief, bacteria were pre-cultured in 5 mL of 869 rich medium at 30 °C and 160 rpm on an orbital shaker until OD_{660} nm = 1.0. Cells were pelleted by centrifugation (4000 g for 20 min), washed three times with 10 mM $MgSO_4$ and incubated overnight at 30 °C to allow the bacteria to use all remaining carbon source traces (starvation). Subsequently, 750 μL of Bushnell and Haas medium supplemented with 50 μL DCPIP solution (100 μg ml^{-1}) were added to a 2-mL sterile microcentrifuge tube. Subsequently, 80 μL of cell suspension and 5 μL filter sterilized diesel were added. Cells were cultivated in the dark to avoid photodegradation of the redox dye (30 °C; 120 rpm) for 1 week. The colour of the reaction medium was compared with three controls: two negative controls with respectively no diesel and no bacteria and a positive control with *Pseudomonas aeruginosa* WatG strain [17]. Strains were scored as positive for microbial hydrocarbon degradation capability if the solution looked clear, and negative if it persisted in blue.

2.8. Analysis of Naphthalene Degradation Using High-Resolution Proton Transfer Reaction Time-of-Flight Mass Spectrometry (PTR-TOF-MS)

The capabilities of the strains to use naphthalene as sole carbon source was assessed by measuring naphthalene degradation using PTR-TOF-MS. 20 mL sterile glass vials with 1.5 g of sterilised sand were inoculated with 200 μL of bacteria suspension (O.D.=1) in Bushnell Haas medium without carbon source. Vials were spiked with 0.3 μL of naphthalene against the inner glass wall to saturate the head-space and incubated for 1 week at 25 °C. A set of vials with the same conditions but no bacteria was used as a control. All experiments were carried out in triplicate.

For PTR-TOF-MS analyses, the headspace of the vials was sampled through the Teflon septum with a 50-mL glass syringe (Z314587, Fortuna Optima, Sigma-Aldrich) equipped with a three-way stopcock attached to the syringe Luer lock tip. The sample was injected into the air inlet of the PTR-TOF-MS instrument (PTR-TOF 8000, Ionicon, Innsbruck, Austria). The instrument drift tube was operated at a field density ratio (E/N) of ≈130 Td, resulting from 2.2 mbar pressure, 80 °C temperature, and 530 V of electric potential. In short, the air sample met a rich mixture of H_3O^+ that protonated the volatiles contained in the air sample. The reaction rate coefficient between the hydronium ion (H_3O^+) and naphthalene was assumed 2.45×10^{-9} cm^3 s^{-1} [36].

2.9. In Vitro Plant Growth Promotion Activity of the Bacterial Strains

Bacteria were screened in vitro for plant growth promoting traits including 1-aminocyclopropane-1-carboxylate (ACC)-deaminase, siderophore production, acetoin, organic acid, and indole-3-acetic acid (IAA) production, phytate mineralization and nitrogen fixation. The production of ACC-deaminase was estimated by monitoring the amount of α-ketobutyrate generated by the enzymatic hydrolysis of ACC [37]. Siderophore release was evaluated via Chrome Azurol S (CAS) assay [38]. The production of the phytohormone IAA was estimated by a colorimetric assay using Salkowski's reagent [39]. The production of the volatile plant growth promoting compound acetoin was assessed using the Voges-Proskauer assay [40]. The evaluation of the results was based on the observations of the colorimetric reactions after 5 days of incubation at 30 °C. Phytate mineralization was evaluated after 12 days of incubation by observing the halo produced around the colonies growing on solid medium supplemented with Na-phytate [41]. Organic acids produced by plant-associated bacteria facilitate the solubility of nutrients in soil, thus facilitating the uptake of nutrients by the plant. The method of Cunningham and Kuiack was used to check their production [42]. Bacteria able to fixate nitrogen possess the enzyme nitrogenase which catalyses the reduction of atmospheric nitrogen to ammonium that can be detected by a colour change on semisolid medium without any nitrogen source [43].

2.10. Pseudomona VI4.1 and VI4T1 Motility

Bacterial motility was tested in sterile glass tubes previously filled with 869 semi-solid agar medium. Inoculation was performed with a sterile straight wire, making a single stab down the centre of the tube to about half the depth of the medium. Subsequently, the tubes were incubated at 30 °C and observed for bacterial growth at regular intervals (1 until 7 days). Non-motile bacteria generally grow into the stab-line, have sharply defined margins and leave the surrounding medium clearly transparent while motile bacteria typically grow by diffusing throughout the medium, rendering it slightly opaque.

2.11. Sequence Database Accession Numbers

The genome sequencing projects have been deposited at DDBJ/EMBL/GenBank with strain VI4T1, Accession Number MULN00000000 and BioProject PRJNA369437, and for strain VI4.1, Accession Number NZ_MULM00000000 and BioProject PRJNA224116. The Genbank accession numbers for the sequenced naphthalene dioxygenase genes are MN030640-MN030657 with associated partial 16S rRNA gene sequences of the isolates: MN006582-MN006610. The other bacterial isolates with 16S rRNA gene sequence identification were also deposited in NCBI Genbank with accession numbers: MN030267-MN030332.

3. Results and Discussion

3.1. Genotypic and Functional Characterization of the Cultivable Hydrocarbon-Degrading Bacterial Community from the Oilfield of Bóbrka

In search for novel hydrocarbon and plant growth promoting strains, we performed an isolation and selective enrichment approach to identify and characterize novel microorganisms capable of degrading oil-related pollutants. Seventy bacterial strains were isolated by using the sandwich diffusion system which had incubated for one month in polluted soil from Bóbrka with crude oil (Figure 2A). 16S rRNA Sanger sequencing showed a dominance of *Pseudomonas* spp. (35%) followed by *Achromobacter* (18%), *Mycobacterium* (12%), *Caulobacter* (6%), *Burkholderia* (4%), *Enterobacter* (3%), *Ralstonia* (3%), and *Stenotrophomonas* spp. (3%) (Figure 2A). A high percentage (60%) of the strains were able to use diesel as sole carbon source. A majority of the strains produced ACC-deaminase (53%). ACC-deaminase is an important trait of many soil and plant-associated bacteria as they can facilitate growth and development of the plant by lowering stress ethylene levels [44]. 33% of the strains produced the hormone indole-3-acetic acid. Strains producing IAA can have a phyto-stimulation effect on the host plant or pathogenic, depending on the concentration [45]. Bacteria use this hormone to interact with the host plants as part of their colonization strategy and to circumvent the basal plant defence mechanisms. It can also serve as signalling molecule between bacteria. Siderophore production and organic acid production was present in 18% and 36% of the isolated strains, respectively (Figure 2B). This can aid in the solubilization of iron and phosphorous in soil, making it more available to microbial and plant uptake. Only 6% of the strains were scored as positive for acetoin production. This ketone showed to enhance plant growth by stimulating root development and increasing resistance against pathogens and drought stress. An overview of the traits per species is shown in Figure 2C. The PGP and diesel use as sole carbon source traits were diversely spread across the different taxa. Strains with multiple PGP traits are most interesting for in vivo plant growth promotion tests, such as *Pseudomonas veronii* YOB01, *Burkholderia cepacia* YOAG01, *Mycobacterium* YOBG04, and *Achromobacter* YOBB03, amongst others.

3.2. BOX-Fingerprinting

BOX-fingerprinting was performed on the isolates that scored positive on the indole/indigo test and thus possess dioxygenases that can have activity against naphthalene and toluene. Based on the differences in the fingerprint patterns and partial 16S rRNA gene identification, eight groups could be discriminated. Subsequently, those strains which gave a positive amplification band and blast hit to an aromatic compound dioxygenase were down-selected (in squares). A multiple sequence alignment of the Rieske subunit of the naphthalene dioxygenases showed a high similarity with the naphthalene dioxygenase large subunit (*nahAc*) of *Pseudomonas putida*, encoded on plasmid pAK5, and verified experimentally to use naphthalene (Figure S2). Based on this information, we chose to sequence two representative strains, one of the *Pseudomonas* sp. group (VI4.1) and *Pseudomonas veronii* group (VI4T1), Figure 3.

3.3. General Features of the Draft Genomes of Strain VI4T1 and VI4.1

The 7,150,343-bp genome of strain VI4T1 (GC 60.54%) contains 7962 genes, of which 7809 are Coding Sequences (CDSs), 9 are rRNAs (7, 1, 1 for 5S, 16S, 23S), and 60 are tRNAs comprising 20 different tRNA types. The 7,346,306-bp genome of *Pseudomonas* sp. VI4.1 (GC 60.02%) contains 8082 genes, of which 7922 are CDS, 9 are rRNAs (7, 1, 1 for 5S, 16S, and 23S, respectively), and 58 are tRNAs composed of 21 different tRNA types. The MaGe server classified 73.65% and 74.38% of the CDSs in at least one Cluster of Orthologous Group (COG) for VI4T1 and VI4.1, respectively. In addition, 83.90% and 82.87% of the CDSs are classified in at least one Evolutionary genealogy of genes: Non-supervised Orthologous Groups (EGGNOG) for VI4T1 and VI4.1 (Figure S3). The most abundant EGGNOG categories for VI4T1 are, amino acid and transport, transcription, inorganic ion transport and metabolism, energy production, while for VI4.1, the most dominant categories were secondary metabolite biosynthesis, inorganic ion transport, lipid transport and metabolism, and coenzyme transport and metabolism (Figure S3). In strain VI4.1, 21 CRISPR sequences and nine transposases were found, while in VI4T1, seven CRISPRs, two Cas (CRISPR-associated), and 10 transposases were present. CRISPR, clustered regularly interspaced short palindromic repeats, along with Cas are a bacterial defence system against bacteriophage predation. This all indicates that both strains have experienced extensive and complex genetic alterations and acquired new genetic elements by bacteriophages and exchanges by lateral gene transfer.

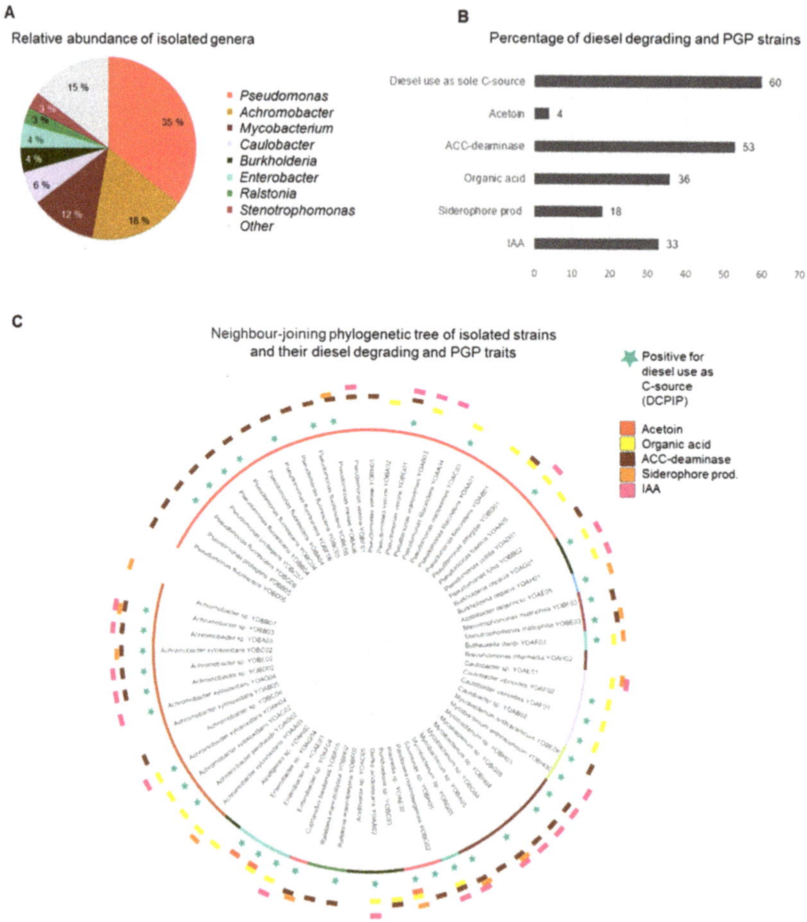

Figure 2. (**A**) Relative abundance of isolated oil tolerant bacteria using the diffusion sandwich system. (**B**) Percentage of strains displaying plant-growth promotion (PGP) properties, specifically Acetoin, Organic acid, 1-aminocyclopropane-1-carboxylate (ACC)-deaminase, siderophore and indole-3-acetic acid (IAA) productions. (**C**) neighbour-joining phylogenetic tree of all isolated taxa and their PGP and diesel-degrading properties. Coloured blocks and coloured stars indicate that the strain scored positive for the tests, non-coloured (empty) blocks and stars indicate the strain should not produce/utilize the compound evaluated by the corresponding assay.

The RAST Server indicated that the most closely related strains of VI4.1 were *Pseudomonas fluorescens* Pf0-1 (205922.3) and *Pseudomonas* sp. GM18, while for VI4T1 these were *Pseudomonas fluorescens* SBW25 and *Pseudomonas extremaustralis* 14-3. Phylogenetic analyses based on the full-length 16S rRNA gene sequences shows the most closely related strains based on the Ribosomal database (Figure 4). *Pseudomonas* sp. VI4.1 clusters with *Pseudomonas fluorescens* (AY538264), while VI4T1 is closely related to the *Pseudomonas veronii* type strain AF064460. The DDH between VI4.1 and its closest genome sequenced relative, *Pseudomonas silesiensis* sp. nov. strain A3T (CP014870), shows a distance of 61.90% (probability DDH >70%, 49%), suggesting that VI4.1 is a new species. The DDH estimate between VI4T1 and *Pseudomonas veronii* 1YdBTEX2 (LT599583) was 87.9%, with a probability of DDH >70% of 95.03%, indicating that VI4T1 belongs to the *Pseudomonas veronii* group.

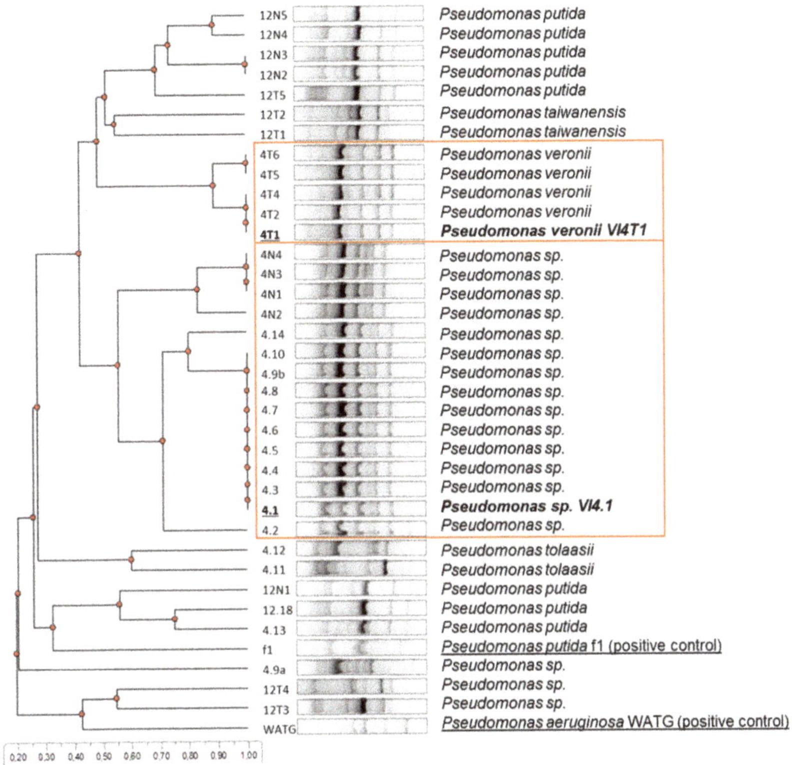

Figure 3. Cluster analyses of BOX-fingerprint PCR products of indole-positive isolates. Strains VI4.1 and VI4T1 were chosen as representative isolates with positive naphthalene dioxygenase large subunit (*nahAc*) gene PCR amplification. *Pseudomonas putida* f1 and *Pseudomonas aeruginosa* WatG were included as reference strains with naphthalene dioxygenase activity.

To gain insights into the genome organisation, the genome assemblies were visualised in Bandage (Figure S4A and Figure S4b). Because of the draft genome completeness, it is still difficult to determine whether one or two chromosomes are present, though both strains contain one plasmid: VI4T1 has a plasmid of 40,201 bp and VI4.1 one of 81,970 bp. Comparative analyses show matches of VI4T1 plasmid with a conjugative plasmid in *Pseudomonas putida* pBI709 and in *Pseudomonas fragi* strain NMC25. The plasmid of VI4.1 shows high similarity with *Pseudomonas fluorescens* strain PC20 plasmid pNAH20, *Pseudomonas putida* NCIB 9816-4, *Pseudomonas frederiksbergensis* AS1, and *Pseudomonas putida* ND6 plasmid pND6-1, all encoding a naphthalene degradation operon (Figure S4).

The 40 kb plasmid of VI4T1 (pVI4T1-1) contains 48 CDSs, while the 81 kb plasmid of VI4.1 (pVI4.1-1) encodes 103 CDSs (Figure S4). pVI4.1-1 has the full aerobic naphthalene degradation operon (*nah1: nahAaAbAcAdBFCED*) which converts naphthalene to salicylaldehyde and salicylate (Figure 5). The enzymes involved are naphthalene 1,2-dioxygenase (*nahAa-d*), which hydroxylates the two-ring compound, and then *cis*-1;2-dihydro-1,2-dihydroxynaphthalene-1,2-dehydrogenase (*nahB*) catalyses the transformation into naphthalene-1,2-diol, with oxygenation by 1,2-dihydroxynaphthalene dioxygenase (*nahC*). Decisive are the actions of the isomerase (*nahD*) and hydratase-aldolase (*nahE*) to obtain salicylaldehyde further dehydrogenated by the enzyme salicylaldehyde (*nahF*). At this point the molecule is metabolized via the catechol cleavage pathway encoded by the lower naphthalene operon (*nah2: nahGTHINLOMKJ*). Salicylate monooxygenase (*nahG*) oxidises salicylate to catechol, and then further transformations are catalysed by the enzymes: ferredoxin (*nahT*), catechol-2, 3-dioxygenase (*nahH*), 2-hydroxymuconic semialdehyde dehydrogenase (*nahI*), hydrolase (*nahN*), hydratase (*nahL*), acetaldehyde dehydrogenase (*nahO*), aldolase (*nahM*), oxalocrotonate decarboxylase (*nahK*), and oxalocrotonate tautomerase (*nahJ*). Furthermore, the typical elements of a conjugative plasmid are found such as the transfer genes (*traA-D*), replication site (*repA*), and a complete type IV secretion system (*mpfABCDEFGHIJ*). These latter components form the secretion machinery including the membrane proteins (VirB6, VirB8, VirB10, VirB9 and VirB7), with three ATPases that form the power unit (VirD4, VirB11, VirB4), and VirB11 and VirB4 are also required for the biogenesis of the T4S pilus. T4SS can mediate in this case the conjugative transfer of the naphthalene operon to other bacteria. Conjugation is an important strategy employed by bacteria to promote bacterial genome plasticity, and to spread a variety of functions, such as the degradation of anthropogenic toxic compounds or the detoxification of heavy metals, but also bacteriocin and toxin production to ward off predators. Having part of the degradation pathways encoded on a plasmid is an interesting property, as the strain can transfer its degradative plasmid via horizontal gene transfer to native soil or plant-associated bacteria, a strategy that has been utilised by our group to clean-up groundwater polluted with toluene and trichloroethylene.

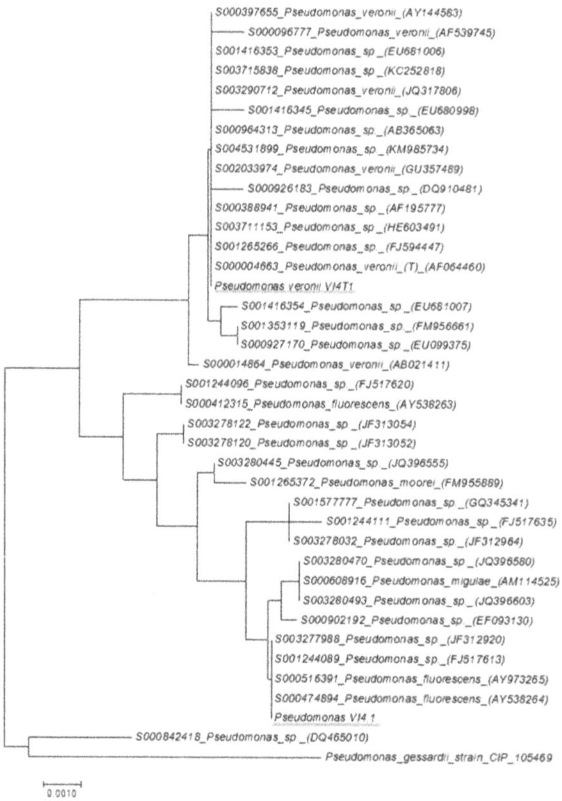

Figure 4. 16S rDNA phylogenetic tree of *Pseudomonas veronii* VI4T1 and *Pseudomonas* sp. VI4.1 compared to closely-related strains and with *Pseudomonas gessardii* as an outgroup. Closely-related strains were evaluated against the Ribosomal Database and the phylogenetic tree was built in MEGA7. The evolutionary distances were computed using the Maximum Composite Likelihood method and are in the units of the number of base substitutions per site.

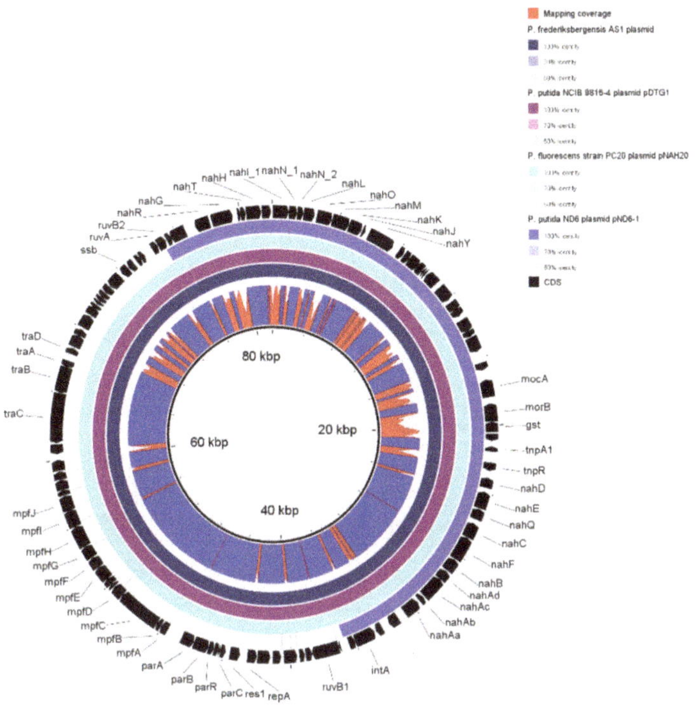

Figure 5. pVI4.1-1 80 kb plasmid encoding a full naphthalene operon, *nahAaAbAcAdBFCED* to convert naphthalene to salicylate, and the lower *nahGTHINLOMKJY* operon to convert salicylate via catechol to tricarboxylic acid (TCA) cycle intermediates.

The 40-kb pVI4T1 plasmid does not contain a naphthalene nor aromatic compound degradation operon, though it does contain partial T4SS proteins (MpfA-E), and ParA and B which are bacterial proteins involved in plasmid replication and partitioning, which suggests this was a functional conjugative plasmid but it might have lost some essential genes (Figure S5). For the many CDS annotated as "hypothetical proteins", we ran a conserved domain search and this revealed the presence of: amidase expression regulating protein, glutathione permease protein (GsiD), histidine/lysine/arginine/ornithine ABC transporter, transposases, cation efflux system (CzcD), transporters (MntH), and components involved in secretion and vesicular transport. It is a subject of speculation as to whether this plasmid is a rudimentary copy of the pVI4.1-1, or a part was not assembled; additional long-read sequencing can confirm this.

3.4. Physiological and Biochemical Properties

Strains VI4T1 and VI4.1 are aerobic, heterotrophic, motile, Gram-negative, and non-sporulating strains. The FAME profile of *P. veronii* VI4T1 consisted of 50.39% 16:1 7c/16:1 w6c, 0.15% 18:0 ante/18:2 w6,9c, 0.11% 19:1 w7c/19:1 w6c, 25.51% 18:1 w7c, 0.54% 14:0, 22.6% 16:0, 0.54% 18:0, and 0.15% 18:1 w7c 11-methyl. For VI4.1, the FAME profile was 42.73% 16:1 w7c/16:1 w6c, 0.61% 18:0 ante/18:2 w6,9c, 3.92% 18:1 w7c, 0.55% 10:0 3OH, 4.56% 14:0, 45.39% 16:0, 1.53% 18:1 w9c, and 0.72% 18:0.

The morphology of stain VI4.1 is smooth, creamy and round, while VI4T1 grows as slightly yellow, smooth and irregular colonies on 869 rich medium. The optimal growth temperature was 30 °C, but the strains showed also good growth at 4 °C. The strains grew preferably on nutrient-rich agar, but they proliferate as well in mineral BH medium and supplemented with diesel oil as the sole carbon source. The strains could grow in 869 medium with NaCl concentrations in the range of 1–4% (w/v) showing salt stress tolerance, and they survived in medium with slight acidic pH, at 5–6.

The ability to capitalize on a variety of carbon sources is an important feature for soil and plant-associated microorganisms. Strain VI4T1 and VI4.1 respired numerous carbon sources as reported in Supplementary Table S1.

Chemical sensitivity of the strains was also tested with the GEN-III array, and this showed tolerance of VI4T1 to grow at pH 5-6, in the presence of 1% NaCl, 1% sodium lactate, fusidic acid, D-serine, troleandomycin, rifamycin SV, minocycline, lincomycin, niaproof 4, vancomycin, tetrazolium violet, tetrazolium blue, nalidixic acid, potassium tellurite, aztreonam, sodium butyrate, and sodium bromate. Strain VI4.1 was also capable of growing at pH 5-6, up to 4% NaCl, in the presence of 1% sodium lactate, fusidic acid, D-serine, troleandomycin, rifamycin SV, minocycline, lincomycin, guanidine HCl, niaproof 4, vancomycin, tetrazolium violet, tetrazolium blue, nalidixic acid, lithium chloride, potassium tellurite, aztreonam, sodium butyrate, and sodium bromate.

3.5. Degradation of Hydrocarbons

The genome of VI4T1 encodes 45 dioxygenases and 20 monooxygenases, VI4.1 has 50 di- and 27 mono-oxygenases. The dioxygenases comprised naphthalene, biphenyl, phenylpropionate, nitropropane, halobenzoate, and catechol dioxygenase, while for monooxygenases, alkanesulfonate, cyclohexane, alkanal, nitrilotriacetate, and l-ornithine were present. The high number and diversity of dioxygenases and monooxygenases suggests that the strains have versatile metabolic capabilities and hydrocarbon degradation potential. Other strains such *Paraburkholderia aromaticivorans* BN5, recently isolated from an oil polluted site in South Korea, have shown to harbour multiple aromatic ring hydroxylating enzymes, facilitating the use of naphthalene and BTEX as sole nutrient source [46]. For efficient phytoremediation activities, it is important to understand the underlying genetics structures of the many degradation pathways to better assess their activity and use in the field. In the following paragraphs we therefore focus on the bioinformatic prediction of degradation pathways of naphthalene, BTEX, and aliphatic hydrocarbons, and their experimental validation, followed by PGP prediction, motility, and emergent contaminant degradation.

Previous studies have shown that aerobic naphthalene degradation in *Pseudomonas* spp. can occur via (1*R*,2*S*)-1,2-dihydronaphthalene-1,2-diol, salicylaldehyde, and salicylate, to be converted to catechol and further degraded to tricarboxylic acid (TCA) cycle intermediates. Bioinformatics analyses showed at least two copies of the upper naphthalene operon *nahAaAbAcAdBFCED* for *P. veronii* VI4T1, similar to *P. veronii* YdBTEX2, and one copy for strain VI4.1, encoded on the pVI4.1-1 plasmid (Figure 6). This confirms that strain VI4.1 has obtained the degradation capabilities of naphthalene through plasmid transfer. Also for VI4T1 one copy of the *nah1* operon seems to be plasmid encoded, since plasmid stabilization proteins and transposases are found flanking the operon. For the downstream *nah2* operon (*nahGTHINLOMKJ*) one complete cluster is present in each strain, based salicylate hydroxylase (*nahG*, in green) and 2.3-catechol *nahH* (in blue) as the key enzymes (Figure 7). Organization of this operon between the strains differs considerably, and many of the enzymes are present in multiple copies, such as 2-hydroxymuconic semialdehyde dehydrogenase (*nahI*). The upper and lower pathways are controlled by a LysR family transcriptional regulator (Figures 6 and 7).

Aerobic toluene degradation can occur via five different pathways as described in MetaCyc. Strain VI4.1 harbours the *tomAo12345tomB* operon to degrade toluene via o-cresol (Figure S6). A comparative analysis shows the similarity of the operon structure with strain *Burkholderia cepacia*. Strain VI4.1 has a partial *tbuA1A2BCUVE* monooxygenase operon (missing *tbuU* and *tbuV*) and also a partial *tomABCDE* operon, lacking a homologue for *tomB* (BLAST E-value <0.001, >50% query coverage). Strain VI4T1 does not have homologues to any of the *tom* operons, but seems to use route IV and V. Strain VI4T1 harbours the complete toluene monooxygenase operon *xylMBCXYZL* in two copies to convert toluene and xylene to catechol, and downstream *meta*-cleavage of catechol via *xylEGHFIJK* to actyl-coA which enters the TCA cycle. The operon structures are very similar to the *xyl* operon of the reference strain *Pseudomonas veronii* 1YDBTEX2. The lower catechol ortho-cleavage pathway contains the genes (Figure S7B) catechol 2,3-dioxygenase (*xylE*), 2-hydroxymuconic semialdehyde dehydrogenase (*xylG*), 4-oxalocrotonate isomerase (*xylH*), 4-oxalocrotonate decarboxylase (*xylI*), 2-oxopent-4-enoate hydratase (*xylJ*), 2-oxo-4-hydroxypentanoate aldolase (*xylK*), and an acetaldehyde dehydrogenase (*todI*). The toluene degradation V route (*todC1C2ABCDEF*) to catalyse toluene degradation via toluene-cis-diol is also present in both strains (Figure S8).

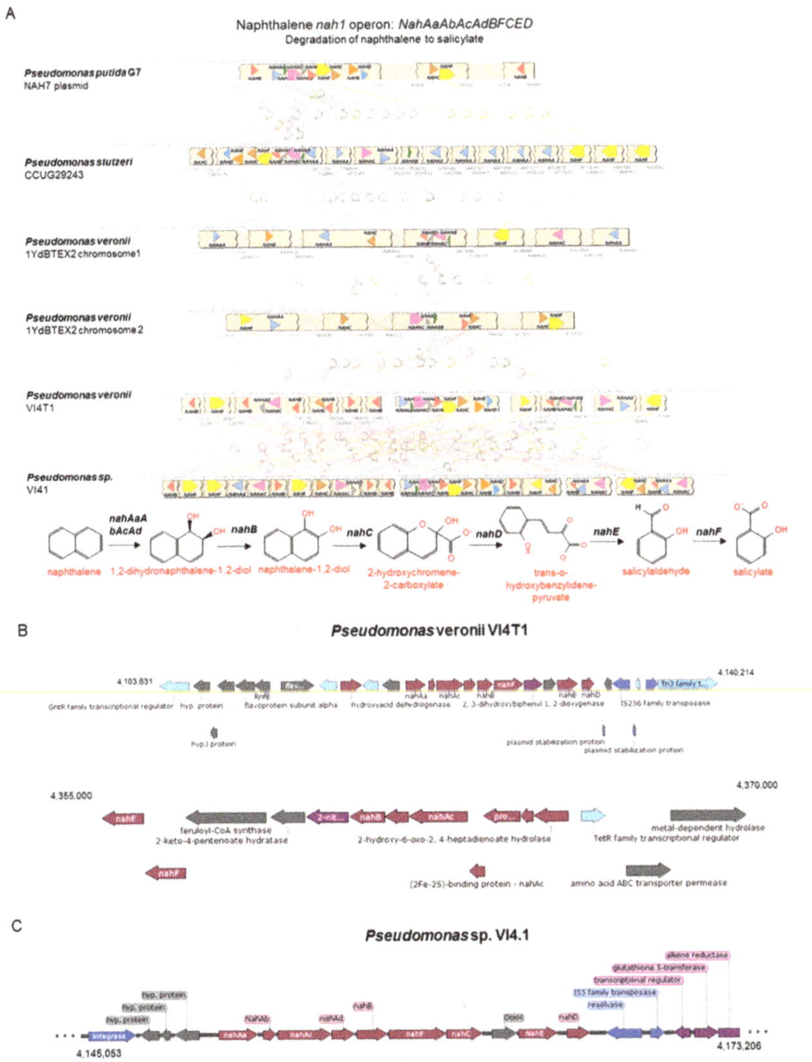

Figure 6. Comparative analyses of the upper naphthalene operon, physical map of the degradation genes and proposed biochemical pathways for naphthalene degradation for strains VI4T1 and VI4.1.

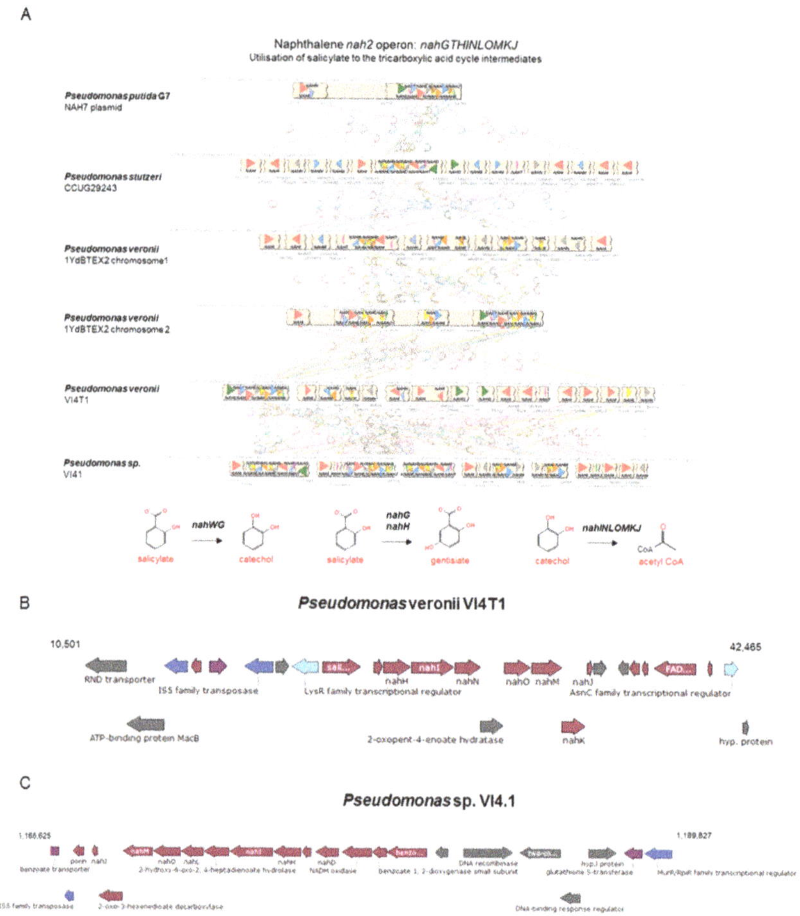

Figure 7. Comparative analyses of the lower naphthalene operon for strains VI4T1 and VI4.1, physical map of the degradation genes, and proposed biochemical pathways for salicylate degradation to catechol or to gentisiate and further breakdown to TCA cycle intermediates.

Strain VI4T1 encodes also the genes to degrade xylene (*xylBCM*), VI4.1 is lacking the *xylM* homologue (e-value <0.001, coverage of 50%) (Figure S7A). Both strains carry the genes to degrade benzene (*bedC1C2AB*) (Figure S9). Strain VI4T1 has additionally an anthranilate degradation operon *antAaAbAc*, downstream of a tryptophan 2,3-dioxygenase gene, as anthranilate is an important intermediate of tryptophan metabolism [47].

Furthermore, genes encoding for enzymes involved in the linear hydrocarbon degradation were investigated. The complete *alkB* operon was present for VI4T1 (Figure 8) but VI4.1 lacks this alkB operon. The 1-alkane hydrolase subunit (*alkBGT*) introduces molecular oxygen in the terminal carbon atom of the hydrocarbons at the expense of NADH to yield primary alcohols, and these are further catabolized by an octanol dehydrogenase (*alkJ*), an aldehyde dehydrogenase (*alkH*) and a medium-chain acyl-CoA synthetase (*alkK*). Finally, the octanoyl-CoA enters the beta-oxidation-cycle and can be utilized as a carbon and energy source (Figure 8). Instead of the complete *AlkB* operon, VI4.1 encodes a homologue of the two-domain *AlkB* system with rubredoxin and rubredoxin-NAD(+) reductase. Both are essential electron transfer components needed for alkane hydroxylation by AlkB. In fact, AlkB-type alkane hydroxylases fused to rubredoxin protein have been shown to hydroxylate n-alkanes with chain lengths up to C32. VI4.1 and VI4T1 also have a homolog of the cytochrome P450 CYP153 family, this is another type of alkane hydroxylase for the degradation of short- and medium-chain-length *n*-alkanes. Other enzymes important in aliphatic hydrocarbon degradation are the alkanesulfonate genes, of which 11 are present in VI4T1 and 15 in VI4.1.

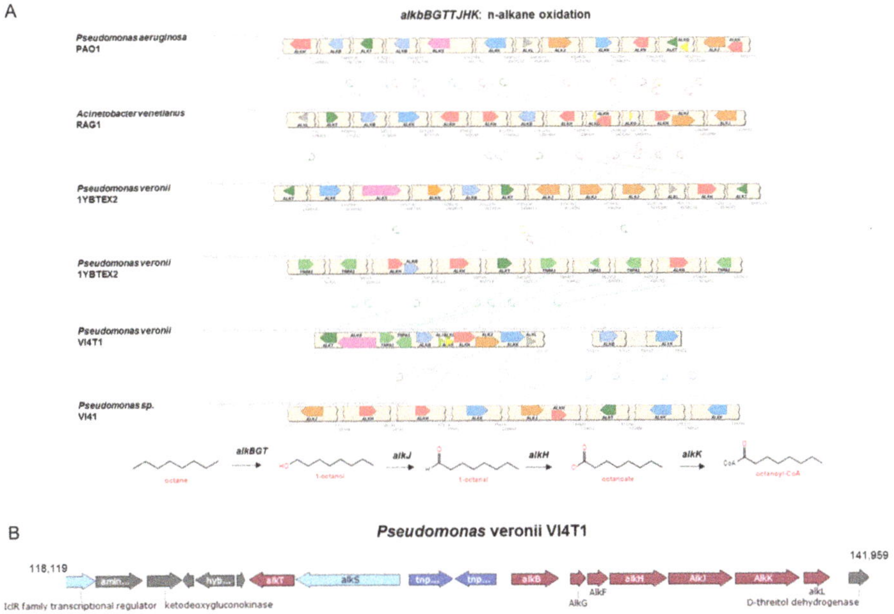

Figure 8. Comparative analyses of the pathways involving alkane-monooxygenase in different *Pseudomonas* strains included *Pseudomonas veronii* VI4T1.

To confirm the degradation potential, experimental evidence of aromatic hydrocarbon compound degradation was performed with PTR-TOF-MS. VI4T1 and VI4.1 were both capable of respiring 15 µg/L naphthalene and 3 µg/L each of benzene, toluene, and xylene as sole carbon source within seven days (Figure 9). Metabolites such as salicylic acid or catechol were not identified with PTR-TOF-MS in the headspace.

3.6. Motility and Chemotaxis

Motility of a strain in its environment influences its survival and competence to colonize to find nutrient sources and colonize plant root surfaces [48]. Additionally, motility allows bacteria to move away from high concentrations of toxic compounds. Bacterial motility was experimentally tested in sterile glass tubes previously filled with 869 semi-solid agar medium. Bacteria were inoculated with a straight wire to about half the depth of the medium. Tubes were incubated at 30 °C and observed for bacterial growth at regular intervals (one until seven days). Non-motile bacteria generally grow into the stab-line, while motile bacteria, such as strains VI4T1 and VI4.1, typically grow by differently diffusing through the medium (Figure S10).

Genomic analyses confirmed the presence of genes coding for the flagellar motor complex: *motA*, *motB*, *flhA*, *flhB*, *fliH*, *fliI*, *fliJ*, *fliO*, *flipP*, *fliQ*, *fliR*, *flgB*, *flgC*, *flgG*, *flgH*, *flgI*, *fliE*, *fliF*, *fligA*, *flgD*, *flgN*, *fliK*, *fliS*, *flgE*, *flgK*, *flgL*, *fliC*, *fliD*, *fliG*, *fliM*, *fliN*, *flgJ*, *flhF*, and *fliL*. Furthermore, chemotaxis proteins CheA and CheY are coded. They are involved in the transmission of sensory signals from the chemoreceptors to the flagellar motors.

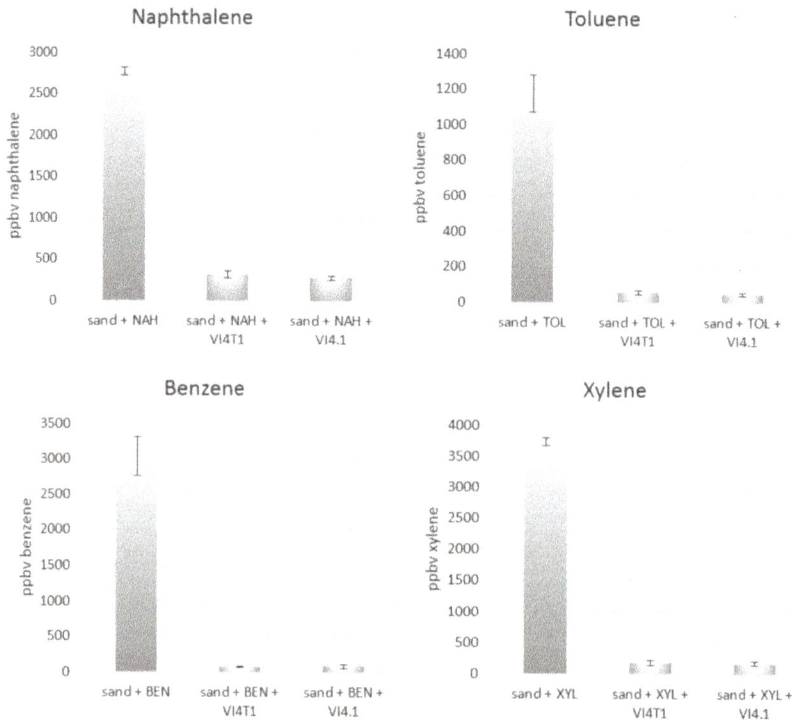

Figure 9. Concentration of naphthalene in the gas phase remaining after seven days of incubation in vials inoculated with *Pseudomonas veronii* VI4T1 and *Pseudomonas* sp. VI4.1. Negative controls are vials containing sterile sand. All vials were spiked in the headspace with 15 μg/L naphthalene or with 3.5 μL/L benzene, toluene, or xylene. 200 μL of bacteria suspension (OD_{600} = 1) were added to the all the vials except for the negative controls. Error bars represent standard error.

3.7. Plant Growth Promotion Potential

In vitro tests showed that VI4T1 and VI4.1 scored positive for indole-3-acetate (IAA) production, siderophore production, acetoin, organic acid and phytate mineralization, but negative for ACC-deaminase and nitrogen fixation.

In both genomes, indole-3-acetamide hydrolase was identified converting indole-3-acetamide to IAA. This pathway is operative in several genera of plant-associated bacteria amongst which many *Pseudomonas* spp. [49]. Additionally, genes of the tryptophan biosynthesis gene cluster *trp* were present: *trpD, trpG, trpE, trpC, trpF, trpD, trpB, trpA*. Strain VI4.1 has also the ability to produce IAA via an additional nitrilase-driven pathway converting indole-3-acetonitrile in IAA. The ability to produce siderophores was confirmed by the presence of the pyoverdin pathway genes, 48 genes for 4T1 and 29 for VI4.1. We found also the fundamental genes for the enzymes converting pyruvate in (R)-acetoin, with acetolactate synthase (*bud*B) and diacetyl reductase (*bud*A). Strain VI4.1 also carries the two genes responsible for the (S)-acetoin biosynthesis (*bud*C). The absent production of some organic acids such as citric, lactic, succinic, gluconic, itaconic, acetic, propionic, tartaric, malonic, malic, oxalic, fumaric, and lactobionic acid was confirmed after the genome annotation.

Growth and development of both, plants and bacteria require macro- and micro-nutrients, some of which are available only from external sources like the soil in which they grow. Phosphorous is such an essential macronutrient, and is usually absorbed and utilized by the plants in the form of phosphate. The role of genes as Purple Acid Phosphatases (PAPs) in strain VI4T1 has not yet been fully clarified. Experimental evidence suggests that the enzyme is induced by a lack of phosphate and excreted from bacterial cells, suggesting that it may be involved in phosphate acquisition. The presence of the gene coding for the acid phosphatase confirmed the in vitro potential of strain VI4T1 to convert a phosphate monoester in more bioavailable phosphate in the presence of water.

No enzymes were detected involved in the biological fixation of atmospheric nitrogen such as *nif* genes or nodulation genes (*nod*), confirming the negative response on the nitrogen fixation test. We did not detect any homologue for 1-aminocyclopropane-1-carboxylate deaminase; that outcome confirmed the results of the corresponding in vitro test.

The aminopeptidase gene (*pepA*) that is involved in seed germination process was detected in both genomes [50]. The enzyme hydrolyses peptide bonds in tissues as the aleurone layer of the endosperm, the scutellum and the growing tissues of the seedling. The activity of these proteins is reported for several plant species.

Furthermore, both strains can synthesize jasmonic acid a member of the jasmonate class of plant hormones which is involved in the regulation of quite a number of processes as embryo and generative organs development, ageing, sex determination, seed germination, root growth, adaptation to stress factors [51].

3.8. Tolerance to Abiotic Stress

Plants are often exposed to abiotic stresses such as heat, drought, metal pollution, high salinity and acidic pH. In such circumstances, inoculating plants with stress alleviating microorganisms may offer a biological alternative to the existing agrochemicals in agriculture [52]. Both strains carry the *speA* gene involved in the survival of the bacterium in acidic conditions via the arginine-dependent pathway, conferring potential acid resistance by exchanging external arginine for internal agmatine [53]. This effectively consumes protons within the cytoplasm, raising the pH.

Abiotic stress can create osmotic deficiencies in plant cells. In this context, the presence of trehalose can act as an osmoprotectant, and strains VI4T1 and VI4.1 possess several genes coding for proteins involved in trehalose anabolism, including pathway IV (trehalose synthase converting b-maltose in trehalose), and pathways VI and VII (trehalose synthase producing trehalose from ADP-alpha-D-glucose and glucose) [54]. Trehalose accumulation may act as a biosurfactant and in this way enhancing the biodegradation of hexachlorocyclohexane.

Interestingly, both strains carry genes for salt tolerance including *betA* choline dehydrogenase, alcohol dehydrogenase (*yiaY*, *adhA*), and betaine aldehyde dehydrogenase (*betB*). Glycine betaine (N,N,N-trimethylglycine) is a very efficient osmolyte found in a wide range of bacteria and plants, where it is accumulated at high cytoplasmic concentrations in response to osmotic stress, to act as an osmoprotectant [55].

All organisms living in an aerobic environment are exposed to reactive oxygen species (ROS). Strains VI4T1 and VI4.1 carry genes coding for superoxide dismutases (*sodB*, *sodM*) and catalases (*katE*, *katB*) converting superoxide molecules into oxygen via hydrogen peroxide formation. Furthermore, the genomes encode for glutathione reductase (*gor*) converting two molecules of glutathione in glutathione disulfide with a co-production of reduced glutaredoxin. Oxidative damage to proteins often results in the formation of mixed disulfides within the polypeptides. A primary defence against this damage is mediated by the action of GSH-dependent thiol-disulfide oxidoreductases, also called thioltransferases and best known as glutaredoxins (Grx). These proteins reduce the protein disulfide groups back to their native form.

Heavy metal transporters are involved in acquisition, metal absorption and detoxification. Both strains possess the genes coding for a cadmium-transporting ATPase and a copper-transporting ATPase (*copA*). Copper is an important element that participates in a high number of enzymatic reactions; in photorespiration, electron transport, in ethylene signalling and many other metabolic processes that have copper-containing enzymes catalysing various reactions.

Furthermore, an arsenic resistance system was detected: arsenical resistance protein ArsH; HTH ArsR-type DNA-binding domain; arsenical-resistance protein Acr3/Arsenical pump membrane protein/ArsBArsenate reductase. In addition, several cation efflux system proteins are present: cation efflux system protein (CusF), cation efflux system protein (CusA), cation efflux system protein (CusC), Cation efflux system protein (CusB).

3.9. Tolerance to Micro-Pollutants and Emerging Contaminants

During the last decades, there has been an increasing concern about the presence of micro-pollutants including pharmaceutical, human health care products, medicines, endocrine disruptors, fluorinated chemicals, and microplastics that are found in the soil and waters in increasing concentrations [56]. There are 2700 such compounds listed by the Joint Research Centre of the European Commission.

Although the novelty of these investigations translates to a lack of information in prokaryotic databases, the genome sequencing of our two *Pseudomonas* spp. allows us to investigate the presence of genes involved in the degradation of endocrine disruptors. We identified in *Pseudomonas* strain VI4.1 the genes androsterone 3-deydrogenase and testosterone dehydrogenase, respectively involved in the transformation of androsterone and testosterone in the common intermediate metabolite androst-4-ene-3,17-dione [57]. This compound links this catabolic pathway to the one of the androstenedione degradations until mineralization via the propanoyl CoA degradation pathway I or TCA cycle. *Comamonas testosterone* was the first strain characterized for its capability to degrade steroids, and its capacity was compared with *Pseudomonas* spp. after incubation in testosterone sewage in the work of Chen et al. (2016) [57].

Concerning fluorinated chemicals, strain VI4.1 carries the gene coding for the enzyme haloacetate dehalogenase, the only known enzyme that can specifically hydrolyse the carbon-fluorine bond. The enzyme was first described from a *Pseudomonas* spp. in 1965 and since then has been described in multiple bacterial strains. Fluorinated organic compounds have widespread applications as pesticides, herbicides, pharmaceuticals, flame-retardants, refrigerants and foam-blowing agents, and are consequently accumulating in the environment. Only a handful of carbon-fluorine bonds from biological origin are known, such as fluoroacetate, a toxin found in the leaves and seeds of a variety of tropical plants, often in high concentrations. The extreme toxicity of fluoroacetate stems from its similarity to acetate. Fluoroacetate combines with coenzyme A to form fluoroacetyl-CoA, which can substitute for acetyl CoA in the tricarboxylic acid cycle. Fluoroacetyl-CoA reacts with citrate synthase to produce 2-fluorocitrate, a metabolite of which then binds very tightly to aconitase, stopping the cycle.

Interestingly, strain VI4.1 possesses the genes terephthalate 1,2-dioxygenase (*tphA*) and terephthalate dihydrodiol dhydrogenase (*tphA*) converting terephthalate to 3,4 dihydroxybenzoate. Terephthalate is the major precursor for polyester fibres and coatings. Fibres from clothing constitute a major problem in our waterways, also polluting drinking water [58]. Strains able to degrade these fibres are an important resource for the plastics recycling/upcycling industry. In relation to this, *Pseudomonas* VI4T1 carries genes (*bph*A) coding for the enzyme biphenyl dioxygenase subunit A. This enzyme is involved in the first step that synthesizes 2-hydroxy-2,4-pentadienoate and benzoate from biphenyl. Biphenyl is an aromatic hydrocarbon and an important precursor for the production of polychlorinated biphenyls (PCBs).

4. Conclusions

Genome analysis of *Pseudomonas veronii* strain VI4T1 and *Pseudomonas* sp. VI4.1 not only revealed the presence of key genes involved in the catabolism of aromatic and aliphatic hydrocarbons but also a diverse set of genes involved in plant growth promotion, stress regulation, and adaptation. Experimental evidence for both hydrocarbon degradation and plant growth promotion confirmed the degradative and versatile metabolic properties. These features make both sequenced strains promising candidates for testing and designing different plant–bacteria systems to perform phytoremediation experiments, and contribute in this way to the development of a cost-effective and eco-friendly method to remediate polluted environments.

Supplementary Materials: The following are available online at http://www.mdpi.com/2073-4425/10/6/443/s1, Figure S1: Side and top view of the custom-made sandwich diffusion system to isolate hydrocarbon-degrading bacteria, Figure S2: Multiple sequence alignment of the Rieske unit of the *nahAc* gene of bacterial isolates from the oil-polluted soil in Bóbrka. Reference strain is *Pseudomonas putida* plasmid with *nahAc* encoded on plasmid pAK5, Figure S3: Circular genome views of strains VI4T1 and VI4.1 in addition to the major EGGNOG categories distribution (in %), Figure S4: Visualizations of the SPAdes genome assembly and plasmidSPAdes assemblies of strains VI4T1 and VI4.1. Closed circular plasmids with coverage (in red), are indicated and results of the BLAST searches are shown in the tables, Figure S5: Rudimentary conjugative 40 kb plasmid of pVI4T1-1, showing some T4SS genes (*mpfABCDEH*), and other genes involved in conjugation (*parA, parB, fmdA, hbpA*), Figure S6: Toluene degradation operon to o-cresol for strain VI4.1, Figure S7: (**A**) Comparative analyses of upper toluene/m-xylene degradation operon to generate *m*-toluate. (**B**) *meta*-cleavage pathway of catechol to acetyl Co-A degradation and C. *ortho*-cleavage pathway, Figure S8: Comparative analyses of the toluene V degradation operon (*todABC1C2DFGHI*), and proposed biochemical pathways for toluene degradation to acetyl-CoA, Figure S9: Comparative analyses of the benzene degradation operon (*bedABC1C2D*), and proposed biochemical pathways, Figure S10: Example of positive result to the motility assay, Table S1: The list of carbon sources catabolized by VI4T1 and VI4.1 tested with the GEN-III array (Biolog).

Author Contributions: V.I. and S.T. executed the main part of the experimental work and wrote the manuscript. Y.D. assisted with the isolation of the hydrocarbon degrading strains and genome annotation. B.M.M. genome sequenced the bacteria strains under supervision of J.V.H., S.W.G. and J.V., experts in phyllo- and phyto-remediation, critically reviewed the manuscripts. M.P-E. performed the PTR-TOF analyses at UAntwerp and revised the manuscript.

Funding: V.I. was supported by a BOF PhD grant from Hasselt University and by Grant G0D0916N. S.T. was supported by a PhD grant from FWO-Flanders and the Methusalem project 08M03VGRJ. M.P.-E. was supported by the Methusalem funding of the Flemish Community through the Research Council of the University of Antwerp and by the Flemish Science Foundation (FWO, Brussels).

Acknowledgments: We thank the director of the The Ignacy Łuksiewicz's Museum of Oil and Gas Industry in Bóbrka for granting access to the Bóbrka forest sampling site. We thank Łukasz Kowalkowski for assisting with the sampling.

Conflicts of Interest: No conflict of interest.

References

1. Aguilera, F.; Méndez, J.; Pásaro, E.; Laffon, B. Review on the effects of exposure to spilled oils on human health. *J. Appl. Toxicol.* **2010**, *30*, 291–301. [CrossRef] [PubMed]
2. Rubailo, A.I.; Oberenko, A.V. *Polycyclic Aromatic Hydrocarbons as Priority Pollutants*; Siberian Federal University: Krasnoyarsk, Russia, 2008.
3. Gurska, J.; Wang, W.; Gerhardt, K.E.; Khalid, A.M.; Isherwood, D.M.; Huang, X.-D.; Glick, B.R.; Greenberg, B.M. Three year field test of a plant growth promoting Rhizobacteria enhanced phytoremediation system at a land farm for treatment of hydrocarbon waste. *Environ. Sci. Technol.* **2009**, *43*, 4472–4479. [CrossRef] [PubMed]
4. Das, S. *Microbial Biodegradation and Bioremediation*; Elsevier: Amsterdam, The Netherlands, 2014.
5. Gkorezis, P.; Daghio, M.; Franzetti, A.; Van Hamme, J.D.; Sillen, W.; Vangronsveld, J. The Interaction between Plants and Bacteria in the Remediation of Petroleum Hydrocarbons: An Environmental Perspective. *Front. Microbiol.* **2016**, *7*, 1836. [CrossRef] [PubMed]
6. Germaine, K.J.; Keogh, E.; Ryan, D.; Dowling, D.N. Bacterial endophyte-mediated naphthalene phytoprotection and phytoremediation. *FEMS Microbiol. Lett.* **2009**, *296*, 226–234. [CrossRef] [PubMed]
7. Ramos, J.L.; Duque, E.; van Dillewjin, P.; Daniels, C.; Krell, T.; Espinosa-Urgel, M.; Ramos-González, M.I.; Rodríguez, S.; Matilla, M.; Wittich, R.; et al. Removal of Hydrocarbons and Other Related Chemicals via the Rhizosphere of Plants. In *Handbook of Hydrocarbon and Lipid Microbiology*; Timmis, K., Ed.; Springer: Berlin/Heidelberg, Germany, 2010; pp. 2575–2581. [CrossRef]
8. Barac, T.; Taghavi, S.; Borremans, B.; Provoost, A.; Oeyen, L.; Colpaert, J.V.; Vangronsveld, J.; van der Lelie, D. Engineered endophytic bacteria improve phytoremediation of water-soluble, volatile, organic pollutants. *Nat. Biotechnol.* **2004**, *22*, 583–588. [CrossRef]

9. Taghavi, S.; Barac, T.; Greenberg, B.; Borremans, B.; Vangronsveld, J.; van der Lelie, D. Horizontal gene transfer to endogenous endophytic bacteria from poplar improves phytoremediation of toluene. *Appl. Environ. Microbiol.* **2005**, *71*, 8500–8505. [CrossRef]
10. Van Hamme, J.D.; Singh, A.; Ward, O.P. Recent Advances in Petroleum Microbiology. *Microbiol. Mol. Biol. Rev.* **2003**, *67*, 503–549. [CrossRef]
11. Kawasaki, A.; Watson, E.R.; Kertesz, M.A. Indirect effects of polycyclic aromatic hydrocarbon contamination on microbial communities in legume and grass rhizospheres. *Plant Soil* **2011**, *358*, 169–182. [CrossRef]
12. Gonzalez, E.; Pitre, F.E.; Pagé, A.P.; Marleau, J.; Nissim, W.G.; St-Arnaud, M.; Labrecque, M.; Joly, S.; Yergeau, E.; Brereton, N.J. Trees, fungi and bacteria: Tripartite metatranscriptomics of a root microbiome responding to soil contamination. *Microbiome* **2018**, *6*, 53. [CrossRef]
13. Bushnell, L.; Haas, H. The utilization of certain hydrocarbons by microorganisms. *J. Bacteriol.* **1941**, *41*, 653.
14. Mergeay, M.; Nies, D.; Schlegel, H.G.; Gerits, J.; Charles, P.; Van Gijsegem, F. *Alcaligenes eutrophus* CH34 is a facultative chemolithotroph with plasmid-bound resistance to heavy metals. *J. Bacteriol.* **1985**, *162*, 328–334.
15. Nagayama, H.; Sugawara, T.; Endo, R.; Ono, A.; Kato, H.; Ohtsubo, Y.; Nagata, Y.; Tsuda, M. Isolation of oxygenase genes for indigo-forming activity from an artificially polluted soil metagenome by functional screening using *Pseudomonas putida* strains as hosts. *Appl. Microbiol. Biotechnol.* **2015**, *99*, 4453–4470. [CrossRef] [PubMed]
16. Zylstra, G.J.; McCombie, W.R.; Gibson, D.T.; Finette, B.A. Toluene degradation by *Pseudomonas putida* F1: Genetic organization of the tod operon. *Appl. Environ. Microbiol.* **1988**, *54*, 1498–1503. [PubMed]
17. Ueno, A.; Hasanuzzaman, M.; Yumoto, I.; Okuyama, H. Verification of degradation of n-alkanes in diesel oil by *Pseudomonas aeruginosa* strain WatG in soil microcosms. *Curr. Microbiol.* **2006**, *52*, 182–185. [CrossRef] [PubMed]
18. Frank, J.A.; Reich, C.I.; Sharma, S.; Weisbaum, J.S.; Wilson, B.A.; Olsen, G.J. Critical Evaluation of Two Primers Commonly Used for Amplification of Bacterial 16S rRNA Genes. *Appl. Environ. Microbiol.* **2008**, *74*, 2461–2470. [CrossRef] [PubMed]
19. Rademaker, J.L. Characterization and classification of microbes by rep-PCR genomic fingerprinting and computer assisted pattern analysis. In *DNA Markers: Protocols, Applications, and Overviews*; Wiley & Sons: Hoboken, NJ, USA, 1997; pp. 151–171.
20. Baldwin, B.R.; Nakatsu, C.H.; Nies, L. Detection and enumeration of aromatic oxygenase genes by multiplex and real-time PCR. *Appl. Environ. Microbiol.* **2003**, *69*, 3350–3358. [CrossRef] [PubMed]
21. Nurk, S.; Bankevich, A.; Antipov, D.; Gurevich, A.; Korobeynikov, A.; Lapidus, A.; Prjibelsky, A.; Pyshkin, A.; Sirotkin, A.; Sirotkin, Y.; et al. Assembling Genomes and Mini-metagenomes from Highly Chimeric Reads. Springer: Berlin/Heidelberg, Germany, 2013; pp. 158–170.
22. Antipov, D.; Hartwick, N.; Shen, M.; Raiko, M.; Lapidus, A.; Pevzner, P.A. plasmidSPAdes: Assembling plasmids from whole genome sequencing data. *Bioinformatics* **2016**, *32*, 3380–3387. [CrossRef]
23. Overbeek, R.; Olson, R.; Pusch, G.D.; Olsen, G.J.; Davis, J.J.; Disz, T.; Edwards, T.A.; Gerdes, S.; Parrello, B.; Shukla, M.; et al. The SEED and the Rapid Annotation of microbial genomes using Subsystems Technology (RAST). *Nucleic Acids Res.* **2014**, *42*, D206–D214. [CrossRef]
24. Tatusova, T.; DiCuccio, M.; Badretdin, A.; Chetvernin, V.; Nawrocki, E.P.; Zaslavsky, L.; Lomsadze, A.; Pruitt, K.D.; Borodovsky, M.; Ostell, J. NCBI prokaryotic genome annotation pipeline. *Nucleic Acids Res.* **2016**, *44*, 6614–6624. [CrossRef]
25. Vallenet, D.; Labarre, L.; Rouy, Z.; Barbe, V.; Bocs, S.; Cruveiller, S.; Lajus, A.; Pascal, G.; Scarpelli, C.; Médigue, C. MaGe: A microbial genome annotation system supported by synteny results. *Nucleic Acids Res.* **2006**, *34*, 53–65. [CrossRef]
26. Tatusov, R.L.; Galperin, M.Y.; Natale, D.A.; Koonin, E.V. The COG database: A tool for genome-scale analysis of protein functions and evolution. *Nucleic Acids Res.* **2000**, *28*, 33–36. [CrossRef] [PubMed]
27. Kanehisa, M.; Goto, S.; Sato, Y.; Furumichi, M.; Tanabe, M. KEGG for integration and interpretation of large-scale molecular data sets. *Nucleic Acids Res.* **2012**, *40*, D109–D114. [CrossRef] [PubMed]
28. Caspi, R.; Billington, R.; Fulcher, C.A.; Keseler, I.M.; Kothari, A.; Krummenacker, M.; Latendresse, M.; Midford, P.E.; Ong, Q.; Ong, W.K.; et al. The MetaCyc database of metabolic pathways and enzymes. *Nucleic Acids Res.* **2017**, *46*, D633–D639. [CrossRef] [PubMed]
29. Veltri, D.; Wight, M.M.; Crouch, J.A. SimpleSynteny: A web-based tool for visualization of microsynteny across multiple species. *Nucleic Acids Res.* **2016**, *44*, W41–W45. [CrossRef] [PubMed]

30. Meier-Kolthoff, J.P.; Auch, A.F.; Klenk, H.-P.; Göker, M. Genome sequence-based species delimitation with confidence intervals and improved distance functions. *BMC Bioinform.* **2013**, *14*, 60. [CrossRef] [PubMed]
31. Meier-Kolthoff, J.P.; Hahnke, R.L.; Petersen, J.; Scheuner, C.; Michael, V.; Fiebig, A.; Rohde, C.; Rohde, M.; Fartmann, B.; Goodwin, L.A.; et al. Complete genome sequence of DSM 30083T, the type strain (U5/41T) of *Escherichia coli*, and a proposal for delineating subspecies in microbial taxonomy. *Stand. Genom. Sci.* **2014**, *9*, 2. [CrossRef] [PubMed]
32. Wick, R.R.; Schultz, M.B.; Zobel, J.; Holt, K.E. Bandage: Interactive visualization of de novo genome assemblies. *Bioinformatics* **2015**, *31*, 3350–3352. [CrossRef]
33. Stecher, G.; Kumar, S.; Tamura, K. MEGA7: Molecular Evolutionary Genetics Analysis Version 7.0 for Bigger Datasets. *Mol. Biol. Evol.* **2016**, *33*, 1870–1874. [CrossRef]
34. Xu, X.; Reid, N. On the robustness of maximum composite likelihood estimate. *J. Stat. Plan. Inference* **2011**, *141*, 3047–3054. [CrossRef]
35. Hanson, K.; Desai, J.D.; Desai, A.J. A rapid and simple screening technique for potential crude oil degrading microorganisms. *Biotechnol. Tech.* **1993**, *7*, 745–748. [CrossRef]
36. Cappellin, L.; Karl, T.; Probst, M.; Ismailova, O.; Winkler, P.M.; Soukoulis, C.; Aprea, E.; Märk, T.D.; Gasperi, F.; Biasioli, F. On quantitative determination of volatile organic compound concentrations using proton transfer reaction time-of-flight mass spectrometry. *Environ. Sci. Technol.* **2012**, *46*, 2283–2290. [CrossRef] [PubMed]
37. Belimov, A.A.; Hontzeas, N.; Safronova, V.I.; Demchinskaya, S.V.; Piluzza, G.; Bullitta, S.; Glick, B.R. Cadmium-tolerant plant growth-promoting bacteria associated with the roots of Indian mustard (*Brassica juncea L. Czern.*). *Soil Biol. Biochem.* **2005**, *37*, 241–250. [CrossRef]
38. Crowley, D.E.; Reid, C.P.; Szaniszlo, P.J. Utilization of microbial siderophores in iron acquisition by oat. *Plant Physiol.* **1988**, *87*, 680–685. [CrossRef] [PubMed]
39. Salkowski, E. Ueber das Verhalten des Skatolcarbonsaure im Organismus. *Z. Physiol. Chem.* **1885**, *9*, 23–33.
40. Barritt, M.M. The intensification of the Voges-Proskauer reaction by the addition of α-naphthol. *J. Pathol. Bacteriol.* **1936**, *42*, 441–454. [CrossRef]
41. Jorquera, M.A.; Hernández, M.T.; Rengel, Z.; Marschner, P.; de la Luz Mora, M. Isolation of culturable phosphobacteria with both phytate-mineralization and phosphate-solubilization activity from the rhizosphere of plants grown in a volcanic soil. *Biol. Fertil. Soils* **2008**, *44*, 1025–1034. [CrossRef]
42. Cunningham, J.E.; Kuiack, C. Production of citric and oxalic acids and solubilization of calcium phosphate by *Penicillium bilaii*. *Appl. Environ. Microbiol.* **1992**, *58*, 1451–1458.
43. Xie, G.H.; Cui, Z.; Yu, J.; Yan, J.; Hai, W.; Steinberger, Y. Identification of *nif* genes in N_2-fixing bacterial strains isolated from rice fields along the Yangtze River Plain. *J. Basic Microbiol.* **2006**, *46*, 56–63. [CrossRef]
44. Glick, B.R. Bacteria with ACC deaminase can promote plant growth and help to feed the world. *Microbiol. Res.* **2014**, *169*, 30–39. [CrossRef]
45. Spaepen, S.; Vanderleyden, J.; Remans, R. Indole-3-acetic acid in microbial and microorganism-plant signaling. *FEMS Microbiol. Rev.* **2007**, *31*, 425–448. [CrossRef]
46. Lee, Y.; Lee, Y.; Jeon, C.O. Biodegradation of naphthalene, BTEX, and aliphatic hydrocarbons by *Paraburkholderia aromaticivorans* BN5 isolated from petroleum-contaminated soil. *Sci. Rep.* **2019**, *9*, 860. [CrossRef] [PubMed]
47. Liu, X.; Dong, Y.; Li, X.; Ren, Y.; Li, Y.; Wang, W.; Wang, L.; Feng, L. Characterization of the anthranilate degradation pathway in *Geobacillus thermodenitrificans* NG80-2. *Microbiology* **2010**, *156*, 589–595. [CrossRef] [PubMed]
48. Barret, M.; Morrissey, J.P.; O'Gara, F. Functional genomics analysis of plant growth-promoting rhizobacterial traits involved in rhizosphere competence. *Biol. Fertil. Soils* **2011**, *47*, 729–743. [CrossRef]
49. Mohite, B. Isolation and characterization of indole acetic acid (IAA) producing bacteria from rhizospheric soil and its effect on plant growth. *J. Soil Sci. Plant. Nutr.* **2013**, *13*, 638–649. [CrossRef]
50. Gonzales, T.; Robert-Baudouy, J. Bacterial aminopeptidases: Properties and functions. *FEMS Microbiol. Rev.* **1996**, *18*, 319–344. [CrossRef] [PubMed]
51. Gimenez-Ibanez, S.; Chini, A.; Solano, R. How Microbes Twist Jasmonate Signaling around Their Little Fingers. *Plants* **2016**, *5*, 9. [CrossRef] [PubMed]
52. Meena, K.K.; Sorty, A.M.; Bitla, U.M.; Choudhary, K.; Gupta, P.; Pareek, A.; Singh, D.P.; Prabha, R.; Sahu, P.K.; Gupta, V.K.; et al. Abiotic Stress Responses and Microbe-Mediated Mitigation in Plants: The Omics Strategies. *Front. Plant Sci.* **2017**, *8*, 172. [CrossRef] [PubMed]

53. Castanie-Cornet, M.-P.; Treffandier, H.; Francez-Charlot, A.; Gutierrez, C.; Cam, K. The glutamate-dependent acid resistance system in *Escherichia coli*: Essential and dual role of the His–Asp phosphorelay RcsCDB/AF. *Microbiology* **2007**, *153*, 238–246. [CrossRef] [PubMed]
54. Iturriaga, G.; Suárez, R.; Nova-Franco, B. Trehalose metabolism: From osmoprotection to signaling. *Int. J. Mol. Sci.* **2009**, *10*, 3793–3810. [CrossRef] [PubMed]
55. Ashraf, M.; Foolad, M. Roles of glycine betaine and proline in improving plant abiotic stress resistance. *Environ. Exp. Bot.* **2007**, *59*, 206–216. [CrossRef]
56. Ben, W.; Zhu, B.; Yuan, X.; Zhang, Y.; Yang, M.; Qiang, Z. Occurrence, removal and risk of organic micropollutants in wastewater treatment plants across China: Comparison of wastewater treatment processes. *Water Res.* **2018**, *130*, 38–46. [CrossRef] [PubMed]
57. Chen, Y.-L.; Wang, C.-H.; Yang, F.-C.; Ismail, W.; Wang, P.-H.; Shih, C.-J.; Wu, Y.-C.; Chiang, Y.-R. Identification of *Comamonas testosteroni* as an androgen degrader in sewage. *Sci. Rep.* **2016**, *6*, 35386. [CrossRef] [PubMed]
58. Webb, H.; Arnott, J.; Crawford, R.; Ivanova, E. Plastic degradation and its environmental implications with special reference to poly (ethylene terephthalate). *Polymers* **2013**, *5*, 1–18. [CrossRef]

© 2019 by the authors. Licensee MDPI, Basel, Switzerland. This article is an open access article distributed under the terms and conditions of the Creative Commons Attribution (CC BY) license (http://creativecommons.org/licenses/by/4.0/).

genes

MDPI

Article

Genomic Analysis of γ-Hexachlorocyclohexane-Degrading *Sphingopyxis lindanitolerans* WS5A3p Strain in the Context of the Pangenome of *Sphingopyxis*

Michal A. Kaminski [1], Adam Sobczak [1,2], Andrzej Dziembowski [1,2] and Leszek Lipinski [1,2,*]

1 Institute of Biochemistry and Biophysics, Polish Academy of Sciences, Pawinskiego 5a, 02-106 Warsaw, Poland

2 Institute of Genetics and Biotechnology, Faculty of Biology, University of Warsaw, Pawinskiego 5a, 02-106 Warsaw, Poland

* Correspondence: lechu@ibb.waw.pl

Received: 29 July 2019; Accepted: 2 September 2019; Published: 6 September 2019

Abstract: *Sphingopyxis* inhabit diverse environmental niches, including marine, freshwater, oceans, soil and anthropogenic sites. The genus includes 20 phylogenetically distinct, valid species, but only a few with a sequenced genome. In this work, we analyzed the nearly complete genome of the newly described species, *Sphingopyxis lindanitolerans*, and compared it to the other available *Sphingopyxis* genomes. The genome included 4.3 Mbp in total and consists of a circular chromosome, and two putative plasmids. Among the identified set of *lin* genes responsible for γ-hexachlorocyclohexane pesticide degradation, we discovered a gene coding for a new isoform of the LinA protein. The significant potential of this species in the remediation of contaminated soil is also correlated with the fact that its genome encodes a higher number of enzymes potentially involved in aromatic compound degradation than for most other *Sphingopyxis* strains. Additional analysis of 44 *Sphingopyxis* representatives provides insights into the pangenome of *Sphingopyxis* and revealed a core of 734 protein clusters and between four and 1667 unique proteins per genome.

Keywords: *Sphingopyxis lindanitolerans*; pesticide; complete genome sequence; pangenome; γ-HCH degradation; *lin* genes

---

1. Introduction

Intensive agriculture entails the extensive use of pesticides to increase yields. One commonly used insecticide has been γ-hexachlorocyclohexane (γ-HCH, also known as lindane or γ-BHC)—a potential carcinogen that is highly toxic to humans, causing neurological diseases, and endocrine disruption [1]. This hydrophobic substance was classified as a persistent organic pollutant [2] and banned in 2009 under the Stockholm Convention. Although the natural degradation of lindane takes a long time, there are known microorganisms that actively degrade it, and can significantly increase the rate of bioremediation [3,4]. Most of the γ-HCH degrading microorganisms belong to *Sphingomonadaceae* (Alphaproteobacteria phylum)—mainly *Sphingobium* sp., *Novosphingobium* sp. and *Sphingomonas* sp. Data showing the γ-HCH degradation activity for the representatives belonging to the *Sphingopyxis* genus has not been published yet, despite the fact that three strains, *Sphingopyxis flava* R11H, *Sphingopyxis indica* DS15 and *Sphingopyxis terrae* subsp. *ummariensis* DSM 24316, were isolated from the environment contaminated with this pesticide [5–8].

The main γ-HCH degradation pathway was described in detail for *Sphingobium japonicum* UT26S and *Sphingobium indicum* B90A [9–11]. These two microorganisms contain the *lin* genes encoding a set of enzymes responsible for lindane conversion to β-ketoadipate, which is further degraded and utilized in other metabolic pathways. The first and key enzyme from the catabolic pathway is dehydrochlorinase LinA [12]. It catalyzes the transformation of γ-HCH to 1,3,4,6-tetrachloro-1,4-cyclohexadiene (1,4-TCDN) via γ-pentachlorocyclohexane (γ-PCCH) by sequential removal of two chloride atoms [13]. This protein is unique among other known dehydrochlorinases as does not require any cofactors [14,15]. LinA is 154 amino acids long and acts as a trimer, where each protomer forms a cone-shaped α + β barrel fold and D25, H73, and R129 residues were shown to be essential for its activity [16,17]. Subsequently, two chloride atoms are removed from 1,4-TCDN by LinB dehalogenase [18] that converts 1,4-TCDN into 2,5-dichloro-2,5-cyclohaxadiene-1,4-diol (2,5-DDOL). LinB requires only H_2O as a cofactor, and unlike LinA acts on a broad range of halogenated substrates—also on a very stable β-HCH isomer [19]. The last enzyme from upper degradation pathway is LinC [20] that converts 2,5-DDOL to 2,5-dichlorohydroquinone (2,5-DCHQ). LinC belongs to the short-chain alcohol dehydrogenase family and requires NAD^+ cofactor. Further reactions from downstream degradation pathway involve further removal of chloride atoms from 2,5-DCHQ to hydroquinone (HQ) by reductive dechlorinase LinD [21] and next ring cleavage by ring cleavage oxygenase LinE [22]. It is also known that 2,5-DCHQ can be transformed directly by LinE or LinEb (enzyme identical to LinE in 53% of amino acids) to maleylacetate (MA) through a yet unknown unstable product [10,11]. Next, MA is converted by maleylacetate reductase LinF, to β-ketoadipate and later LinG; LinH and LinJ proteins (acyl-CoA transferase and thiolase) transform β-ketoadipate to succinyl-CoA and acetyl-CoA [11].

Here, we present a detailed genomic analysis of the γ-HCH degrader—*Sphingopyxis lindanitolerans* WS5A3p [23]. As described previously, WS5A3p was selectively enriched using the γ-HCH as a sole carbon source. The genome of this new microorganism was sequenced and analyzed, leading to the identification of 13 *lin* genes involved in the γ-HCH degradation pathway. Additionally, we provide insights into *Sphingopyxis* based on the analysis of 44 genomes, with a description of the core pangenome and the unique proteins for each representative.

2. Materials and Methods

2.1. Genome Sequencing and Assembly

For genomic DNA isolation, all steps were performed as described previously [23]. Two libraries were prepared for genome sequencing: (1) a paired-end library with an insert size of 500 bp, (2) a Nextera® Mate Pair library (Illumina Inc., San Diego, CA, USA) with an average insert size of 8 kbp. Sequencing was performed on the Illumina MiSeq platform (Illumina Inc.) with a 300 bp read length resulting in 311,675 and 1,376,608 raw paired reads respectively. Next, a number of bioinformatics tools were used to analyze the raw reads. Adapters from the raw reads were trimmed using Cutadapt [24], and quality filtered with Sickle [25] (quality at least Q30). The mate-paired libraries were processed using NxTrim script v0.4.0 provided by the manufacturer. Assembly was performed using the SPAdes 3.9.1 software [26]. The assembly was manually checked by mapping sequencing reads to obtained contigs using Genious 6.1.6 [27]. This Whole Genome Shotgun project has been deposited at DDBJ/ENA/GenBank under the accession PHFW00000000. The version described in this paper is version PHFW01000000. The genome was annotated using the NCBI Prokaryotic Genome Annotation Pipeline [28]. Genes with signal peptides were identified with SignalP 4.1 [29]. Association of genes and proteins to COG categories was done using COGNIZER software [30] and MicroScope platform [31]. Analysis of secondary metabolite biosynthesis gene clusters was done with antiSMASH 5.0 (https://antismash.secondarymetabolites.org) [32]; prophage sequences in WS5A3p genome were identified with PHASTER (http://phaster.ca/) [33]; and CRISPRCasFinder software (http://crispr.i2bc.paris-saclay.fr) [34] with default parameters was used to identify CRISPR-related sequences.

2.2. Average Nucleotide Identity Based on BLAST (ANIb) and GGDC Analyses

The similarity of the sequenced genome of *S. lindanitolerans* WS5A3p to other public genomes of closely related *Sphingopyxis* species [8,23,35–44] was determined based on the average nucleotide identity using both BLASTn (ANIb) and MUMer (ANIm) algorithms with the help of the pyani Python module [45] and average amino acid identity (AAI), which was calculated using CompareM (available at: https://github.com/dparks1134/CompareM).

2.3. Pangenomic Analyses

For pangenomic analyses, nucleotide sequences of *Sphingopyxis* sp. genomes were downloaded from the NCBI database on December 2017. Due to low quality and completeness, genomes retrieved from metagenomes and single-cell sequencing were excluded from the analysis. To unify annotation, open reading frames (ORFs) were predicted for all genomes using Prokka [46].

Phylogenetic analysis of *Sphingopyxis* genomes was performed using Phylophlan [47], and ggtree R package [48] was used to display the results. Phylophlan software identifies 400 ubiquitous genes based on their amino acid content, selects the most discriminative positions in each sequence and concatenates them into a single long sequence which is used for ML tree construction by RAxML [49]. From the Phylophlan generated protein alignment we ran RAxML to generate 500 bootstrap replicates (ML search with PMB substitution matrix, as the best-scoring AA model).

For pangenomic analysis of amino acid sequences obtained from annotated ORFs, all genomes were merged into one set. Next, the clustering process based on sequence identity was performed using CD-HIT [50]. Parameters for CD-HIT were as follows: s parameter (length difference cutoff) = 0.9, and c parameter (sequence identity threshold) = 0.8. Using in-house Python and R scripts processing CD-HIT output (scripts collected into Pangenome Analysis Pipeline (PAPi) available at https://github.com/michkam89/PAPi), we identified clusters that had protein representatives present in all genomes (core pangenome) but also clusters specific for each genome included in the analysis (unique clusters). For the unique clusters of each particular genome, we performed additional blastp alignment against other *Sphingopyxis* genomes to exclude false-positive sequences.

Sphingopyxis amino acid sequences obtained from annotated ORFs of the analyzed dataset were also searched for the presence of proteins responsible for aromatic compound degradation. For this purpose, we performed a blastp search of all ORFs against in-house protein database developed by Kato and colleagues [51] filtering reads with parameters: e-value $<1 \times 10^{-5}$ and alignment percent identity >50%. This database covers the metabolic paths of aerobic and anaerobic degradation routes of 3CB, phenanthrene, biphenyl, and carbazole (phenanthrene-phthalate-protocatechuate route, biphenyl-benzoate route, carbazole-anthranilate route, naphthalene-salicylate route, catechol 1,2- and 2,3-dioxygenation route, gentisate route, anaerobic polycyclic aromatic hydrocarbon (naphthalene) route, and anaerobic benzoate and toluene route) [51].

Alignment of protein sequences for analyzed Lin proteins was performed using MUSCLE [52], phylogenetic relationship and alignment visualization was done in Geneious [27].

3. Results and Discussion

The WS5 soil sample, from which the WS5A3p strain was isolated, had been contaminated with a high concentration of organochlorine pesticides including γ-HCH (lindane), dichlorodiphenyltrichloroethane (DDT) and methoxychlor. WS5A3p was isolated through an 8-week enrichment under constant γ-HCH pressure at 5 mg/L [23]. Its activity was confirmed by the ability to create clear zones on γ-HCH agar plates. Moreover, the decrease of γ-HCH was observed in a liquid medium assay where the presence of small amounts of 2,5-dichlorophenol was identified (Figure S1).

The WS5A3p strain was cultured in LB medium and genomic DNA was isolated and sequenced on the Illumina MiSeq platform (see Materials and Methods section) with genome coverage estimated at 112× [23]. As a result of genomic sequencing and assembly, three DNA scaffolds were obtained. General WS5A3p genome statistics are presented in Table S1. The size of the assembled genome sequence is 4,372,786 bp, and contains 65.1% G + C and is 90.4% coding. The assembly consists of one circular scaffold, representing a single chromosome of 4.14 Mbp with three unsequenced gaps of the total estimated length of 305 nucleotides. Two remaining scaffolds of length 181,517 bp and

42,040 bp having separate partition and transfer modules were classified as plasmids. There are a total of 4184 predicted genes, from which 4,133 are protein-coding and 75.1% have assigned putative function. The WS5A3p genome has 51 RNA genes (three rRNA as one ribosomal operon, 45 tRNA and three ncRNA). Observed substitution of lysine 88 by arginine in S12 protein of the 30S ribosomal subunit (RpsL) suggests natural resistance of this strain to streptomycin, what is considered as a feature of *Sphingomonadaceae* and *Erythrobacteraceae* [43]. The functional categorization of genes into COGs (clusters of orthologous groups) for WS5A3p is presented in Table 1. Further analysis of WS5A3p genome in a context of phage-related sequences using PHASTER [33], revealed four prophage regions: one intact region (score = 110) according to the tool's completeness core, one questionable (score = 90) and two incomplete regions (score = 60 and 50). The intact region (16.1 kb in length) localized on chromosome sequence shows the highest similarity to genome fragment of *Sphingopyxis granuli* TFA [43]. No CRISPR-related sequences were identified in WS5A3p genome using CRISPRCasFinder application [34]. Like other *Sphingopyxis* genomes, this one contains ectoine synthesis genes *ectABCD*, Aspartate kinase Ask_Ect and Ectoine/proline transporter ProP [43]. Secondary metabolite biosynthesis gene clusters analysis using antiSMASH 5.0 also showed the presence of regions potentially involved in the biosynthesis of carotenoids (one region with 75% similarity to astaxanthin dideoxyglycoside cluster) or natural antimicrobials such as β-lactones (one region), lenthipeptides (one region), lasso peptides (three regions), which may give advantage to *S. lindanitolerans* in an oligotrophic environment.

Table 1. Gene counts associated with the general COGs (clusters of orthologous groups) functional categories for the genome of *Sphingopyxis lindanitolerans* WS5A3p, obtained from COGNIZER software [30].

Code	Description	WS5A3p	
		Value	**%**
J	Translation, ribosomal structure and biogenesis	241	6.8%
K	Transcription	287	8.0%
L	Replication, recombination and repair	319	8.9%
D	Cell cycle control, cell division, chromosome partitioning	38	1.1%
V	Defense mechanisms	81	2.3%
T	Signal transduction mechanisms	179	5.0%
M	Cell wall/membrane/envelope biogenesis	232	6.5%
N	Cell motility	57	1.6%
W	Extracellular structures	8	0.2%
U	Intracellular trafficking, secretion, and vesicular transport	209	5.9%
O	Posttranslational modification, protein turnover, chaperones	189	5.3%
C	Energy production and conversion	381	10.7%
G	Carbohydrate transport and metabolism	218	6.1%
E	Amino acid transport and metabolism	512	14.3%
F	Nucleotide transport and metabolism	91	2.5%
H	Coenzyme transport and metabolism	175	4.9%
I	Lipid transport and metabolism	481	13.5%
P	Inorganic ion transport and metabolism	481	13.5%
Q	Secondary metabolites biosynthesis, transport and catabolism	302	8.5%
R	General function prediction only	651	18.2%
S	Function unknown	277	7.8%

We compared the WS5A3p genome to 46 other *Sphingopyxis* genomes deposited in the NCBI database. Phylogenetic placement based on 400 highly conserved protein sequences showed that WS5A3p is placed on a distinct branch on the phylogenetic tree with 100% bootstrap support (Figure 1A). We also saw that *Sphingopyxis baekryungensis* DSM 16222 and LPB0140 isolate, previously assigned to *Sphingopyxis*, clustered with outgroup genomes analyzed in the dataset (containing representatives of *Sphingobium*, *Novosphingobium*, and *Sphingorhabdus*). To extend these findings we also performed genomic analysis of whole DNA sequences (average nucleotide identity based on Mummer—ANIm, average nucleotide identity based on BLAST—ANIb) and amino acid content of all protein sequences (average amino acid Identity—AAI). These tests showed that WS5A3p is unique among available genomes. We also identified that values for MWB1 isolate in all of the three tests were outlying the most from other species, therefore we decided to treat this isolate also as an outgroup. Average values (excluding outgroup and LPB0140, MWB1 and DSM 16222 genomes) for similarity among *Sphingopyxis* genomes to WS5A3p based on ANIb, ANIm and AAI were 82.6 ± 0.98%, 85.8 ± 0.63%, and 81.8 ± 1.35%, respectively. The ranges were 80.1–83.7% for ANIb, 84.5–86.8% for ANIm and 79.7–83.8% for AAI. Results of the tests mentioned above are illustrated in Figure 1B. All of the analyses distinguished two groups of genomes highly similar to each other. Group H1 consisting of isolates: H005, H012, H038, H053, H077, H080, H085, H093 and group H2 consisting of isolates: H057, H067, H071, H073, H081, H100, H107 that are genomic clones with almost identical sequences. This result provides additional evidence that exclusion of *S. baekryungensis* DSM 16222 from *Sphingopyxis* genus and reclassification of isolates LPB0140 and MWB1 should be considered in the near future.

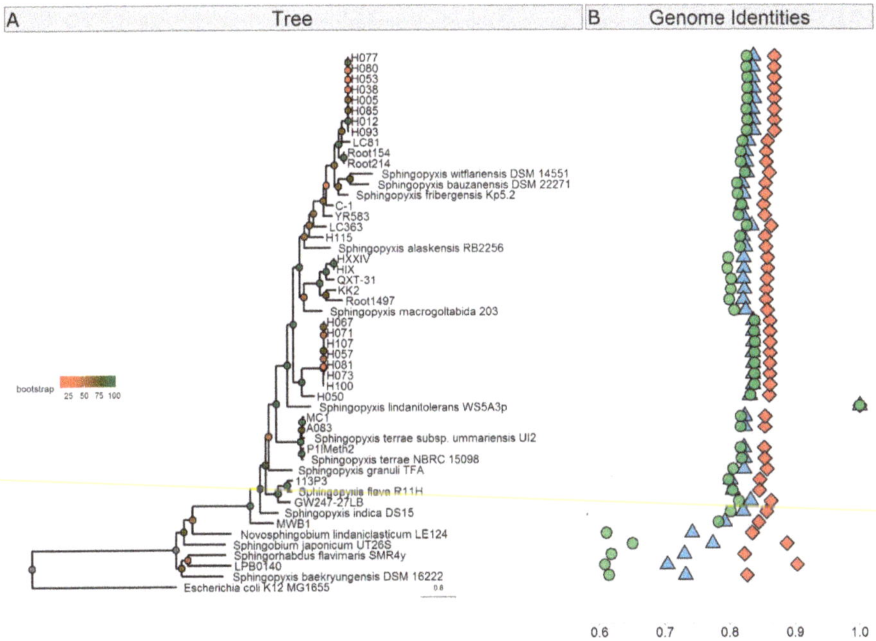

Figure 1. Phylogenetic analyses of *Sphingopyxis* genomes. (**A**) Phylogenetic relationship of *Sphingopyxis* genomes based on a maximum likelihood analysis of 400 conserved protein sequences. Branches are colored according to bootstrap values, which are presented on the nodes. The bar represents 0.8 substitutions per site. The WS5A3p branch is indicated with a blue diamond. The blue shaded box contains outgroup genomes included in the analysis. (**B**) Sequence similarity of analyzed *Sphingopyxis* genomes comparing to WS5A3p strain based on ANIb (blue triangle), ANIm (red diamond) and AAI (green circle).

To analyze the uniqueness of WS5A3p genome, we first performed the pangenomic comparison of protein sequences present in all *Sphingopyxis* representatives. From these tests, we excluded genomes of LB0140, MWB1 and DSM 16222, as based on the previous phylogenetic analysis they clustered closely to outgroup genomes (LB0140, DSM 16222) and had AAI values below 80% (LB0140, DSM 16222 and MWB1). Table S2 presents general information about all genomes included in the analysis. Representatives had been isolated from diverse water environments (drinking water, seawater, ground, lake, and cave water) and soils (mainly contaminated with organic pollutants) as well as anthropogenic sites like mines or wastewater treatment plants. Isolates derived from geographically distant regions in Europe, Asia, North America and one from Australia. The median size of the *Sphingopyxis* genome is 4.3 Mbp and ranged from 3.4 Mbp for *Sphingopyxis alaskensis* RB2256 and 5.7 Mbp for *Sphingopyxis macrogoltabida* 203. The average *Sphingopyxis* genome contains 4236 genes with a minimal number of 3287 and maximal of 5572 for *S. alaskensis* RB2256 and *S. macrogoltabida* 203, respectively.

Pangenomic analyses usually focus on pathogenic bacteria at the species level [53]. Few studies have described generic bacterial pangenomes beyond *Streptococcus* [54] or *Bifidobacterium* [55]. The first attempt to describe the *Sphingopyxis* pangenome was performed by Garcia-Romero [33] and Parthasarathy [56]. They analyzed seven *Sphingopyxis* genomes and established the core pangenome size at 1371 gene families or 1515 single copy orthologues. In our pangenomic analysis, we grouped proteins into clusters based on their amino acid sequence, differentiating 734 core clusters present in all of the analyzed genomes, which represent less than 20% of the average *Sphingopyxis* genome. Compared to the previous reports, our dataset increased from seven to 44 genomes and allowed us to reduce the size of the core *Sphingopyxis* pangenome by almost half. It is only 1.5× larger than the minimal genome content of the smallest free-living microorganism (470 coding regions) [57]. We also calculated the size of the entire pangenome with the median size of 28,914 clusters (IQR = 2645). Accumulation plots of the pangenome and core pangenome are presented in Figure S2. While core genome seems to reach its saturation and addition of new genomes should not affect it significantly, the pangenome median curve started to flatten but it is not saturated yet and the pangenome still should be considered as "open".

Next, we assigned representatives of core clusters to COGs. Nineteen percent and 20% of clusters were assigned to cellular processes and signaling (Figure 2A), and information storage and processing functions (Figure 2B), respectively. Almost 45% of all core clusters belonged to metabolic processes (Figure 2C). The enrichment in housekeeping functions like translation and ribosome biogenesis is clearly visible here, but amino acid and nucleotide metabolism were also very abundant processes.

We also identified and compared unique protein clusters for each genome included in the analysis (Table S2), and showed significant differences among the number of unique clusters; from four to 1667. The number of unique clusters positively correlate with phylogenetic distance to closest relatives (Spearman rank correlation p-value $<2.2 \times 10^{-16}$, Figure S3). As expected, closely related genomes contained fewer unique clusters (previously mentioned groups H1 and H2, but also genomes of Root154 and Root214) than more distant relatives.

In the pangenomic analysis, we found that the WS5A3p strain has 1274 unique protein clusters (one of the highest number among species analyzed in our dataset—Table S2), representing almost 31% of all clusters identified in its genome. Representatives of 1087 of unique protein clusters were successfully assigned to COGs, with 286 (26%) of them assigned to groups of general function prediction or unknown function (Figure 2D). However, within the general function group, we identified numerous aromatic ring-cleaving dioxygenases, monooxygenases, hydrolases and reductases that may participate in xenobiotic decomposition as well as members of the major facilitator superfamily (MFS) known to be involved in membrane transport. Here, we also found enzymes responsible for lindane turnover. Among poorly characterized COGs, the most abundant clusters were assigned to metabolic processes (369 clusters) and among them, 86 clusters were involved with inorganic ion transport and metabolism (Figure 2C). These clusters represent various iron-related proteins in high amounts, with the largest fraction as TonB-dependent receptors associated with the uptake and transport of large substrates such as iron siderophore complexes and vitamin B12 [58]. We also identified 123 clusters involved in transcription processes as well as clusters from replication, recombination and repair group (65 clusters) (Figure 2B). These were mainly transposases, integrases, and methyltransferases, indicating a high potential for recombinational processes. Analyzing the former, we detected transcriptional regulators assigned to numerous transcriptional regulator families like TerR, HxlR, GntR, LysR, IclR, AcrR, MarR, LuxR, ArsR, and AsnC.

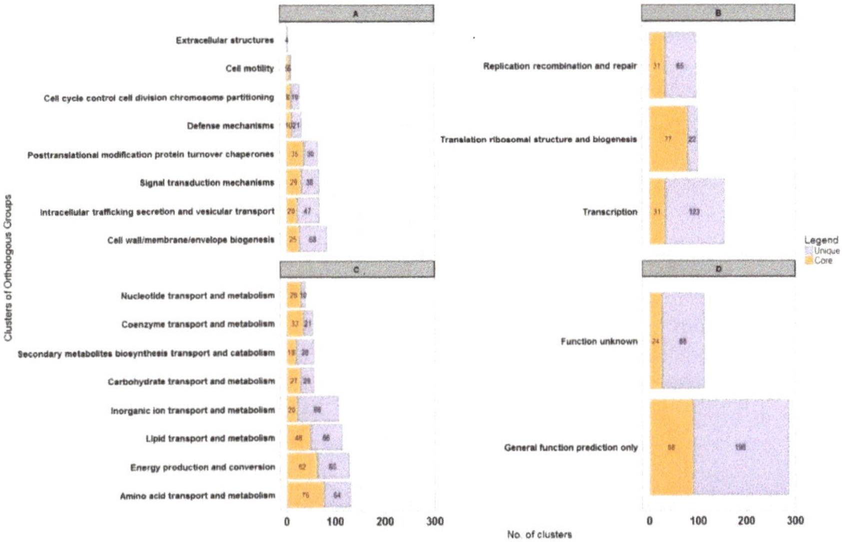

Figure 2. Genome-scale analysis of protein functions for *Sphingopyxis lindanitolerans* WS5A3p in context of *Sphingopyxis* pangenome. Bar chart illustrating protein clusters participation in different COGs functional categories for *Sphingopyxis* core pangenome (orange) and WS5A3p unique clusters (purple). (**A**) Cellular processes and signaling; (**B**) information storage and processing functions; (**C**) metabolism process; (**D**) unknown functions.

For many *Sphingomonadaceae* members, a high number of genes putatively involved in aromatic compound degradation pathways have been annotated. Thus we also searched *Sphingopyxis* proteomes for the presence of proteins potentially involved in aromatic compound degradation. To reduce the number of calculations we selected H005 and H057 as representatives of H1 and H2 groups of nearly identical genomes respectively. We analyzed 31 proteomes and obtained 314 hits in total. The median number of identified proteins per proteome was 12 (IQR = 7), whereas the highest number of hits were found for *S. bauzanensis* DSM2271 ($n = 38$), strain LC363 ($n = 29$), *S. lindanitolerans* WS5A3p ($n = 27$) and strain H050 ($n = 26$) (Figure S4A,B). DSM2271 and WS5A3p strains were isolated from soils that were contaminated with either aromatic hydrocarbons or pesticides and might have acquired more genes related to the metabolism of aromatic compounds than other microorganisms analyzed. On the other hand, LC363 and H050 strains were isolated from water environments—cave and drinking water respectively—that are rather oligotrophic. For *S. lindanitolerans* WS5A3p we identified more enzymes potentially involved in aromatic compound degradation than for other strains isolated also from a γ-HCH contaminated environment like *S. flava* R11H ($n = 15$), *S. indica* DS15 ($n = 9$) and *S. terrae* subsp. *ummariensis* UI2 ($n = 9$). Majority of blastp hits (62%) were unique per genome, but also multiple copies of some proteins were identified (duplicates 31%, triplicates 4%, quadruplicates and more 2%). The most frequently identified proteins (that occurred more than 30 times) were dienelactone hydrolase (clCinC), 2-keto-4-pentanoate hydratase (CexF), 3-oxoadipate CoA-transferase (CinE), cytochrome P450 (CYP) and 4-hydroxy-2-oxovalerate aldolase (CexG). The highest frequency of the aforementioned enzymes can be explained by the fact that they conduct more distant reactions in their metabolic pathways and those substrates can participate in other pathways. The most frequently identified enzymes belonged to catechol 1,2-dioxygenation and catechol 2,3-dioxygenation route. Only for *S. bauzanensis* DSM2271 all seven proteins from catechol 2,3-dioxygenation route and only for strain KK2 all four proteins from catechol 1,2-dioxygenation route (for chlorocatechol) were identified (Figure S4C). Those results support the thesis that *Sphingopyxis*, like other sphingomonads, are specialized in the degradation of a particular compound rather than degrading a wide range of substrates [59]. Since the catechol is the main intermediate in the monoaromatic hydrocarbons biodegradation pathways, WS5Ap3 could be involved in degradation of benzene, phenol, benzoic acid or derivatives thereof. Catechol is also an intermediate product in the naphthalene and biphenyl pathways, so these compounds could also support growth of this strain, which may also confirm the presence of trans-o-hydroxybenzylidenepyruvate hydrolase-aldolase (NE), salicylaldehyde dehydrogenase (NF) (naphthalene-salicylate route) and biphenyl-2,3-diol 1,2-dioxygenase (BpC), 2,6-dioxo-6-phenylhexa-3-enoate hydrolase (BpD), benzolate 1,2-dioxygenase (BpG), dihydroxycyclohexadiene carboxylate dehydrogenase (BpH) (biphenyl-benzoate route) coding sequences in the genome (Figure S4C). However, the effect of individual aromatic compounds on the growth of this strain requires further study.

Plasmids present in bacteria isolated from extreme environments often contain catabolic genes (e.g., for xenobiotic degradation) that are essential for survival in harsh conditions [60,61]. A thorough analysis of the WS5A3p genome led to the identification of two potential plasmid sequences. Their comparison to other known catabolic plasmid sequences derived from *Sphingomonadaceae* family is shown in Figure 3. Both of them have a mosaic structure of segments identified in other plasmids. The fact that the segments of high similarity are flanked by transposases suggests that they were acquired by fusion with ancestral plasmids [62]. The sequence of the larger plasmid—pSPMK1—has regions of high similarity to plasmids pMI1, pMI2 and pMI3 isolated from *Sphingobium* sp. MI1205—another γ-HCH degrader [59]. One region of pSPMK1 has exclusive similarity to plasmid pHSL3 from *Sphingomonas* sp. JJ-A5 (Genbank: CP018224.1), a chlorinated pesticide (alachlor) degrading bacterium. This region contains mostly conjugal transfer module, in particular, type IV secretion system genes. The shorter plasmid—pSPMK2—has the largest region similar to plasmids also derived from γ-HCH degrading microorganisms like pISP3 from *Sphingomonas* sp. MM-1 [63] and pSRL3 from *S. indicum* B90A [64,65]. Except for the first 8 kbp this region covers almost the entire sequence. Homologous regions include the stabilization and transfer module also present in pSA4 from non-γ-HCH-degrading strain *Novosphingobium resinovorum* [66]. The first 8 kbp of pSPMK2 is present only in the pMI1 plasmid (however in a different part of pMI1 than the sequence present in contig 1), and a small part in pCHQ1 from *S. japonicum* UT26S.

Since WS5A3p created clear zones on LB plates covered with γ-HCH and showed its degradation ability in a liquid medium, the genome sequence was analyzed for the presence of genes coding proteins involved in xenobiotic degradation pathways. pSPMK1 contains fragments of high similarity to the genes encoding a major pathway for the aerobic degradation of HCH isomers - the Lin pathway: *linA*, *linB*, *linX*, *linEb*, *linF*, and *linGHIJ* genes, whereas pSPMK2 contains *linC* and *linDER* genes (Table S3). Likely a consequence of numerous acquisitions, loss and structural rearrangements of mobile elements among different species, we can distinguish different localizations of *lin* genes. The location of *lin* genes ranges from almost solely chromosomally encoded in *S. japonicum* UT26S, to highly dispersed on multiple plasmids as in *Sphingobium* sp. MI1205 or MM-1 strains [62,67]. Numerous lines of evidence indicate that the accumulation of these genes is an ongoing process, as there are microorganisms identified with partial sets of *lin* genes in their genomes [68]. In *S. lindanitolerans* WS5A3p, the genes comprising the entire γ-HCH degradation pathway were detected exclusively on the two plasmids. The evolution and horizontal transfer of *lin* genes is connected with transposons [69]. By searching the IS FINDER database [70] we identified multiple insertion sequences (IS) in the WS5A3p genome, with IS*6* (n = 12) and IS*21* (n = 8) as the major IS families. The IS*6100* involved in the distribution of *linA* to *linF* genes in sphingomonad strains [12,42], belonging to the family of IS*6* insertion sequences were detected solely on pSPMK1 and pSPMK2 plasmids, not the chromosome (Figure S5). On pSPMK1 we identified seven complete IS*6100* and two partial matches on opposite edges of scaffold sequence, whereas on pSPMK2 only one complete IS*6100* and two partial matches. On pSPMK1 all of the *lin* genes are associated with at least one IS*6100* in close proximity. Arrangement of *linA* and *linX* genes resembles mostly the pattern detected in pMI2 where one of the flanking IS*6100* is localized distantly from *linA*. On the other hand, the *linB* gene is surrounded directly by IS*6100* sequences like on *Sphingobium* sp. TKS chromosome. Differently, than in other sphingomonad's plasmids, *linC* gene is not flanked by two IS*6100* sequences [42]. In the case of pSPMK2, only partial IS*6100* sequence downstream *linC* gene was identified (Figure S5).

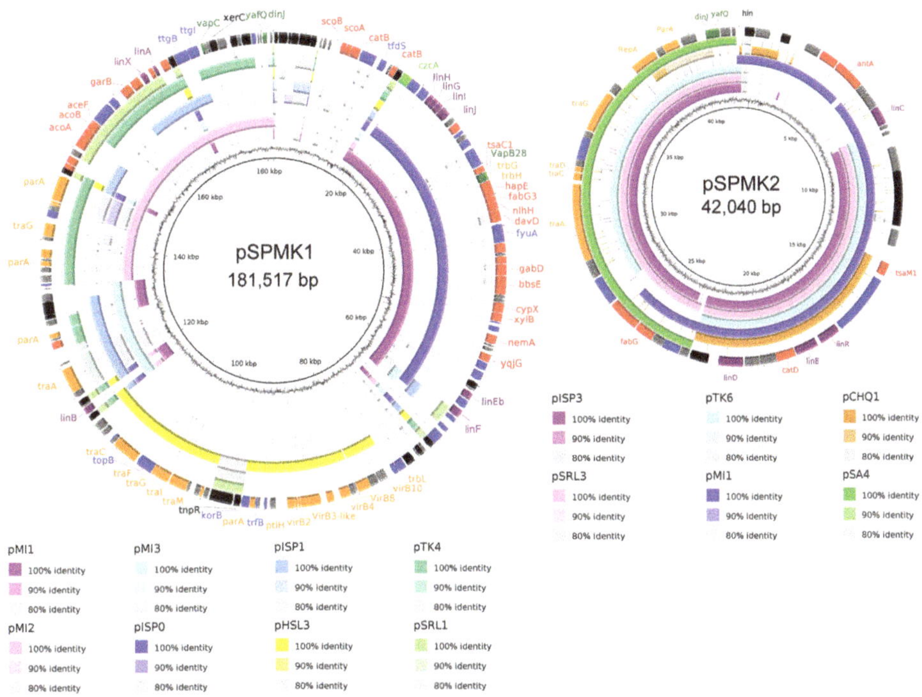

Figure 3. Comparative map of plasmid pSPMK1 and plasmid pSPMK2. Sequence similarity between pSPMK1 and pSPMK2 compared to selected plasmids derived from representative *Sphingomonadaceae*. The first ring represents GC content, rings 2–9 (for pSPMK1) and 2–7 (for pSPMK2) represents the similarity between chosen reference sequences. The last ring for both plasmids indicates open reading frames identified in the sequence, colored according to their function: black—transposases, integrases; purple—*lin* genes; orange—transfer, conjugation, partition genes; red—other catabolic genes; dark green—toxin/anti-toxin; light green—heavy metal resistance; blue—other genes, grey—hypothetical proteins.

Protein sequences of lindane upper degradation pathway proteins LinA and LinB are characterized by a high level of polymorphism and this alters their enantio- and stereoselectivity [71]. Sequences of Lin proteins identified in WS5A3p are almost identical to those from the archetypal strain UT26. The most differentiated in amino acid composition is an enzyme from the γ-HCH upper degradation pathway— LinA (named here LinA type-1a), which shows 98.7% identity with a LinA type 1 (also described as LinA2) [12] that was described to be more efficient in γ-HCH degradation than LinA1 from *S. indicum* B90A [72] to which it has a 90.3% identity (Figure 4). Also, three other variants of the LinA protein have been proposed in the literature [16,73–75]. To distinguish them, we use the nomenclature LinA3 (from *Sphingobium* sp. HDIPO4) and LinA type-2 and LinA type-3 (from the soil metagenomes). Their identities with WS5A3p LinA type-1a protein are 93.6% and 86.6%, respectively. LinA type-1a protein has conserved D25 and H73 residues necessary for its specific enzymatic activity. It differs from LinA type-1 and type-3 by A23G and I35V, which is unique for WS5A3p. While G23 is involved in protein thermostability [16], V35 have not been yet identified as accountable for specific protein features and need to be further investigated. Other LinA type-1a residues are the same as in

LinA type-1. Based on the presence of K20, L96, and A131 residues we can predict that LinA type-1a will exhibit preference to the (−) enantiomers [76].

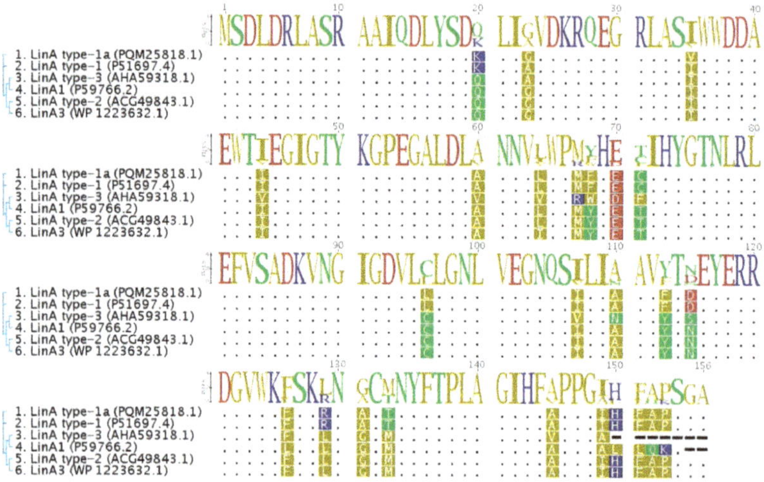

Figure 4. Multiple sequence alignment of LinA protein variants. The amino acid sequence of LinA type-1a from WS5A3p strain (1) compared to previously known variants of this protein (LinA type-1 from *Sphingobium japonicum* UT26 (2), LinA1 from *Sphingobium indicum* B90A (4), LinA3 from *Sphingobium sp.* HDIPO4 (6), LinA type-2 (5) and LinA type-3 from soil metagenomes (3)). Sequences were aligned and visualized using Geneious [27]. Dots correspond to the identical amino acids, dashes indicate the gaps. Phylogenetic relationships are indicated next to the sequence names and accession numbers.

The genes encoding putative ABC transporter (*linKLMN*) necessary for γ-HCH utilization, reported for *Sphingobium* sp. were not identified in the WS5A3p genome. The highest amino acid identity of one of the WS5A3p ABC-transporter proteins to products of those genes were: 61% to LinK, 68% to LinL, 56% to LinM and 50% to LinN. The fact that homologs of the mentioned WS5A3p chromosomally encoded, putative ABC-transporter system are widely distributed among *Sphingopyxis* sp. with low identity levels (e.g., 42–75% identities at aa level for linKLMN identified in UT26 and MM-1 strains [62]) seems consistent with the hypothesis that several variants of these proteins underwent convergent evolution to facilitate lindane utilization [73] and more, this divergence roughly reflects the phylogenetic relationship between the strains [62].

4. Conclusions

In this work, we analyzed the genome of *Sphingopyxis lindanitolerans* WS5A3p derived from long-term pesticide-contaminated soil and capable of degrading xenobiotics. As predicted by its nucleotide sequence, WS5A3p degrades γ-HCH and may play an important role in the soil microbial community contaminated with this pesticide. This ability has been acquired by the process of horizontal gene transfer of *lin* genes. We also identified a new variant of LinA type-1 protein, namely LinA type-1a. This is the first well-documented communication of horizontal gene transfer of lindane degrading machinery to *Sphingopyxis* sp. Furthermore, a core protein set present in all sequenced representatives of this genus was described, as well as the unique protein sets identified for each genome, providing data for further investigations.

Supplementary Materials: The following are available online at http://www.mdpi.com/2073-4425/10/9/688/s1, Figure S1: γ-HCH degrading abilities of *Sphingopyxis lindanitolerans* WS5A3p, Figure S2: Accumulation plots of the *Sphingopyxis pangenome*, Figure S3: Correlation between minimal phylogenetic distance to the closest neighbour and number of unique protein clusters in the genome, Figure S4: Identification of enzymes potentially involved in aromatic compound degradation, Figure S5: Organization of lin genes (blue) and IS6100 (green) identified on *Sphingopyxis lindanitolerans* WS5A3p plasmids, Table S1: General statistics of the *Sphingopyxis lindanitolerans* WS5A3p genome, Table S2: General information about 44 *Sphingopyxis* genomes used in this work, Table S3: Genome coordinates and locus tags of *lin* genes identified in WS5A3p genome sequence.

Author Contributions: L.L., A.S. conceived and directed the studies. DNA isolation, DNA libraries preparation, genome sequencing, and bioinformatics analysis were performed by M.A.K. with the support of L.L. The manuscript was written by M.A.K. and L.L., consulted and corrected by A.S., A.D. Funding for this work was provided by L.L. and A.D. All authors read and approved the final manuscript.

Funding: This work was supported by the European Union's European Regional Development Fund through the Innovative Economy Operational Program, 2007–2013 (project support agreement POIG.01.01.02-14-054/09-00).

Conflicts of Interest: The authors declare that the research was conducted in the absence of any commercial or financial relationships that could be construed as a potential conflict of interest.

References

1. Nolan, K.; Kamrath, J.; Levitt, J. Lindane toxicity: A comprehensive review of the medical literature. *Pediatr. Dermatol.* **2012**, *29*, 141–146. [CrossRef] [PubMed]
2. Vijgen, J.; Abhilash, P.C.; Li, Y.F.; Lal, R.; Forter, M.; Torres, J.; Singh, N.; Yunus, M.; Tian, C.; Schaffer, A.; et al. Hexachlorocyclohexane (HCH) as new Stockholm Convention POPs—A global perspective on the management of Lindane and its waste isomers. *Environ. Sci. Pollut. Res.* **2011**, *18*, 152–162. [CrossRef] [PubMed]
3. Manickam, N.; Pathak, A.; Saini, H.S.; Mayilraj, S.; Shanker, R. Metabolic profiles and phylogenetic diversity of microbial communities from chlorinated pesticides contaminated sites of different geographical habitats of India. *J. Appl. Microbiol.* **2010**, *109*, 1458–1468. [CrossRef] [PubMed]
4. Böltner, D.; Moreno-morillas, S.; Ramos, J. 16S rDNA phylogeny and distribution of lin genes in novel hexachlorocyclohexane-degrading *Sphingomonas* strains. *Environ. Microbiol.* **2005**, *7*, 1329–1338. [CrossRef] [PubMed]
5. Verma, H.; Rani, P.; Singh, A.K.; Kumar, R.; Dwivedi, V.; Negi, V.; Lal, R. *Sphingopyxis flava* sp. nov., isolated from a hexachlorocyclohexane (HCH)-contaminated soil. *Int. J. Syst. Evolut. Microbiol.* **2015**, *65*, 3720–3726. [CrossRef] [PubMed]
6. Jindal, S.; Dua, A.; Lal, R. *Sphingopyxis indica* sp. nov., isolated from a high dose point hexachlorocyclohexane (HCH)- contaminated dumpsite. *Int. J. Syst. Evolut. Microbiol.* **2013**, *63*, 2186–2191. [CrossRef] [PubMed]
7. Sharma, P.; Verma, M.; Bala, K.; Nigam, A.; Lal, R. *Sphingopyxis ummariensis* sp. nov., isolated from a hexachlorocyclohexane dump site. *Int. J. Syst. Evolut. Microbiol.* **2010**, *60*, 780–784. [CrossRef] [PubMed]
8. Feng, G.D.; Wang, D.D.; Yang, S.Z.; Li, H.P.; Zhu, H.H. Genome-based reclassification of *Sphingopyxis ummariensis* as a later heterotypic synonym of *Sphingopyxis terrae*, with the descriptions of *Sphingopyxis terrae* subsp *terrae* subsp. nov. and *Sphingopyxis terrae* subsp. *ummariensis* subsp. nov. *Int. J. Syst. Evolut. Microbiol.* **2017**, *67*, 5279–5283.
9. Lal, R.; Dadhwal, M.; Kumari, K.; Sharma, P.; Singh, A.; Kumari, H.; Jit, S.; Gupta, S.K.; Nigam, A.; Lal, D.; et al. *Pseudomonas* sp. to *Sphingobium indicum*: A journey of microbial degradation and bioremediation of Hexachlorocyclohexane. *Indian J. Microbiol.* **2008**, *48*, 3–18. [CrossRef] [PubMed]
10. Endo, R.; Kamakura, M.; Miyauchi, K.; Ohtsubo, Y.; Tsuda, M.; Fukuda, M.; Nagata, Y.I. Identification and characterization of genes involved in the downstream degradation pathway of γ -hexachlorocyclohexane in *Sphingomonas paucimobilis* UT26. *J. Bacteriol.* **2005**, *187*, 847–853. [CrossRef] [PubMed]
11. Nagata, Y.; Endo, R.; Ito, M.; Ohtsubo, Y.; Tsuda, M. Aerobic degradation of lindane (γ-hexachlorocyclohexane) in bacteria and its biochemical and molecular basis. *Appl. Microbiol. Biotechnol.* **2007**, *76*, 741–752. [CrossRef] [PubMed]
12. Imai, R.; Nagata, Y.; Fukuda, M.; Takagi, M.; Yano, K. Molecular cloning of a *Pseudomonas paucimobilis* gene encoding a 17-kilodalton polypeptide that eliminates HCl molecules from γ-hexachlorocyclohexane. *J. Bacteriol.* **1991**, *173*, 6811–6819. [CrossRef] [PubMed]
13. Nagata, Y.; Futamura, A.; Miyauchi, K.; Takagi, M. Two different types of dehalogenases, LinA and LinB, involved in γ- hexachlorocyclohexane degradation in *Sphingomonas paucimobilis* UT26 are localized in the periplasmic space without molecular processing. *J. Bacteriol.* **1999**, *181*, 5409–5413. [PubMed]

14. Nagata, Y.; Hatta, T.; Imai, R.; Kimbara, K.; Fukuda, M.; Yano, K.; Takagi, M. Purification and characterization of γ -hexachlorocyclohexane (γ -HCH) dehydrochlorinase (LinA) from *Pseudomonas paucimobilis*. *Biosci. Biotechnol. Biochem.* **1993**, *57*, 1582–1583. [CrossRef]
15. Nagata, Y.; Mori, K.; Takagi, M.; Murzin, A.G.; Damborsk, J. Identification of protein fold and catalytic residues of γ-hexachlorocyclohexane dehydrochlorinase LinA. *Proteins Struct. Funct. Genet.* **2001**, *45*, 471–477. [CrossRef] [PubMed]
16. Macwan, A.S.; Kukshal, V.; Srivastava, N.; Javed, S.; Kumar, A.; Ramacheandran, R. Crystal structure of the hexachlorocyclohexane dehydrochlorinase (LinA-Type2): Mutational analysis, thermostability and enantioselectivity. *PLoS ONE* **2012**, *7*. [CrossRef]
17. Okai, M.; Kubota, K.; Fukuda, M.; Nagata, Y.; Nagata, K.; Tanokura, M. Crystallization and preliminary X-ray analysis of- hexachlorocyclohexane dehydrochlorinase LinA from *Sphingobium japonicum* UT26. *Acta Crystallogr. Sect. F Struct. Biol. Cryst. Commun.* **2009**, *65*, 822–824. [CrossRef]
18. Nagata, Y.; Nariya, T.; Ohtomo, R.; Fukuda, M.; Yano, K.; Takagi, M. Cloning and sequencing of a dehalogenase gene encoding an enzyme with hydrolase γ-hexachlorocyclohexane involved in the degradation of y-hexachlorocyclohexane. *J. Bacteriol.* **1993**, *175*, 6403–6410. [CrossRef]
19. Kmunícek, J.; Hynková, K.; Jedlicka, T.; Nagata, Y.; Negri, A.; Gago, F.; Wade, R.C.; Damborsky, J. Quantitative analysis of substrate specificity of haloalkane dehalogenase LinB from *Sphingomonas paucimobilis* UT26. *Biochemistry* **2005**, *44*, 3390–3401. [CrossRef]
20. Nagata, Y.; Ohtomo, R.; Miyauchi, K.; Fukuda, M.; Yano, K.; Takagi, M. Cloning and sequencing of a 2,5-dichloro-2,5-cyclohexadiene-1,4-diol dehydrogenase gene involved in the degradation of γ-hexachlorocyclohexane in *Pseudomonas paucimobilis*. *J. Bacteriol.* **1994**, *176*, 3117–3125. [CrossRef]
21. Miyauchi, K.; Suh, S.K.; Nagata, Y.; Takagi, M. Cloning and sequencing of a 2,5-dichlorohydroquinone reductive dehalogenase gene whose product is involved in degradation of γ-hexachlorocyclohexane by *Sphingomonas paucimobilis*. *J. Bacteriol.* **1998**, *180*, 1354–1359. [PubMed]
22. Miyauchi, K.; Adachi, Y.; Nagata, Y. Cloning and sequencing of a novel meta-cleavage dioxygenase gene whose product is involved in degradation of γ -Hexachlorocyclohexane in *Sphingomonas paucimobilis*. *J. Bacteriol.* **1999**, *181*, 6712–6719. [PubMed]
23. Kaminski, M.A.; Sobczak, A.; Spolnik, G.; Danikiewicz, W.; Dziembowski, A.; Lipinski, L. *Sphingopyxis lindanitolerans* sp. nov. strain WS5A3pT enriched from a pesticide disposal site. *Int. J. Syst. Evol. Microbiol.* **2018**, *68*, 1–7. [CrossRef] [PubMed]
24. Martin, M. Cutadapt removes adapter sequences from high-throughput sequencing reads. EMBnet. *Journal* **2011**, *17*, 10–12.
25. Joshi, N.; Fass, J. Sickle: A sliding-window, adaptive, quality-based trimming tool for FastQ files [Internet]. 2011. Available online: https://github.com/najoshi/sickle (accessed on 1 December 2017).
26. Bankevich, A.; Nurk, S.; Antipov, D.; Gurevich, A.; Dvorkin, M.; Kulikov, A.S.; Lesin, V.M.; Nikolenko, S.I.; Pham, S.; Prjibrlski, A.D.; et al. SPAdes: A new genome assembly algorithm and its applications to single-cell sequencing. *J. Comput. Biol.* **2012**, *19*, 455–477. [CrossRef] [PubMed]
27. Kearse, M.; Moir, R.; Wilson, A.; Stones-Havas, S.; Cheung, M.; Sturrock, S.; Buxton, S.; Cooper, A.; Markowitz, S.; Duran, C.; et al. Geneious Basic: An integrated and extendable desktop software platform for the organization and analysis of sequence data. *Bioinformatics* **2012**, *28*, 1647–1649. [CrossRef]
28. Tatusova, T.; Dicuccio, M.; Badretdin, A.; Chetvernin, V.; Nawrocki, P.; Zaslavsky, L.; Lomsadze, A.; Pruitt, K.D.; Borodovsky, M.; OSTELL, J. NCBI prokaryotic genome annotation pipeline. *Nucleic Acid. Res.* **2016**, *44*, 6614–6624. [CrossRef] [PubMed]
29. Petersen, T.N.; Brunak, S.; Heijne, G.V.; Nielsen, H. SignalP 4.0: Discriminating signal peptides from transmembrane regions. *Nat. Method* **2011**, *8*, 785–786. [CrossRef]
30. Bose, T.; Haque, M.M.; Reddy, C.; Mande, S.S. COGNIZER: A framework for functional annotation of metagenomic datasets. *PLoS ONE* **2015**, *10*, 1–16. [CrossRef]
31. Calteau, A.; Gachet, M.; Gautreau, G.; Josso, A.; Langlois, J.; Médigue, C.; Cruveiller, S.; Lajus, A.; Pereira, H.; Planel, P.; et al. MicroScope-an integrated resource for community expertise of gene functions and comparative analysis of microbial genomic and metabolic data. *Brief Bioinform.* **2017**, 1–14. [CrossRef]
32. Blin, K.; Shaw, S.; Steinke, K.; Villebro, R.; Ziemert, N.; Lee, S.Y.; Medema, M.H.; Weber, T. antiSMASH 5.0: Updates to the secondary metabolite genome mining pipeline. *Nucleic Acids Res.* **2019**, *47*, W81–W87. [CrossRef] [PubMed]

33. Arndt, D.; Grant, J.R.; Marcu, A.; Sajed, T.; Pon, A.; Liang, Y.J.; Wishart, D.S. PHASTER: A better, faster version of the PHAST phage search tool. *Nucleic Acids Res.* **2016**, *44*, W16–W21. [CrossRef] [PubMed]
34. Grissa, I.; Vergnaud, G.; Pourcel, C. CRISPRFinder: A web tool to identify clustered regularly interspaced short palindromic repeats. *Nucleic Acids Res.* **2007**, *35*, W52–W57. [CrossRef] [PubMed]
35. Lauro, F.M.; Dougald, D.M.; Thomas, T.; Williams, T.J.; Egan, S.; Rice, S.; DeMaere, M.Z.; Ting, L.; Ertan, H.; Johnson, J.; et al. The genomic basis of trophic strategy in marine bacteria. *Proc. Natl. Acad. Sci.USA* **2009**, *106*, 15527–15533. [CrossRef] [PubMed]
36. Ohtsubo, Y.; Nonoyama, S.; Nagata, Y.; Numata, M.; Tsuchikane, K.; Hosoyama, A.; Yamazoe, A.; Tsuda, M.; Fujita, N.; Kawai, F. Complete genome sequence of *Sphingopyxis terrae* strain 203-1 (NBRC 111660), a polyethylene glycol degrader. *Genome Announc.* **2016**, *4*. [CrossRef] [PubMed]
37. Ohtsubo, Y.; Nagata, Y.; Numata, M.; Tsuchikane, K.; Hosoyama, A.; Yamazoe, A.; Tsuda, M.; Fujita, N.; Kawai, F. Complete genome sequence of polyvinyl alcohol-degrading strain *Sphingopyxis* sp. 113P3 (NBRC 111507). *Genome Announc.* **2015**, *3*. [CrossRef] [PubMed]
38. Gan, H.Y.; Gan, H.M.; Tarasco, A.M.; Busairi, N.I.; Barton, H.A.; Hudson, A.O.; Savka, M.A. Whole-genome sequences of five oligotrophic bacteria isolated from deep within Lechuguilla Cave, New Mexico. *Genome Announc.* **2014**, *2*. [CrossRef] [PubMed]
39. Okano, K.; Shimizu, K.; Maseda, H.; Kawauchi, Y.; Utsumi, M.; Itayama, T.; Zhang, Z.Y.; Sugiura, N. Whole-genome sequence of the microcystin-degrading bacterium *Sphingopyxis* sp. strain C-1. *Genome Announc.* **2015**, *3*. [CrossRef]
40. Kaminski, M.A.; Furmanczyk, E.M.; Dziembowski, A.; Sobczak, A.; Lipinski, L. Draft genome sequence of the type strain *Sphingopyxis witflariensis* DSM 14551. *Genome Announc.* **2017**, *5*. [CrossRef]
41. Kaminski, M.A.; Furmanczyk, E.M.; Dziembowski, A.; Sobczak, A.; Lipinski, L. Draft genome sequence of the type strain *Sphingopyxis bauzanensis* DSM 22271. *Genome Announc.* **2017**, *5*. [CrossRef]
42. Oelschlägel, M.; Rückert, C.; Kalinowski, J.; Schmidt, G.; Schlömann, M.; Tischler, D. *Sphingopyxis fribergensis* sp. nov., a soil bacterium with the ability to degrade styrene and phenylacetic acid. *Int. J. Syst. Evolut. Microbiol.* **2015**, *65*, 3008–3015. [CrossRef] [PubMed]
43. García-Romero, I.; Pérez-Pulido, A.J.; González-Flores, Y.E.; Reyes-Ramírez, F.; Santero, E.; Floriano, B. Genomic analysis of the nitrate-respiring *Sphingopyxis granuli* (formerly *Sphingomonas macrogoltabida*) strain TFA. *BMC Genom.* **2016**, *17*, 93. [CrossRef]
44. Ohtsubo, Y.; Nonoyama, S.; Nagata, Y.; Numata, M.; Tsuchikane, K.; Hosoyama, A.; Yamazoe, A.; Tsuda, M.; Fujita, N.; Kawai, F. Complete genome sequence of *Sphingopyxis macrogoltabida* strain 203N (NBRC 111659), a polyethylene glycol degrader. *Genome Announc.* **2016**, *4*. [CrossRef] [PubMed]
45. Pritchard, L.; Glover, R.H.; Humphris, S.; Elphinstone, J.G.; Toth, I.K. Genomics and taxonomy in diagnostics for food security: Soft-rotting enterobacterial plant pathogens. *Anal. Methods* **2016**, *8*, 12–24. [CrossRef]
46. Seemann, T. Prokka: Rapid prokaryotic genome annotation. *Bioinformatics.* **2014**, *30*, 2068–2069. [CrossRef]
47. Segata, N.; Börnigen, D.; Morgan, X.C.; Huttenhower, C. PhyloPhlAn is a new method for improved phylogenetic and taxonomic placement of microbes. *Nat. Commun.* **2013**, *4*, 2304. [CrossRef]
48. Yu, G.; Smith, D.K.; Zhu, H.; Guan, Y.; Lam, T.T.Y. Ggtree: An r package for visualization and annotation of phylogenetic trees with their covariates and other associated data. *Methods Ecol. Evolut.* **2017**, *8*, 28–36. [CrossRef]
49. Stamatakis, A. RAxML Version 8: A tool for phylogenetic analysis and post-analysis of large phylogenies. *Bioinformatics* **2014**, *30*, 1312–1313. [CrossRef]
50. Fu, L.; Niu, B.; Zhu, Z.; Wu, S.; Li, W. CD-HIT: Accelerated for clustering the next-generation sequencing data. *Bioinformatics* **2012**, *28*, 3150–3152. [CrossRef]
51. Kato, H.; Mori, H.; Maruyama, F.; Toyoda, A.; Oshima, K.; Endo, R.; Fuchu, G.; Miyakoshi, M.; Dozono, A.; Ohtsubo, Y. Time-series metagenomic analysis reveals robustness of soil microbiome against chemical disturbance. *DNA Res.* **2015**, *22*, 413–424. [CrossRef]
52. Edgar, R.C. MUSCLE: Multiple sequence alignment with high accuracy and high throughput. *Nucleic Acids Res.* **2004**, *32*, 1792–1797. [CrossRef] [PubMed]
53. Vernikos, G.; Medini, D.; Riley, D.R.; Tettelin, H. Ten years of pan-genome analyses. *Curr Opin Microbiol.* **2015**, *23*, 148–154. [CrossRef] [PubMed]
54. Lefébure, T.; Stanhope, M.J. Evolution of the core and pan-genome of *Streptococcus*: Positive selection, recombination, and genome composition. *Genome Biol.* **2007**, *8*, 71. [CrossRef] [PubMed]

55. Bottacini, F.; Medini, D.; Pavesi, A.; Turroni, F.; Foroni, E.; Riley, D.; Giubellini, V.; Tettelin, H.; Sinderen, D.V.; Ventura, M. Comparative genomics of the genus *Bifidobacterium*. *Microbiology* **2010**, *156*, 3243–3254. [CrossRef] [PubMed]
56. Parthasarathy, S.; Azam, S.; Lakshman Sagar, A.; Narasimha Rao, V.; Gudla, R.; Parapatla, H.; Yakkala, H.; Vemuri, S.G.; Siddavattam, D. Genome-guided insights reveal organophosphate-degrading *Brevundimonas diminuta* as *Sphingopyxis wildii* and define its versatile metabolic capabilities and environmental adaptations. *Genome Biol. Evolut.* **2017**, *9*, 77–81. [CrossRef] [PubMed]
57. Fraser, C.M.; Gocayne, J.D.; White, O.; Adams, M.D.; Clayton, R.A.; Fleischmann, R.D.; Bult, C.J.; Kerlavage, A.R.; Sutton, G.; Kelley, J.M.; et al. The minimal gene complement of *Mycoplasma genitalium*. *Science* **1995**, *270*, 397–403. [CrossRef]
58. Koebnik, R.; Locher, K.P.; Van Gelder, P. Structure and function of bacterial outer membrane proteins: Barrels in a nutshell. *Mol. Microbiol.* **2000**, *37*, 239–353. [CrossRef]
59. Tabata, M.; Ohhata, S.; Nikawadori, Y.; Sato, T.; Kishida, K.; Ohtsubo, Y.; Tsuda, M.; Nagata, Y. Complete genome sequence of a γ-hexachlorocyclohexane-degrading bacterium, *Sphingobium* sp. strain MI1205. *Genome Announc.* **2016**, *4*. [CrossRef]
60. Kumar, S.; Mukerji, K.G.; Lai, R. Molecular aspects of pesticide degradation by microorganisms. *Crit. Rev. Microbiol.* **1996**, *22*, 1–26. [CrossRef]
61. Ramakrishnan, B.; Venkateswarlu, K.; Sethunathan, N.; Megharaj, M. Local applications but global implications: Can pesticides drive microorganisms to develop antimicrobial resistance? *Sci. Total Environ.* **2019**, *654*, 177–189. [CrossRef]
62. Tabata, M.; Ohhata, S.; Nikawadori, Y.; Kishida, K.; Sato, T.; Kawasumi, T.; Kato, H.; Ohtsubo, Y.; Tsuda, M.; Nagata, Y. Comparison of the complete genome sequences of four γ-hexachlorocyclohexane-degrading bacterial strains: Insights into the evolution of bacteria able to degrade a recalcitrant man-made pesticide. *DNA Res.* **2016**, *23*, 581–599. [CrossRef]
63. Tabata, M.; Ohtsubo, Y.; Ohhata, S.; Tsuda, M.; Nagata, Y. Complete genome sequence of the γ-hexachlorocyclohexane-degrading bacterium *Sphingomonas* sp. strain MM-1. *Genome Announc.* **2013**, 1. [CrossRef] [PubMed]
64. Anand, S.; Sangwan, N.; Lata, P.; Kaur, J.; Dua, A.; Singh, A.K.; Verma, M.; Kaur, J.; Khurana, J.P.; Khurana, P.; et al. Genome sequence of *Sphingobium indicum* B90A, a hexachlorocyclohexane-degrading bacterium. *J Bacteriol.* **2012**, *194*, 4471–4472. [CrossRef] [PubMed]
65. Verma, H.; Bajaj, A.; Kumar, R.; Kaur, J.; Anand, S.; Nayyar, N.; Puri, A.; Singh, Y.; Khurana, J.P.; Lal, R. Genome organization of *Sphingobium indicum* B90A: An archetypal hexachlorocyclohexane (HCH) degrading genotype. *Genome Biol. Evolut.* **2017**, *9*, 2191–2197. [CrossRef] [PubMed]
66. Hegedus, B.; Kos, P.B.; Balint, B.; Maroti, G.; Gan, H.M.; Perei, K.; Rákhely, G. Complete genome sequence of *Novosphingobium resinovorum* SA1, a versatile xenobiotic-degrading bacterium capable of utilizing sulfanilic acid. *J. Biotechnol.* **2017**, *241*, 76–80. [CrossRef] [PubMed]
67. Tabata, M.; Endo, R.; Ito, M.; Ohtsubo, Y.; Kumar, A.; Tsuda, M.; Nagata, Y. The *lin* genes for γ-hexachlorocyclohexane degradation in *Sphingomonas* sp. MM-1 proved to be dispersed across multiple plasmids. *Biosci. Biotechnol. Biochem.* **2011**, *75*, 466–472. [CrossRef] [PubMed]
68. Pearce, S.L.; Oakeshott, J.G.; Pandey, G. Insights into ongoing evolution of the hexachlorocyclohexane catabolic pathway from comparative genomics of ten *Sphingomonadaceae* strains. *G3 (Bethesda).* **2015**, *5*, 1081–1094. [CrossRef] [PubMed]
69. Dogra, C.; Raina, V.; Pal, R.; Suar, M.; Lal, S.; Gartemann, K.H.; Holliger, C.; Meer, J.R.V.D.; Lal, R. Organization of lin genes and IS6100 among different strains of hexachlorocyclohexane-degrading *Sphingomonas paucimobilis*: Evidence for horizontal gene transfer. *J. Bacteriol.* **2004**, *186*, 2225–2235. [CrossRef]
70. Siguier, P.; Perochon, J.; Lestrade, L.; Mahillon, J.; Chandler, M. ISfinder: The reference centre for bacterial insertion sequences. *Nucleic Acids Res.* **2006**, *34*, 32–36. [CrossRef]
71. Sharma, P.; Pandey, R.; Kumari, K.; Pandey, G.; Jackson, C.J.; Russell, R.J.; Oakeshott, J.G.; Lal, R. Kinetic and sequence-structure-function analysis of known LinA variants with different hexachlorocyclohexane isomers. *PLoS ONE.* **2011**, *6*. [CrossRef]
72. Kumari, R.; Subudhi, S.; Suar, M.; Dhingra, G.; Raina, V.; Dogra, C.; Lal, S.; Meer, J.R.V.D.; Holliger, C.; Lal, R. Cloning and characterization of *lin* genes responsible for the degradation of hexachlorocyclohexane isomers by *Sphingomonas paucimobilis* strain B90. *Appl. Environ Microbiol.* **2002**, *68*, 6021–6028. [CrossRef] [PubMed]

73. Verma, H.; Kumar, R.; Oldach, P.; Sangwan, N.; Khurana, J.P.; Gilbert, J.A.; Lal, R. Comparative genomic analysis of nine *Sphingobium* strains: Insights into their evolution and hexachlorocyclohexane (HCH) degradation pathways. *BMC Genomics* **2014**, *15*, 1014. [CrossRef] [PubMed]
74. Macwan, A.S.; Javed, S.; Kumar, A. Isolation of a novel thermostable dehydrochlorinase (LinA) from a soil metagenome. *3Biotech* **2011**, *1*, 193–198. [CrossRef] [PubMed]
75. Shrivastava, N.; Prokop, Z.; Kumar, A. Novel LinA type 3 δ-hexachlorocyclohexane dehydrochlorinase. *Appl. Environ. Microbiol.* **2015**, *81*, 7553–7559. [CrossRef] [PubMed]
76. Shrivastava, N.; Macwan, A.S.; Kohler, H.P.E.; Kumar, A. Important amino acid residues of hexachlorocyclohexane dehydrochlorinases (LinA) for enantioselective transformation of hexachlorocyclohexane isomers. *Biodegradation* **2017**, *28*, 171–180. [CrossRef] [PubMed]

© 2019 by the authors. Licensee MDPI, Basel, Switzerland. This article is an open access article distributed under the terms and conditions of the Creative Commons Attribution (CC BY) license (http://creativecommons.org/licenses/by/4.0/).

genes

MDPI

Article

Biphenyl/PCB Degrading *bph* Genes of Ten Bacterial Strains Isolated from Biphenyl-Contaminated Soil in Kitakyushu, Japan: Comparative and Dynamic Features as Integrative Conjugative Elements (ICEs)

Jun Hirose [1,*], Hidehiko Fujihara [2], Takahito Watanabe [3], Nobutada Kimura [4], Hikaru Suenaga [5], Taiki Futagami [6], Masatoshi Goto [7], Akiko Suyama [2] and Kensuke Furukawa [2]

1 Department of Applied Chemistry, Faculty of Engineering, University of Miyazaki, Miyazaki 889-2192, Japan
2 Department of Food and Fermentation Sciences, Faculty of Food and Nutrition Sciences, Beppu University, Beppu 874-8501, Japan; fujihara@nm.beppu-u.ac.jp (H.F.); aksuyama@nm.beppu-u.ac.jp (A.S.); kfurukaw@nm.beppu-u.ac.jp (K.F.)
3 Research Institute for Sustainable Humanosphere, Kyoto University, Uji 611-0011, Japan; takahito@rish.kyoto-u.ac.jp
4 Bioproduction Research Institute, National Institute of Advanced Industrial Science and Technology (AIST), Tsukuba 305-8566, Japan; n-kimura@aist.go.jp
5 Biotechnology Research Institute for Drug Discovery, National Institute of Advanced Industrial Science and Technology (AIST), Tokyo 135-0064, Japan; suenaga-hikaru@aist.go.jp
6 Education and Research Center for Fermentation Studies, Faculty of Agriculture, Kagoshima University, Kagoshima 890-0065, Japan; futagami@chem.agri.kagoshima-u.ac.jp
7 Faculty of Agriculture, Saga University, Saga 840-8502, Japan; mgoto@cc.saga-u.ac.jp
* Correspondence: jhirose@cc.miyazaki-u.ac.jp; Tel.: +81-985-58-7322

Received: 30 March 2019; Accepted: 20 May 2019; Published: 27 May 2019

Abstract: We sequenced the entire genomes of ten biphenyl/PCB degrading bacterial strains (KF strains) isolated from biphenyl-contaminated soil in Kitakyushu, Japan. All the strains were Gram-negative bacteria belonging to β- and γ-proteobacteria. Out of the ten strains, nine strains carried a biphenyl catabolic *bph* gene cluster as integrative conjugative elements (ICEs), and they were classified into four groups based on the structural features of the *bph* genes. Group I (five strains) possessed *bph* genes that were very similar to the ones in *Pseudomonas furukawaii* KF707 (formerly *Pseudomonas pseudoalcaligenes* KF707), which is one of the best characterized biphenyl-utilizing strains. This group of strains carried salicylate catabolic *sal* genes that were approximately 6-kb downstream of the *bph* genes. Group II (two strains) possessed *bph* and *sal* genes similar to the ones in KF707, but these strains lacked the *bphX* region between *bphC* and *bphD*, which is involved in the downstream catabolism of biphenyl. These *bph-sal* clusters in groups I and II were located on an integrative conjugative element that was larger than 110 kb, and they were named $ICE_{bph\text{-}sal}$. Our previous study demonstrated that the $ICE_{bph\text{-}sal}$ of *Pseudomonas putida* KF715 in group II existed both in an integrated form in the chromosome (referred to as $ICE_{bph\text{-}sal}$KF715 (integrated)) and in a extrachromosomal circular form (referred to as $ICE_{bph\text{-}sal}$ (circular)) (previously called pKF715A, 483 kb) in the stationary culture. The $ICE_{bph\text{-}sal}$ was transferred from KF715 into *P. putida* AC30 and *P. putida* KT2440 with high frequency, and it was maintained stably as an extrachromosomal circular form. The $ICE_{bph\text{-}sal}$KF715 (circular) in these transconjugants was further transferred to *P. putida* F39/D and then integrated into the chromosome in one or two copies. Meanwhile, group III (one strain) possessed *bph* genes, but not *sal* genes. The nucleotide sequences of the *bph* genes in this group were less conserved compared to the genes of the strains belonging to groups I and II. Currently, there is no evidence to indicate that the *bph* genes in group III are carried by a mobile element. Group IV (two strains) carried *bph* genes as ICEs (59–61 kb) that were similar to the genes found in Tn*4371* from *Cupriavidus oxalacticus* A5 and ICE_{KKS102}*4677* from the *Acidovorax* sp. strain KKS102. Our study found that *bph* gene islands have integrative functions, are transferred among soil bacteria, and are diversified through modification.

Keywords: biphenyl; *bph* gene; integrative conjugative element; genome sequence

1. Introduction

To date, a number of biphenyl-utilizing bacteria have been isolated and characterized in terms of the degradation of polychlorinated biphenyls (PCBs), which are serious environmental contaminants that are prevalent worldwide [1–3]. These strains include both Gram-negative and Gram-positive bacteria. Biphenyl catabolic enzymes co-metabolize certain PCBs into chlorobenzoic acids. It is well documented that PCB degradation is highly dependent on chlorine substitutions, such as the number and positions of the substituted chlorine [4]. Degradation capabilities are also strain dependent. For the first time, biphenyl catabolic *bph* genes were cloned from *Pseudomonas furukawaii* KF707 [5]. Since then, *bph* genes were cloned from various strains, including both Gram-negative and Gram-positive bacteria, and then they were analyzed in detail [3]. These studies indicated that some strains possessed *bph* genes that were very similar to the ones in KF707 in terms of gene organization and nucleotide sequences, although some strains possessed *bph* gene clusters that were different from KF707 and diversified from each other [3]. Some *bph* genes are located on chromosomes, whereas others are present on plasmids. The *bph* genes of *C. oxalacticus* A5 (formerly *Ralstonia* sp. strain A5) [6] and *Acidovorax* sp. KKS102 [7] are located on the ICEs. Gram-positive *Rhodococcus jostii* RHA1 possesses multiple *bph* genes on large linear plasmids [8,9].

The typical *bph* gene cluster shown in KF707 is composed of *bphRA1A2A3A4BCX0X1X2X3D* (Figure 1) [3]. Briefly, the biphenyl dioxygenase is a multi-component enzyme encoded by *bphA1A2A3A4,* and it catalyzes the initial oxygenation of biphenyl, converting the biphenyl into dihydrodiol, where *bphA1* and *bphA2* encode a large and a small subunit of the terminal dioxygenase, respectively. *bphA3* encodes ferredoxin, and *bphA4* encodes ferredoxin reductase. The dihydrodiol compound is then converted to a dihydroxy-compound by the dehydrogenase encoded by the *bphB*. The dihydroxy-compound is then degraded into 2-hydroxy-6-oxo-6-phenylhexa-2,4-dienoic acid by the ring-cleavage dioxygenase encoded by the *bphC*. Then, the ring *meta*-cleavage compound is degraded into benzoic acid and 2-hydroxypenta-2,4-dienoic acid by the hydrolase (encoded by the *bphD*). BphX1X2X3 is responsible for the further degradation of 2-hydroxypenta-2,4-dienoic acid into acetyl CoA. These structural *bph* genes are regulated by the *bphR* located on the *bph* gene cluster [10,11] Among these *bph* genes, *bphA1* is critically important for substrate specificity, i.e., the biodegradation capability for various aromatic compounds, including PCBs [12–14].

Previously, we isolated more than ten biphenyl-utilizing bacterial strains (KF strains) from biphenyl-contaminated soil in Kitakyushu, Japan [15]. Among these KF strains, we determined the complete nucleotide sequence of the *P. putida* KF715 genome [16], which revealed five replicons: one circular chromosome and four plasmids. Southern blot analysis indicated that the majority of the KF715 cell population carried the *bph-sal* cluster on its chromosome. However, a small population of cells carried the cluster on a huge extrachromosomal circular element called pKF715A (483 kb). In addition, this element carried the *oriT* sequence, the *repA* gene involved in replication, the conjugal transfer gene (*tra*), and the partitioning gene (*par*). In this study, we were interested in how the KF strains isolated from the same location carried *bph* gene clusters along with other catabolic genes. We performed whole genome sequencing of these strains, anticipating that their genome information would shed light on the diversity and evolution of biphenyl-utilizing bacteria. Our results indicated that specific DNA blocks, including the *bph* gene cluster, were integrated within glycine tRNA (tRNA-Gly) genes and that some blocks contained an integrase gene, illustrating that certain *bph* gene islands had integrative functions.

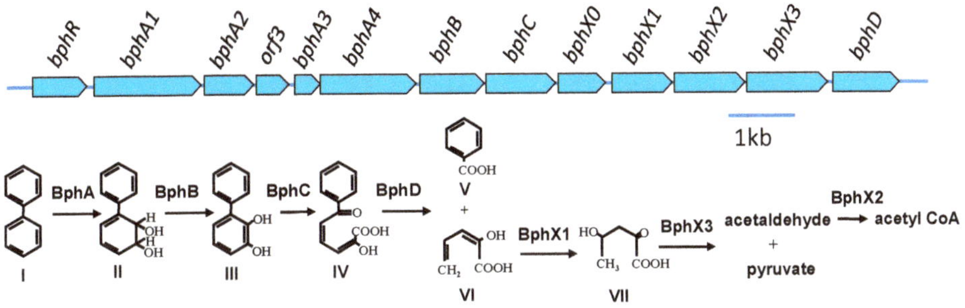

Figure 1. Catabolic pathway of the biphenyl degradation and organization of the *bph* gene cluster in *P. furukawaii* KF707. Compounds: I, biphenyl; II, 2,3-dihydroxy-4-phenylhexa-4,6-diene (dihydrodiol compound); III, 2,3-dihydroxybiphenyl; IV, 2-hydroxy-6-oxo-6-phenylhexa-2,4-dienoic acid (biphenyl *meta*-cleavage compound: HOPD); V, benzoic acid; VI, 2-hydroxypenta-2,4-dienoic acid; VII, 4-hydroxy-2-oxovalerate. Enzymes: BphA1-BphA2-BphA3-BphA4, biphenyl dioxygenase; BphB, dihydrodiol dehydrogenase; BphC, 2,3-dihydroxybiphenyl dioxygenase; BphX0, glutathione *S*-transferase; BphX1, 2-hydroxypenta-2,4-dienoate hydratase; BphX2, acetaldehyde dehydrogenase (acylating); BphX3, 4-hydoxy-2-oxovalerate aldolase; BphD, 2-hydroxy-6-oxo-6-phenylhexa-2,4-dieonic acid hydrolase. The BphR protein, belonging to the GntR family, is a transcriptional regulator involved in the expression of *bphR* and *bphX0X1X2X3D*. The function of *orf3* remains unclear.

2. Materials and Methods

2.1. Bacterial Strains and Cultivation

The bacterial strains (KF strains) used in this study are presented in Table 1. These biphenyl/PCB degrading strains were isolated at the biphenyl-manufacturing factory in Kitakyushu, Japan [15], and were deposited to the National Biological Resource Center (NBRC). Strains KF701, KF703, KF707, KF708, and KF712 were renamed based on the 16S rRNA sequence [17–21]. *P. putida* AC30Bph+ and *P. putida* KT2440Bph+ were obtained through conjugation with *P. putida* KF715, and these two transconjugants grew on biphenyl as a sole source of carbon and energy, as described in Reference [22]. *P. putida* F39/D is a mutant of toluene utilizing *P. putida* F1 [23], in which the *todD* gene is defective. This strain was used as a recipient for the conjugation experiments. The growth of the KF strains and the transconjugants was examined on a basal salt agar medium with biphenyl and various aromatic compounds as described in Reference [5].

Table 1. Biphenyl/PCB degrading KF strains used in this study.

Strain	NBRC Number	DDBJ/EMBL/GenBank Accession Number	References
Pseudomonas abietaniphila KF701	110664	B BQJ01000001-BBQJ01000140	[15,17]
Pseudomonas aeruginosa KF702	110665	B BQK01000001-BBQK01000091	[15,24]
Pseudomonas putida KF703	110666	BBQL01000001-BBQL01000135	[15,18]
Pseudomonas furukawaii KF707	110670	AP014862	[15,19,25]
Cupriavidus basilensis KF708	110671	B BQM01000001-BBQM01000062	[15,20]
Cupriavidus pauculus KF709	110672	BBQN01000001-BBQN01000227	[15,26]
Pseudomonas toyotomiensis KF710	110674	BBQO01000001-BBQO01000029	[15,27]
Comamonas testosteroni KF712	110673	BBQP01000001-BBQP01000097	[15,21]
Pseudomonas putida KF715	110667	AP015029, AP015030-AP015033	[15,16,22,28]
Pseudomonas stutzeri KF716	110668	BBQQ01000001-BBQQ01000030	[29]

2.2. Genome Sequencing and Computational Analysis

The whole genome sequences of the KF strains were determined by the National Institute of Technology and Evaluation (NITE), using a combination of shotgun sequencing on a 454 GS FLX+ system (Roche, Basel, Switzerland) and paired-end sequencing on a HiSeq sequencing system (Illumina,

San Diego, CA, USA) as previously reported in Reference [17]. The reads obtained by the two systems were assembled using the Newbler version 2.8 (Roche). The draft sequence data of the *P. furukawaii* KF707 and *P. putida* KF715 were further completed using the GenoFinisher computer program (http://www.ige.tohoku.ac.jp/joho/gf_e/). Remaining gaps between the contigs were closed using polymerase chain reaction (PCR) amplification and DNA sequencing with standard Sanger technology. The genome sequences were annotated using the RAST (Rapid Annotation using Subsystem Technology) server [30]. The identification of the coding genes was checked using a BLAST search (http://www.ncb.gov/BLAST/). Sequence comparison was performed using EasyFig Ver. 2.1 [31], and the map was generated using drawGeneArrows3 (http://www.ige.tohoku.ac.jp/joho/labhome/tool.html). The whole genome sequences of the 10 strains were deposited to the DDBJ/EMBL/GenBank under the accession numbers presented in Table 1. The nucleotide sequences of the integrative conjugative elements in KF701, KF702, KF703, KF707, KF708, KF710, KF712, and KF716 were deposited separately. Their accession numbers were LC469607, LC469608, LC469609, LC469610, LC469611, LC469612, LC469613, and LC469614, respectively.

2.3. Phylogenetic Analysis and Gene Alignment

The nucleotide sequences were aligned computationally using the ClustalW algorithm as in Reference [32]. Phylogenetic trees were generated using the neighbor-joining method with the Mega 6.0 program [33]. The trees were evaluated through bootstrap resampling (500 replicates).

2.4. Conjugation Experiments

Transfer of the Bph^+ phenotype (*P. putida* AC30Bph+ and *P. putida* KT2440Bph+) by conjugation into the recipient cells (*P. putida* F39/D) was carried out through filter mating as described in Reference [22].

2.5. DNA Manipulation

DNA isolation, Southern blot analysis, PCR, DNA sequencing, and other DNA manipulations were performed according to standard procedures as described in Reference [34]. Pulsed-field gel electrophoresis was performed in accordance with the manufacturer's instructions (Bio-Rad Laboratories, Hercules, CA, USA).

3. Results

3.1. KF Strains and their Genomic Features

In this study, we found that various types of *bph* genes were present in the ten biphenyl/PCB degrading strains isolated from the biphenyl-contaminated soil. Figure 2 shows the 16S rRNA phylogenetic tree of the ten strains. Among the ten strains, seven belonged to the *Pseudomonas* genus, two strains to *Cupriavidus* spp., and one to *Comamonas* spp. These were all Gram-negative bacteria belonging to β-proteobacteria (*Comamonas* and *Cupriavidus*) and γ-proteobacteria (*Pseudomonas*). The complete genome sequences were determined for *P. furukawaii* KF707 and *P. putida* KF715. KF707 possessed one circular chromosome of 6,242,949 bp and one plasmid (pKF707) of 59,819 bp. On the other hand, *P. putida* KF715 possessed one circular chromosome of 6,583,376 bp and four plasmids as previously reported in Reference [16]. The total length of the contigs (total number of contigs, >500 bp) of the remaining strains were as follows: *P. abietaniphila* KF701, 6,886,250 bp (140) [17]; *P. aeruginosa* KF702, 7,167,540 bp (91) [24]; *P. putida* KF703, 6,434,897 bp (135) [18]; *C. basilensis* KF708, 7,826,077 bp (62) [20]; *C. pauculus* KF709, 6,826,799 bp (227) [24]; *P. toyotomiensis* KF710, 5,596,721 bp (29) [27]; *C. testosteroni* KF712, 5,890,323 bp (97) [21]; and *P. stutzeri* KF716, 4,188,013 bp (30) [29].

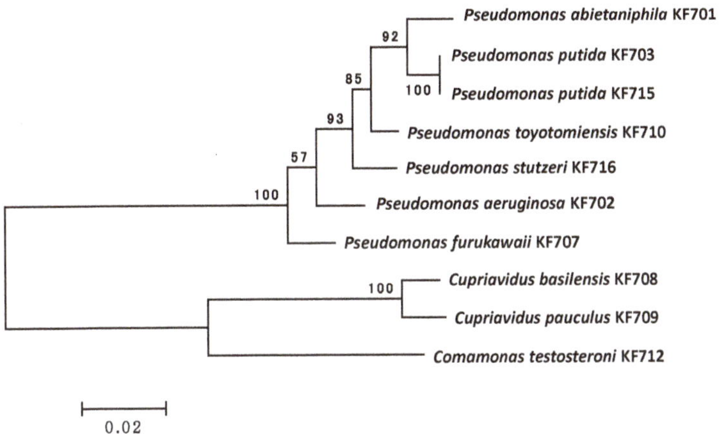

Figure 2. 16S rRNA phylogenetic trees of the ten KF strains. The multiple alignment outputs were used to generate neighbor-joining phylogenetic trees using MEGA 6.0 [33]. The bar indicates expected nucleotide substitutions per site. Numbers indicate the percentage occurrence of the branch in the bootstrapped trees on 500 replicates.

All ten strains possessed the *bph* genes, while seven strains possessed the salicylate catabolic *sal* genes. The *bph-sal* cluster was localized on the chromosome in KF707. The same cluster was located on the chromosome in the majority of the KF715 cells and also existed as an extrachromosomal circular form (483,376 bp) as well in the minor part of the cells in the stationary phase culture [16]. Based on the features of the *bph* genes, we classified the ten strains into four groups. Group I (five strains) possessed the *bph-sal* cluster that was almost identical to that of KF707, where the *sal* genes were localized approximately 6-kb downstream of the *bph* genes. This group included strains KF702, KF703, KF707, KF710, and KF716. The *bph* gene cluster of this group was composed of *bphRA1A2A3A4BCX0X1X2X3D* (11.2 kb), which encodes the catabolic enzymes that degrade biphenyl into benzoate and acetyl CoA via the *meta*-cleavage pathway. Group II (strains KF701 and KF715) possessed the *bph-sal* cluster that was similar to that of KF707, although the *bphX* region (3.5 kb) that is involved in the metabolism of 2-hydroxypenta-2,6-dienoic acid to acetyl CoA (Figure 1, lower pathway of biphenyl metabolism) was deleted as described in Reference [22]. The structural features of the *bphRABCD* (7.8 kb) and the flanking regions, including the *sal* genes (10.2 kb), were almost identical between KF701 and KF715, indicating that the *bph-sal* clusters of these two strains were horizontally transferred to one another. Group III (KF709) possessed the *bph* genes, but not the adjacent *sal* genes as in the case of *Burkholderia xenovorans* LB400 [35], another well-characterized PCB-degrader. Overall alignment of the *bph* gene cluster in KF709 was similar to that of group I, but each *bph* component of the *bph* genes was relatively low. Group IV (KF708 and KF712) possessed *bph* genes similar to the genes found in Tn*4371* from *C. oxalacticus* A5 [6], and the ones in ICE$_{KKS102}$*4677* from the *Acidovorax* sp. strain KKS102 [7]. Some rearrangements of the *bph* gene cluster were observed compared to that in groups I–III as described below. This group of strains did not carry the *sal* genes.

All the strains possessed the catabolic genes of benzoate, the lower pathway intermediate of biphenyl catabolism. Strains KF701, KF702, KF703, KF707, KF710, and KF715 possessed multiple benzoate catabolic genes that encoded both the extradiol cleavage *meta*-pathways (*bza*) and intradiol cleavage *ortho*-pathways (*ben*). KF708, KF709, and KF712 possessed the *box* genes encoding the benzoate catabolic pathway via benzoyl-CoA. Besides the *bph*, *sal*, and benzoate catabolic (*bza*, *ben*, and *box*) genes, these ten strains possessed catabolic genes for various aromatic compounds. Eight strains (KF701, KF703, KF707, KF708, KF710, KF712, KF715, and KF716) possessed a phenol-catabolic *dmp* gene cluster [36]. Five strains (KF702, KF703, KF707, KF710 and KF715) had protocatechuate catabolic *pca* genes. Five strains (KF703, KF707, KF708, KF709 and KF715) possessed phenylacetate catabolic *paa* genes. Other than the aromatic catabolic genes, many strains carried heavy metal-resistant genes. Six strains (KF701, KF702, KF708, KF710, KF712, and KF715) possessed a cluster of *mer* genes that were responsible for the resistance of inorganic mercury [37]. All of the ten strains possessed putative *czc* genes involved in the resistance of cobalt, zinc, and cadmium [38]. In KF708 and KF709, more than ten putative *czc* gene clusters were present. Arsenate-resistant genes were present in all the ten strains. These results, except for the arsenate-resistant genes, are summarized in Table S1.

3.2. Comparison of the bph Genes

The identities (percent) of the nucleotide sequences of the *bph* genes in groups I–IV strains (named as types I–IV *bph* genes, respectively) are shown in Tables S2–S13. The phylogenetic trees of *bphA1*, *bphB*, *bphC*, and *bphD* belonging to types I and II are shown in Figure 3. A comparative analysis of the nucleotide sequences of each gene revealed that all the *bph* genes were almost identical in types I and II (96.7–100%), except for that of the *bphX* region (3.5 kb) between *bphC* and *bphD*, which was missing in the *bph* gene cluster of type II (Figure 4).

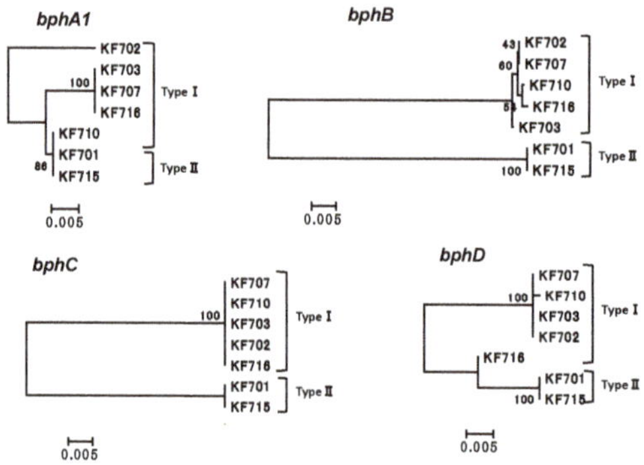

Figure 3. Phylogenetic tree of *bphA1*, *bphB*, *bphC*, and *bphD* of KF strains in groups I and II (types I and II *bph* genes). The multiple alignment outputs were used to generate neighbor-joining phylogenetic trees using MEGA 6.0 [33]. The bar indicates expected nucleotide substitutions per site.

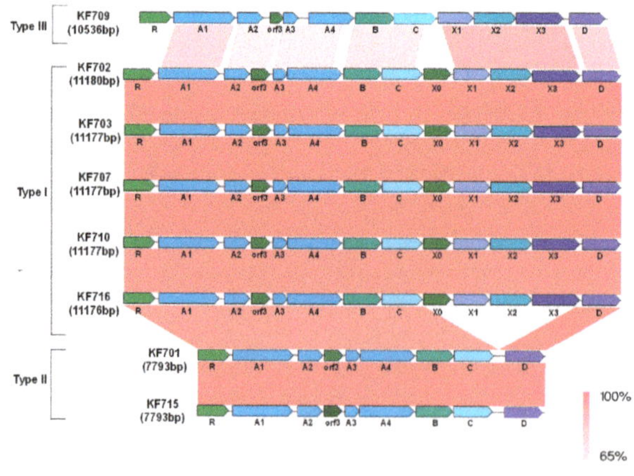

Figure 4. Comparison of the *bph* gene clusters of KF strains belonging to groups I–III (types I–III). The shading in pink to red shows the identity (65–100%) of the gene clusters as indicated at the bottom of the figure. R, *bphR*; A1, *bphA1*; A2, *bphA2*; A3, *bphA3*; A4, *bphA4*; B, *bphB*; C, *bphC*; D, *bphD*; X0, *bphX0*, X1, *bphX1*; X2, *bphX2*; and X3, *bphX3*.

The *bphA1*, *bphA2*, *bphA3*, *bphA4*, *bphB*, and *bphC* genes of KF716 were almost identical to that of the type I *bph* gene. However, the 3′-terminus of the *bphX3* gene (754 bp) and 5′-terminus of the *bphD* gene (260 bp) of KF716 were different from that of type I (Figure S1a,b), and the 5′-terminus of the *bphD* gene (260 bp) of KF716 was almost identical to that of the type II *bph* gene. These genetic features are reflected in the phylogenetic tree of the *bphD* (Figure 3). Thus, the *bph* genes of KF716 showed the structural features of both groups I and II.

The KF709 (group III strain) possessed a type III *bph* gene similar to the one in group I, but the respective *bph* genes were less conserved and rearranged (Figure 4). The nucleotide sequences of the KF709 *bph* genes had less than 72.7% identities compared to types I and II in the *bphABC* region (5.9 kb). However, the *bphX1, X2, X3* genes and *bphD* gene were more identical (88.6–90.6% in *bphX1* and 74.3–84.9% in *bphX2*, *bphX2*, *bphX3*, or *bphD*) (Tables S2–S13). The *bphX0* encoding putative glutathione S-transferase was not present between the *bphC* and *bphX1*, but it was located downstream of the tRNA-Gly gene (data not shown). *orf3* was present between the *bphA2* and *bphA3* of types I–III. Despite the conserved feature of *orf3*, its function remains unclear [3].

The genetic features of KF708 and KF712 in the group IV strain *bph* genes (Type IV *bph* gene) were very different to types I and II, yet they were similar to those of the *Acidovorax* sp. strain KKS102 and *C. oxalacticus* A5 (Figure 5). The *bph* gene cluster of this type was composed of *bphSX1X2X3(V)A1A2A3BCD(W)A4*. Thus, the *bphX1X2X3* region was located upstream of the *bphA1A2A3BCD* region, and *bphA4* was present just downstream of the *bphD* (Figure 5). The insertion sequence (1190 bp) and the transposase gene (IS/tnp) were present between the *bphS* and *bphX1* of KF708 and KKS102 [39], but these were not present in KF712 and *C. oxalacticus* A5. The identity of the KF708 and KF712 *bph* genes of type IV was compared to types I and II (Tables S2–S13). The nucleotide sequences of *bphA1*, *bphA2*, *bphA3*, *bphA4*, *bphB*, *bphC*, and *bphD* were conserved between types I and IV; however, the identities were less than 77%. Two unidentified gene components, *bphV* and *bphW*, were present in all the type IV *bph* genes, but they were not present in types I–III. The type IV *bph* genes possessed a transcriptional regulator, *bphS* [39]. It was reported that *bphS* acts as a repressor, whereas the *bphR* of Type I acts as an activator for biphenyl catabolism. The functions of *bphS* and *bphR* oppose each other, although they belong to the same GntR family [10,39].

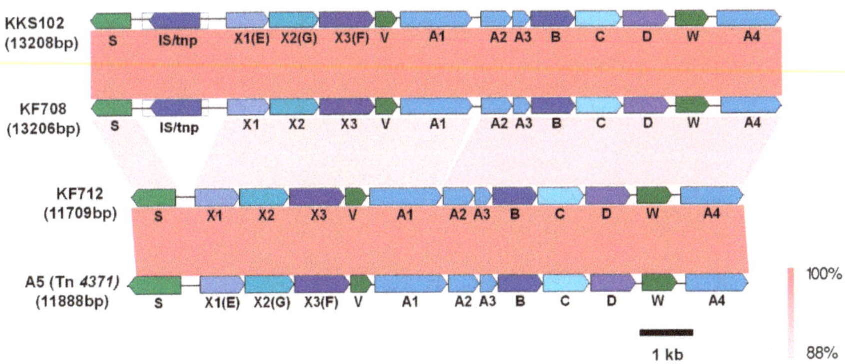

Figure 5. Comparison of the *bph* genes of *Cupriavidus basilensis* KF708 and *Comamonas testosteroni* KF712 belonging to group IV (type IV) to those of *Acidovorax* sp. KKS102 and *Cupriavidus oxalacticus* A5. The identities (88–100%) between the gene clusters are shown by shading in pink to red as indicated at the right bottom of the figure. E, *bphE*; F, *bphF*; G, *bphG*; S, *bphS*; V, *bphV*; W, *bphW*; IS, insertion sequence; tnp, transposase. The *bphE, bphF, and bphG of* KKS102/Tn4371 are homologous genes of *bphX1*, *bphX3*, and *bphX2* of KF708/KF712, respectively. The other signs are defined in the legend in Figure 4.

3.3. ICE$_{bph\text{-}sal}$ in KF Strains Belonging to Groups I and II

The genome sequence analysis of KF701, KF702, KF703, KF707, KF715, and KF716 revealed highly conserved large "genomic islands" that included the *bph-sal* genes adjacent to the tRNA-Gly(CCC) gene (Figure 6). There was an 18 bp direct repeat (5′-TTCCC(T/A)(T/C)(C/T)(G/A)CCCGCTCCA-3′) on the border of the conserved region and the non-conserved region (*attL* and *attR*, Figure 7). The *attL* included 18 bp of the 3′ end of the tRNA-Gly(CCC) gene. These 18 bp direct repeat sequences could be generated by the integration of genomic islands into the chromosome. In strain KF707, *attL* was followed by a phage-related integrase (*int*) gene. The *bph* cluster was located just downstream of the *int* gene, followed by the *sal* gene cluster approximately 6-kb downstream, and by the *bza* gene approximately 49-kb downstream. The *attR* site was located downstream of the *bza* gene. Thus, the genomic island was estimated to be 122.0 kb in size (Figure 6). Genes identified as the VirB4 component (ATPase) and VirD4 component (coupling protein) of the type IV secretory pathway [40] were located 40-kb and 80-kb downstream of the tRNA-Gly(CCC) gene, respectively. The *parA* and *parB* genes encoding the replication partition proteins were present near the right end, and were proposed to act as a stabilization system for the maintenance of mobile elements in the bacterial genomes [41]. These structures corresponded to the common backbone of many integrative conjugative elements (ICEs) [42], such as ICE*clc* from *Pseudomonas knackmussii* B13 [43], which carries the catabolic genes of chlorocatechols, the tyrosine integrase gene, and type IV secretory machinery. We designated it as ICE$_{bph\text{-}sal}$KF707. Within this element, many putative mobile protein genes were present surrounding the *sal* gene cluster. To note, five IS genes at the upstream region and two IS genes at the downstream region were identified. ICE$_{bph\text{-}sal}$ was observed in other KF strains of group I and KF701 of group II. An ICE including the *bph* and *sal* gene clusters of *P. aeruginosa* KF702, named ICE$_{bph\text{-}sal}$KF702, was calculated to be 126.7 kb (Figure 6). Likewise, the ICE$_{bph\text{-}sal}$KF703 of *P. putida* KF703, ICE$_{bph\text{-}sal}$KF710 of *P. toyotomiensis* KF710, and ICE$_{bph\text{-}sal}$KF716 of *P. stutzeri* KF716 were calculated to be 120.8 kb, 130.3 kb, and 117.3 kb, respectively. The ICE$_{bph\text{-}sal}$KF701 of *P. abietaniphila* KF701 in the group II strain was 117.4 kb in size. As in the case of ICE$_{bph\text{-}sal}$KF707, ICE$_{bph\text{-}sal}$KF701, ICE$_{bph\text{-}sal}$KF702, ICE$_{bph\text{-}sal}$KF703, and ICE$_{bph\text{-}sal}$KF710 contained three gene clusters of *bph*, *sal*, and *bza*. ICE$_{bph\text{-}sal}$KF716 contained *bph* and *sal* genes, but not the *bza* genes. Sequence comparison revealed an inversion in ICE$_{bph\text{-}sal}$KF702. The *sal:bza* and *bza:sal* fusion gene clusters were found, in which half parts of the *sal* genes and the *bza* genes were replaced with each other. These fusion genes were likely generated via homologous recombination between the *sal* and the *bza* genes. ICE$_{bph\text{-}sal}$KF710 and ICE$_{bph\text{-}sal}$KF716 contained the 5-kb region encoding putative multidrug efflux pumps, but ICE$_{bph\text{-}sal}$s from the remaining KF strains was deficient of the corresponding region. Thus, ICE$_{bph\text{-}sal}$ in groups I and II contained highly conserved nucleotide sequences larger than 110 kb, which were larger than many other ICEs found in bacterial strains to date [44].

We previously reported that KF715 harbors an approximately 90-kb conjugative *bph-sal* gene cluster in the chromosome [22], and that the cluster could be transferred to *P. putida* AC30 and *P. putida* KT2440 with very high frequency. Whole genome sequencing of the KF715 studied here revealed that the *bph-sal* cluster was located on a huge 483-kb extrachromosomal element, previously designated as pKF715A [16]. An 18 bp DNA sequence identical to the 3′- end portion of a tRNA-Gly(CCC) gene as part of its attachment site (*attP*) was present just upstream of the *int* gene in this element. It was also confirmed that the bacterial integration site (*attB*) was present within the 3′-end portion of a tRNA-Gly gene in the KF715 chromosome. Furthermore, the two *Spe*I digested bands, approximately 300 kb and 200 kb, hybridized with a *bphA1* probe were observed as reported [16]. Since there was only one *Spe*I site in the element, the large, faint band was matched to 314 kb of the *Spe*I digested extrachromosomal element. The second major band of 220 kb was matched to that of the *Spe*I digested chromosome of strain KF715. This suggested that the majority of the KF715 cells carried an integrated ICE$_{bph\text{-}sal}$KF715, whereas fewer cells carried it as an extrachromosomal circular form. Thus, it was likely that the circular form could be obtained by recombining the *attL* site (present at the 3′-end of the tRNA-Gly gene) and *attR* site (present far downstream, 483 kb from the *attL* site), forming the *attP* site.

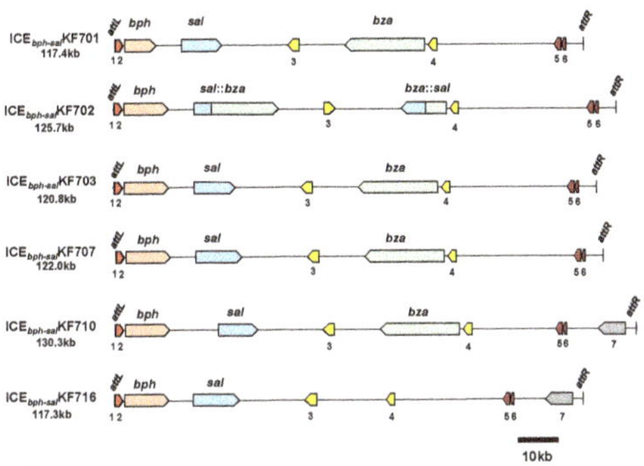

Figure 6. Organization of the ICE$_{bph\text{-}sal}$ in KF strains of group I and KF701 of group II. ICE$_{bph\text{-}sal}$KF701, ICE$_{bph\text{-}sal}$KF702, ICE$_{bph\text{-}sal}$KF703, ICE$_{bph\text{-}sal}$KF707, and ICE$_{bph\text{-}sal}$KF710 carry the *int* gene, *bph* genes, *sal* genes, and *bza* genes. ICE$_{bph\text{-}sal}$KF716 carries the *int* gene, *bph*, and *sal* genes, but not the *bza* genes. The *sal* genes and *bza* genes in ICE$_{bph\text{-}sal}$KF702 are recombined. 1, tRNA-Gly(CCC) genes (partial); 2, *int* genes; 3, VirB4 components of the type IV secretory pathway; 4, VirD4 component of the type IV secretory pathway; 5, *parB* genes; 6, *parA* genes; and 7, putative multidrug efflux pumps. The other, undefined gene components are not shown in the figure.

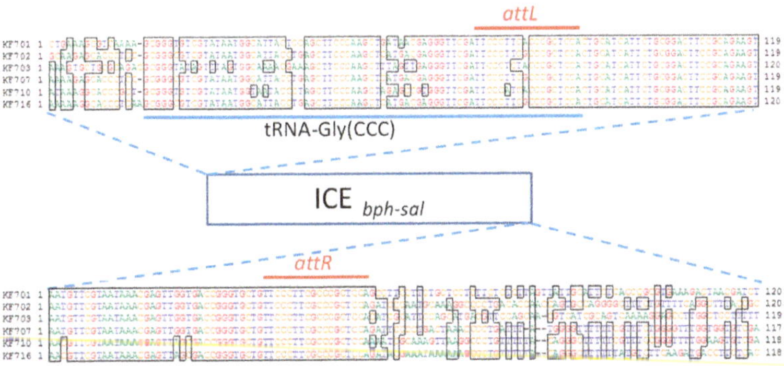

Figure 7. Integration sites of ICE$_{bph\text{-}sal}$ in the KF strains of group I and KF701 of group II. The *attL* (18 bp) site, including 18 bp of the 3′ end of tRNA-Gly gene, and *attR* (18 bp) sites are indicated.

3.4. ICEbph in the Group IV Strains

The *bph* genes of *C. basilensis* KF708 and *C. testosteroni* KF712 were also located on different types of ICEs. ICE$_{bph}$KF708 (61.8 kb), including *bph* genes of KF708, was almost identical to ICE$_{KKS102}$*4677* of the *Acidovorax* sp. strain KKS102 [7] (Figure 8). ICE$_{bph}$KF708 had covalently bound ends of the conserved 5′-GATTTTAAG-3′ sequence (*attL1* and *attR1*). This *att* sequence was identical to that of ICE$_{KKS102}$*4677*. Nine nucleotide substitutions, three nucleotide deletions, and one nucleotide insertion were found in ICE$_{bph}$KF708 when compared to ICE$_{KKS102}$*4677*. The ICE$_{bph}$KF712 (59.4 kb) carrying the *bph* genes of KF712 was almost identical to the Tn*4371* from *C. oxalaticus* A5, being the first ICE found carrying the *bph* gene cluster [6]. ICE$_{bph}$KF712 had the covalently bound ends of the conserved *attL2* and *attR2* sequence (5′-TTTTCAT-3′). This *att* sequence was identical to that of Tn*4371*, but it was different from the *attL1* and *attR1* sequence (5′-GATTTTAAG-3′) of ICE$_{KKS102}$*4677* and KF708. The core part, including the *bph* and *trb* gene of ICE$_{bph}$KF712 in length of 33 kb, was almost identical to that of Tn*4371*, whereas the remaining part flanked by the *attL2* site was less conserved. Major parts of these ICEs shared common structures, including the *bph* gene cluster; genes encoding the replication and partition proteins; *parA*; VirD2 component (relaxase); as well as two conjugative transfer elements, the *trb* gene cluster, and the *tra* genes (Figure 8). ICE$_{bph}$KF708 and ICE$_{KKS102}$*4677* possessed a putative arsenate-resistant gene cluster at 15.5–18.5 kb downstream of the *attL* site, which encoded the transcriptional regulator, arsenate reductase, and arsenite efflux transporter. The gene cluster corresponding to them was not found in ICE$_{bph}$KF712 and Tn*4371*.

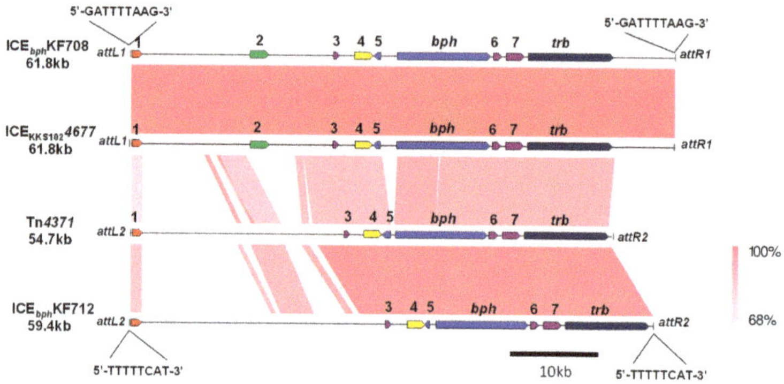

Figure 8. Organization of ICE$_{bph}$KF708 and ICE$_{bph}$KF712 in comparison to ICE$_{KKS102}$*4677* and Tn*4371*. *attL1* and *attR1* represent integration sites for the ICE$_{bph}$KF708 and ICE$_{KKS102}$*4677*. *attL2* and *attR2* represent integration sites for the Tn*4371* or ICE$_{bph}$KF712. 1, integrase (*int*) gene; 2, putative arsenate-resistant gene; 3, *parA* gene; 4, VirD2 component (relaxase); 5, transcriptional regulator *bphS* gene; 6, *traR* gene; and 7, *traG* gene. The shading in pink to red shows the identities (68–100%) of the gene clusters as indicated at the right bottom of the figure.

3.5. Conjugal Transfer of Extrachromosomal ICEbph-salKF715 Into P. putida F39/D

Since *P. putida* AC30 Bph+ and *P. putida* KT2440 Bph+ harbor an extrachromosomal ICE*bph-sal*KF715, from strain KF715, we tried to transfer this element into *P. putida* F39/D through filter mating. The results are shown in Figure 9. Interestingly, the Bph+ transconjugant of F39/D exhibited two bands hybridized to KF715 *bphA1* DNA when KT2440Bph+ was used as a donor strain of pKF715A. On the other hand, only a single band was detected when AC30Bph+ was used as a donor strain. These hybridized bands were all different in size to that of the 310-kb *Spe*I fragment of pKF715A, indicating that the extrachromosomal ICE$_{bph\text{-}sal}$KF715 of the two donor strains were transferred into the recipient strain F39/D and then integrated into the chromosomes at different loci.

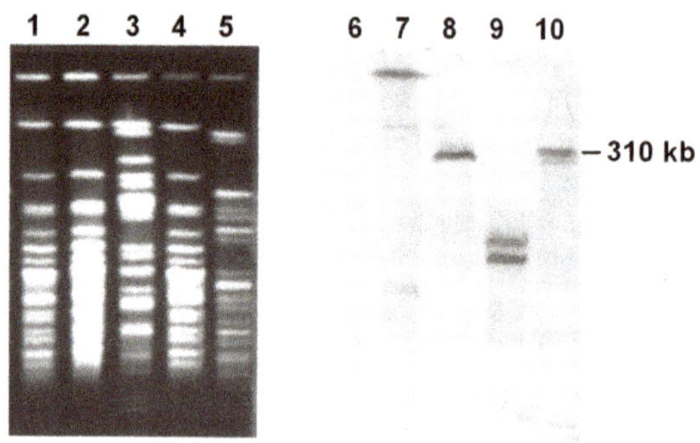

Figure 9. Conjugative transfer of the *bph* genes from *P. putida* AC30Bph+ and *P. putida* KT2440Bph+ into *P. putida* F39/D. Genomic DNA was digested with *Spe*I, applied to pulse field gel electrophoresis, and subjected to Southern blot analysis. The *bphA1* DNA of KF715 was used as a probe. Lanes 1 and 6, F39/D; lanes 2 and 7, F39/DBph+ transconjugant from AC30Bph+; Lanes 3 and 8, AC30Bph+; Lanes 4 and 9, F39/DBph+ transconjugant from KT2440Bph+; and Lanes 5 and 10, KT2440Bph+.

4. Discussion

Bacteria evolve through a variety of genetic events, such as mutations, intergenomic shuffling, and horizontal gene transfer. ICEs are being identified in increasing numbers via bacterial genomic analysis. Genomic islands, including ICEs, are discrete DNA segments and play an important role in bacterial evolution [42,45,46]

In this study, we revealed that *bph* genes differentially existed in ten biphenyl/PCB degrading strains isolated from biphenyl-contaminated soil. The types I and II *bph* genes of KF strains belonging to groups I and II were very similar in terms of gene organization and nucleotide sequences, except that the *bphX* region was missing in the type II *bph* gene cluster. The ICE structures carrying the *bph* genes were identified in several strains. Typical ICE$_{bph\text{-}sal}$ of approximately 120 kb was observed in the group I strains and KF701 of group II, and they were accompanied by the *sal* genes, approximately 6-kb downstream of the *bph* genes. The KF715 strain of group II carried *bph* and *sal* gene clusters in the chromosome and also on a plasmid (pKF715A, 483 kb) [16]. ICEs are thought to reintegrate into the recipient's chromosome immediately after transfer. However, a recent study suggested that certain ICEs, such as ICE*Bs1* from *Bacillus subtilis* [47] and SXT/R391 ICEs from *Vibrio cholerae* [48], are capable of autonomous replication. Transconjugants of *P. putida* AC30Bph+ and *P. putida* KT2440Bph+ carry the *bph-sal* cluster as an extrachromosomal circular form [16]. The mobile element carrying the *bph-sal* cluster replicates autonomously like plasmid, maintained stably, and consists of genes sharing homologies to components of the DNA replication and stabilization machinery.

The first example of ICE-harboring genes for the degradation of xenobiotic compounds is the *clc* element (ICE*clc*, 105 kb) from *P. knackmussii* B13. ICE*clc* encodes for the catabolic pathway involved in 3- and 4-chlorocatechol degradation [43,49–51]. Besides this catabolic gene cluster, ICE*clc* has the core genes, such as type IV secretion system-encoding genes, relaxase, and integrase. ICE*clc* can excise through recombination between short direct repeats at either end (*attL* and *attR*). The excised ICE*clc* can transfer to a new recipient cell through the conjugation apparatus and it integrates into the recipient's chromosome between the 18-bp sequence at the 3′ end of the tRNA-Gly gene on the chromosome (*attB*) and the identical sequence on the excised ICE*clc* (*attP*), thereby restoring the tRNA-Gly gene. Both excision and integration are mediated by the IntB13 integrase. The ICE$_{bph\text{-}sal}$s of group I (strains KF702, KF703, KF707, KF710, and KF716) were related to ICE*clc* in terms of the conjugation apparatus and the core gene set. However, the genetic organizations of ICE$_{bph\text{-}sal}$s are different from those of ICE*clc*. The integrases in the group I ICE$_{bph\text{-}sal}$ were almost identical (~99% in amino acid sequences), but it was as low as 59% between that of ICE$_{bph\text{-}sal}$ and ICE*clc*. Such low identity of the integrases between ICE$_{bph\text{-}sal}$ and ICE*clc* reflected the differences of the insertion sites. The ICE$_{bph\text{-}sal}$s of group I were inserted at the 3′- end of the tRNA-Gly gene (76 bp in length), which carries the CCC anticodon. ICE*clc* inserts into a number of tRNA-Gly genes, but only the genes which carry the GCC anticodon [51]. It should also be noted that the ICE$_{bph\text{-}sal}$ in strains belonging to groups I and II are present in *Pseudomonas* spp., whereas other strains than *Pseudomonas* possess different ICE$_{bph}$ as seen in group IV strains, indicating that ICE$_{bph\text{-}sal}$ have restricted host ranges.

In this study, we found that various types of *bph* genes are present in ten different strains isolated from the same soil sample. Several lines of evidence suggested that many *bph* genes in these strains were present on the chromosome as an integrated form. However, it is also true that certain ICE$_{bph\text{-}sal}$ is present stably as a plasmid. ICE$_{bph\text{-}sal}$KF715 was stably maintained as a circular form in the two transconjugants, *P. putida* AC30Bph+ and *P. putida* KT2440Bph+. On the other hand, ICE$_{bph\text{-}sal}$KF715 (circular) from *P. putida* AC30Bph+ and *P. putida* KT2440Bph+ seemed to integrate differently in another recipient *P. putida* F39/D. When *P. putida* AC30Bph+ was used as a donor strain, the largest *Spe*I DNA fragment of the F39/DBph+ transconjugant was hybridized with the KF715 *bphA1* probe (Figure 9). When *P. putida* KT2440Bph+ was used as a donor strain, two copies of the *bphA1* DNA were detected in the F39/DBph+ transconjugant strain at different positions (Figure 9). Investigations are currently underway to reveal how the ICE$_{bph\text{-}sal}$KF715 is integrated in the genome of the F39/DBph+ transconjugants. The ICE*clc* of the *P. knackmussii* strain B13 was transferred by conjugation and integrated into two nonadjacent sites on the chromosome of toluene utilizing *P. putida* F1 [49]. Our repeated attempts to conjugally transfer the ICE$_{bph\text{-}sal}$ of other strains than KF715 have not been successful. This may be due to a lack of expression of the integrase genes or mutations in certain gene(s) involved in the excision or conjugal transfer.

ICE$_{bph}$KF708 and ICE$_{bph}$KF712 were found in *Cupriavidus* and *Comamonas*, respectively. ICE$_{bph}$KF708 was almost identical to the ICE$_{KKS102}$*4677* from *Acidovorax* sp. KKS102, and they had several nucleotide differences, indicating that this type of ICE$_{bph}$ could be transferred between *Cupriavidus* and *Acidovorax*. The right wing corresponding to the core legion (33 kb) of ICE$_{bph}$KF712 and Tn*4371* is highly conserved, but the left wing is diversified (Figure 8). This indicated that these two ICE$_{bph}$ were rearranged in the left wing. ICE$_{bph\text{-}sal}$ (from the groups I and II strain) and ICE$_{bph}$ (from the group IV strain) are typical ICEs, which possess type IV secretion machinery. However, there are few relationships between ICE$_{bph\text{-}sal}$ and ICE$_{bph}$ in terms of gene organization, nucleotide sequence, and size (Figure 6, Figure 8). No significant identity was detected between the integrases of ICE$_{bph}$ and ICE$_{bph\text{-}sal}$. The region encoding two conjugative transfer components, the *tra*/*trb* genes adjacent to the *bph* gene cluster, was found in ICE$_{bph}$. The corresponding gene clusters were not found in ICE$_{bph\text{-}sal}$. Gene components downstream of the *sal* gene cluster are likely to be involved in conjugative transfer in ICE$_{bph\text{-}sal}$; however, they have not been identified because of the lack of reliable homologous genes that are identified as conjugative transfer components in public databases. Although ICE$_{bph\text{-}sal}$ and ICE$_{bph}$ possess common genetic components involved in biphenyl catabolism, their platforms are different.

The biphenyl degrading bacteria are considered to be responsible for lignin degradation at the final stage. Lignin is a complex compound based on the phenylpropane structure and contains a variety of biphenyl related molecules. Thus, the *bph* genes could be very ancient and distributed across a wide range of soil bacteria. Mobilization of the *bph* genes in soil bacteria can be achieved through various mobile genetic elements, including ICEs, transposons, and plasmids. It is highly conceivable that these genes can be modified and rearranged in different ways in new host cells. The results in this study provide a better understanding as to how soil bacteria exchange genetic islands involved in the catabolism of aromatic compounds, as well as how such genes are rearranged and modified in the natural environment.

Supplementary Materials: The following tables and figure are available online at http://www.mdpi.com/2073-4425/10/5/404/s1. Table S1: Distribution of the catabolic genes for aromatic compounds and the heavy metal resistance genes in the biphenyl/PCB degrading KF strains. Tables S2–S13: Identity (%) of nucleotide sequence of *bphR, bphA1, bphA2, bphA3, bphA4, bphB, bphC, bphX0, bphX1, bphX2, bphX3*, and *bphD*. Figure S1: Comparison of the *bphX3* (a) and *bphD* (b) genes belonging to types I and II.

Author Contributions: K.F. planned the genome sequencing project and performed Southern blot hybridization analysis of the transconjugants. H.F. performed the sequence analysis of KF701 and KF702. H.S. performed the sequence analysis of KF703, KF708, and KF715. N.K. performed the sequence analysis of KF707. T.W. performed the sequence analysis of KF709 and KF710. J.H. participated in the sequence analysis of KF712 and KF716 and annotation of ICE$_{bph\text{-}sal}$/ICE$_{bph}$. A.S., T.F., and M.G. were involved in the whole genome annotation of the KF strains and completed the manuscript. All authors read and approved the final manuscript.

Funding: This work was partly supported by a research grant from the Institute for Fermentation, Osaka (G-2018-3-020 to T.W.). The APC was funded by a research grant from Beppu University (to H.F.).

Acknowledgments: We thank Atsushi Yamazoe and Akito Nishi for their useful discussion and suggestions.

Conflicts of Interest: The authors declare no conflict of interest.

References

1. Furukawa, K. Microbial degradation of polychlorinated biphenyls. In *Biodegradation and Detoxification of Environmental Pollutants*; Chakrabarty, A.M., Ed.; CRC Press: Boca Raton, FL, USA, 1982; pp. 33–57.
2. Abramowicz, D.A. Aerobic and anaerobic biodegradation of PCBs: A review. *Crit. Rev. Biotechnol.* **1990**, *10*, 241–251. [CrossRef]
3. Furukawa, K.; Fujihara, H. Microbial degradation of polychlorinated biphenyls: Biochemical and molecular features. *J. Biosci. Bioeng.* **2008**, *105*, 433–449. [CrossRef] [PubMed]

4. Furukawa, K.; Tonomura, K.; Kamibayashi, A. Effect of chlorine substitution on the biodegradability of polychlorinated biphenyls. *Appl. Env. Microbiol.* **1978**, *35*, 223–227.
5. Furukawa, K.; Miyazaki, T. Cloning of a gene cluster encoding biphenyl and chlorobiphenyl degradation in *Pseudomonas pseudoalcaligenes*. *J. Bacteriol.* **1986**, *166*, 392–398. [CrossRef]
6. Toussaint, A.; Merlin, C.; Monchy, S.; Benotmane, M.A.; Leplae, R.; Mergeay, M.; Springael, D. The biphenyl- and 4-chlorobiphenyl-catabolic transposon Tn*4371*, a member of a new family of genomic islands related to IncP and Ti plasmids. *Appl. Env. Microbiol.* **2003**, *69*, 4837–4845. [CrossRef] [PubMed]
7. Ohtsubo, Y.; Ishibashi, Y.; Naganawa, H.; Hirokawa, S.; Atobe, S.; Nagata, Y.; Tsuda, M. Conjugal transfer of polychlorinated biphenyl/biphenyl degradation genes in *Acidovorax* sp. strain KKS102, which are located on an integrative and conjugative element. *J. Bacteriol.* **2012**, *194*, 4237–4248. [CrossRef]
8. Masai, E.; Sugiyama, K.; Iwashita, N.; Shimizu, S.; Hauschild, J.E.; Hatta, T.; Kimbara, K.; Yano, K.; Fukuda, M. The *bphDEF meta*-cleavage pathway genes involved in biphenyl/polychlorinated biphenyl degradation are located on a linear plasmid and separated from the initial *bphACB* genes in *Rhodococcus* sp. strain RHA1. *Gene* **1997**, *187*, 141–149. [CrossRef]
9. Shimizu, S.; Kobayashi, H.; Masai, E.; Fukuda, M. Characterization of the 450-kb linear plasmid in a polychlorinated biphenyl degrader, *Rhodococcus* sp. strain RHA1. *Appl. Env. Microbiol.* **2001**, *67*, 2021–2028. [CrossRef] [PubMed]
10. Watanabe, T.; Inoue, R.; Kimura, N.; Furukawa, K. Versatile transcription of biphenyl catabolic *bph* operon in *Pseudomonas pseudoalcaligenes* KF707. *J. Biol. Chem.* **2000**, *275*, 31016–31023. [CrossRef]
11. Fujihara, H.; Yoshida, H.; Matsunaga, T.; Goto, M.; Furukawa, K. Cross-regulation of biphenyl- and salicylate-catabolic genes by two regulatory systems in *Pseudomonas pseudoalcaligenes* KF707. *J. Bacteriol.* **2006**, *188*, 4690–4697. [CrossRef]
12. Kumamaru, T.; Suenaga, H.; Mitsuoka, M.; Watanabe, T.; Furukawa, K. Enhanced degradation of polychlorinated biphenyls by directed evolution of biphenyl dioxygenase. *Nat. Biotechnol.* **1998**, *16*, 663. [CrossRef] [PubMed]
13. Suenaga, H.; Goto, M.; Furukawa, K. Emergence of multifunctional oxygenase activities by random priming recombination. *J. Biol. Chem.* **2001**, *276*, 22500–22506. [CrossRef] [PubMed]
14. Suenaga, H.; Mitsuoka, M.; Ura, Y.; Watanabe, T.; Furukawa, K. Directed evolution of biphenyl dioxygenase: Emergence of enhanced degradation capacity for benzene, toluene, and alkylbenzenes. *J. Bacteriol.* **2001**, *183*, 5441–5444. [CrossRef]
15. Furukawa, K.; Hayase, N.; Taira, K.; Tomizuka, N. Molecular relationship of chromosomal genes encoding biphenyl/polychlorinated biphenyl catabolism: Some soil bacteria possess a highly conserved *bph* operon. *J. Bacteriol.* **1989**, *171*, 5467–5472. [CrossRef]
16. Suenaga, H.; Fujihara, H.; Kimura, N.; Hirose, J.; Watanabe, T.; Futagami, T.; Goto, M.; Shimodaira, J.; Furukawa, K. Insights into the genomic plasticity of *Pseudomonas putida* KF715, a strain with unique biphenyl-utilizing activity and genome instability properties. *Environ. Microbiol. Rep.* **2017**, *9*, 589–598. [CrossRef]
17. Fujihara, H.; Yamazoe, A.; Hosoyama, A.; Suenaga, H.; Kimura, N.; Hirose, J.; Watanabe, T.; Futagami, T.; Goto, M.; Furukawa, K. Draft genome sequence of *Pseudomonas abietaniphila* KF701 (NBRC110664), a polychlorinated biphenyl-degrading bacterium isolated from biphenyl-contaminated soil. *Genome Announc.* **2015**, *3*, e00473-15. [CrossRef] [PubMed]
18. Suenaga, H.; Yamazoe, A.; Hosoyama, A.; Kimura, N.; Hirose, J.; Watanabe, T.; Fujihara, H.; Futagami, T.; Goto, M.; Furukawa, K. Draft genome sequence of the polychlorinated biphenyl-degrading bacterium *Pseudomonas putida* KF703 (NBRC 110666) isolated from biphenyl-contaminated soil. *Genome Announc.* **2015**, *3*, e00142-15. [CrossRef]
19. Kimura, N.; Watanabe, T.; Suenaga, H.; Fujihara, H.; Futagami, T.; Goto, M.; Hanada, S.; Hirose, J. *Pseudomonas furukawaii* sp. nov., a polychlorinated biphenyl-degrading bacterium isolated from biphenyl-contaminated soil in Japan. *Int. J. Syst. Evol. Microbiol.* **2018**, *68*, 1429–1435. [CrossRef]
20. Suenaga, H.; Yamazoe, A.; Hosoyama, A.; Kimura, N.; Hirose, J.; Watanabe, T.; Fujihara, H.; Futagami, T.; Goto, M.; Furukawa, K. Draft genome sequence of the polychlorinated biphenyl-degrading bacterium *Cupriavidus basilensis* KF708 (NBRC 110671) isolated from biphenyl-contaminated soil. *Genome Announc.* **2015**, *3*, e00143-15. [CrossRef]

21. Hirose, J.; Yamazoe, A.; Hosoyama, A.; Kimura, N.; Suenaga, H.; Watanabe, T.; Fujihara, H.; Futagami, T.; Goto, M.; Furukawa, K. Draft genome sequence of the polychlorinated biphenyl-degrading bacterium *Comamonas testosteroni* KF712 (NBRC 110673). *Genome Announc.* **2015**, *3*, e01214-15. [CrossRef]
22. Nishi, A.; Tominaga, K.; Furukawa, K. A 90-kilobase conjugative chromosomal element coding for biphenyl and salicylate catabolism in *Pseudomonas putida* KF715. *J. Bacteriol.* **2000**, *182*, 1949–1955. [CrossRef]
23. Zylstra, G.J.; Gibson, D.T. Toluene degradation by *Pseudomonas putida* F1. Nucleotide sequence of the *todC1C2BADE* genes and their expression in *Escherichia coli*. *J. Biol. Chem.* **1989**, *264*, 14940–14946.
24. Fujihara, H.; Yamazoe, A.; Hosoyama, A.; Suenaga, H.; Kimura, N.; Hirose, J.; Watanabe, T.; Futagami, T.; Goto, M.; Furukawa, K. Draft genome sequence of *Pseudomonas aeruginosa* KF702 (NBRC 110665), a polychlorinated biphenyl-degrading bacterium isolated from biphenyl-contaminated soil. *Genome Announc.* **2015**, *3*, e00517-15. [CrossRef]
25. Triscari-Barberi, T.; Simone, D.; Calabrese, F.M.; Attimonelli, M.; Hahn, K.R.; Amoako, K.K.; Turner, R.J.; Fedi, S.; Zannoni, D. Genome sequence of the polychlorinated-biphenyl degrader *Pseudomonas pseudoalcaligenes* KF707. *J. Bacteriol.* **2012**, *194*, 4426–4427. [CrossRef]
26. Watanabe, T.; Yamazoe, A.; Hosoyama, A.; Fujihara, H.; Suenaga, H.; Hirose, J.; Futagami, T.; Goto, M.; Kimura, N.; Furukawa, K. Draft genome sequence of *Cupriavidus pauculus* strain KF709, a biphenyl-utilizing bacterium isolated from biphenyl-contaminated soil. *Genome Announc.* **2015**, *3*, e00222-15. [CrossRef]
27. Watanabe, T.; Yamazoe, A.; Hosoyama, A.; Fujihara, H.; Suenaga, H.; Hirose, J.; Futagami, T.; Goto, M.; Kimura, N.; Furukawa, K. Draft genome sequence of *Pseudomonas toyotomiensis* KF710, a polychlorinated biphenyl-degrading bacterium isolated from biphenyl-contaminated soil. *Genome Announc.* **2015**, *3*, e00223-15. [CrossRef]
28. Suenaga, H.; Yamazoe, A.; Hosoyama, A.; Kimura, N.; Hirose, J.; Watanabe, T.; Fujihara, H.; Futagami, T.; Goto, M.; Furukawa, K. Complete genome sequence of the polychlorinated biphenyl-degrading bacterium *Pseudomonas putida* KF715 (NBRC 110667) isolated from biphenyl-contaminated soil. *Genome Announc.* **2017**, *5*, e01624-16. [CrossRef]
29. Hirose, J.; Yamazoe, A.; Hosoyama, A.; Kimura, N.; Suenaga, H.; Watanabe, T.; Fujihara, H.; Futagami, T.; Goto, M.; Furukawa, K. Draft genome sequence of the polychlorinated biphenyl-degrading bacterium *Pseudomonas stutzeri* KF716 (NBRC 110668). *Genome Announc.* **2015**, *3*, e01215-15. [CrossRef]
30. Aziz, R.K.; Bartels, D.; Best, A.A.; DeJongh, M.; Disz, T.; Edwards, R.A.; Formsma, K.; Gerdes, S.; Glass, E.M.; Kubal, M.; et al. The RAST server: Rapid annotations using subsystems technology. *BMC Genom.* **2008**, *9*, 75. [CrossRef]
31. Sullivan, M.J.; Petty, N.K.; Beatson, S.A. Easyfig: A genome comparison visualizer. *Bioinformatics* **2011**, *27*, 1009–1010. [CrossRef]
32. Li, K.-B. ClustalW-MPI: ClustalW analysis using distributed and parallel computing. *Bioinformatics* **2003**, *19*, 1585–1586. [CrossRef]
33. Tamura, K.; Stecher, G.; Peterson, D.; Filipski, A.; Kumar, S. MEGA6: molecular evolutionary genetics analysis version 6.0. *Mol. Biol. Evol.* **2013**, *30*, 2725–2729. [CrossRef]
34. Sambrook, J.; Russell, D.W. *Molecular Cloning: A Laboratory Manual*, 3rd ed.; CSHL Press: Cold Spring Harbor, NY, USA, 2001.
35. Erickson, B.D.; Mondello, F.J. Nucleotide sequencing and transcriptional mapping of the genes encoding biphenyl dioxygenase, a multicomponent polychlorinated-biphenyl-degrading enzyme in *Pseudomonas* strain LB400. *J. Bacteriol.* **1992**, *174*, 2903–2912. [CrossRef] [PubMed]
36. Shingler, V.; Powlowski, J.; Marklund, U. Nucleotide sequence and functional analysis of the complete phenol/3,4-dimethylphenol catabolic pathway of *Pseudomonas* sp. strain CF600. *J. Bacteriol.* **1992**, *174*, 711–724. [CrossRef]
37. Brown, N.L.; Misra, T.K.; Winnie, J.N.; Schmidt, A.; Seiff, M.; Silver, S. The nucleotide sequence of the mercuric resistance operons of plasmid R100 and transposon Tn*501*: further evidence for *mer* genes which enhance the activity of the mercuric ion detoxification system. *Mol. Gen. Genet.* **1986**, *202*, 143–151. [CrossRef]
38. Intorne, A.C.; de Oliveira, M.V.V.; de M Pereira, L.; de Souza Filho, G.A. Essential role of the *czc* determinant for cadmium, cobalt and zinc resistance in *Gluconacetobacter diazotrophicus* PAl 5. *Int. Microbiol. Off. J. Span. Soc. Microbiol.* **2012**, *15*, 69–78.

39. Ohtsubo, Y.; Delawary, M.; Kimbara, K.; Takagi, M.; Ohta, A.; Nagata, Y. BphS, a key transcriptional regulator of *bph* genes involved in polychlorinated biphenyl/biphenyl degradation in *Pseudomonas* sp. KKS102. *J. Biol. Chem.* **2001**, *276*, 36146–36154. [CrossRef]
40. Christie, P.J.; Whitaker, N.; González-Rivera, C. Mechanism and structure of the bacterial type IV secretion systems. *Biochim. Biophys. Acta* **2014**, *1843*, 1578–1591. [CrossRef] [PubMed]
41. Mavrodi, D.V.; Loper, J.E.; Paulsen, I.T.; Thomashow, L.S. Mobile genetic elements in the genome of the beneficial rhizobacterium *Pseudomonas fluorescens* Pf-5. *BMC Microbiol.* **2009**, *9*, 8. [CrossRef]
42. Johnson, C.M.; Grossman, A.D. Integrative and conjugative elements (ICEs): What they do and how they work. *Annu. Rev. Genet.* **2015**, *49*, 577–601. [CrossRef]
43. Gaillard, M.; Vallaeys, T.; Vorhölter, F.J.; Minoia, M.; Werlen, C.; Sentchilo, V.; Pühler, A.; van der Meer, J.R. The *clc* element of *Pseudomonas* sp. strain B13, a genomic island with various catabolic properties. *J. Bacteriol.* **2006**, *188*, 1999–2013. [CrossRef]
44. Bellanger, X.; Payot, S.; Leblond-Bourget, N.; Guédon, G. Conjugative and mobilizable genomic islands in bacteria: Evolution and diversity. *FEMS Microbiol. Rev.* **2014**, *38*, 720–760. [CrossRef]
45. Juhas, M.; van der Meer, J.R.; Gaillard, M.; Harding, R.M.; Hood, D.W.; Crook, D.W. Genomic islands: Tools of bacterial horizontal gene transfer and evolution. *FEMS Microbiol. Rev.* **2009**, *33*, 376–393. [CrossRef]
46. Wozniak, R.A.F.; Waldor, M.K. Integrative and conjugative elements: Mosaic mobile genetic elements enabling dynamic lateral gene flow. *Nat. Rev. Microbiol.* **2010**, *8*, 552–563. [CrossRef]
47. Lee, C.A.; Babic, A.; Grossman, A.D. Autonomous plasmid-like replication of a conjugative transposon. *Mol. Microbiol.* **2010**, *75*, 268–279. [CrossRef]
48. Carraro, N.; Poulin, D.; Burrus, V. Replication and active partition of integrative and conjugative elements (ICEs) of the SXT/R391 family: The line between ICEs and conjugative plasmids is getting thinner. *PLoS Genet.* **2015**, *11*, e1005298. [CrossRef]
49. Ravatn, R.; Studer, S.; Springael, D.; Zehnder, A.J.B.; van der Meer, J.R. Chromosomal integration, tandem amplification, and deamplification in *Pseudomonas putida* F1 of a 105-kilobase genetic element containing the chlorocatechol degradative genes from *Pseudomonas* sp. strain B13. *J. Bacteriol.* **1998**, *180*, 4360–4369.
50. Sentchilo, V.; Ravatn, R.; Werlen, C.; Zehnder, A.J.B.; van der Meer, J.R. Unusual integrase gene expression on the *clc* genomic island in *Pseudomonas* sp. strain B13. *J. Bacteriol.* **2003**, *185*, 4530–4538. [CrossRef]
51. Sentchilo, V.; Czechowska, K.; Pradervand, N.; Minoia, M.; Miyazaki, R.; van der Meer, J.R. Intracellular excision and reintegration dynamics of the ICE*clc* genomic island of *Pseudomonas knackmussii* sp. strain B13. *Mol. Microbiol.* **2009**, *72*, 1293–1306. [CrossRef]

© 2019 by the authors. Licensee MDPI, Basel, Switzerland. This article is an open access article distributed under the terms and conditions of the Creative Commons Attribution (CC BY) license (http://creativecommons.org/licenses/by/4.0/).

genes

MDPI

Article

Metagenomic Insights into the Bacterial Functions of a Diesel-Degrading Consortium for the Rhizoremediation of Diesel-Polluted Soil

Daniel Garrido-Sanz [1], Miguel Redondo-Nieto [1], María Guirado [2], Oscar Pindado Jiménez [2], Rocío Millán [2], Marta Martin [1] and Rafael Rivilla [1,*]

[1] Departamento de Biología, Facultad de Ciencias, Universidad Autónoma de Madrid, Darwin 2, 28049 Madrid, Spain; daniel.garrido@uam.es (D.G.-S.); miguel.redondo@uam.es (M.R.-N.); m.martin@uam.es (M.M.)

[2] Centro de Investigaciones Energéticas, Medioambientales y Tecnológicas, Avenida Complutense 40, 28040 Madrid, Spain; maria.guirado@ciemat.es (M.G.); oscar.pindado@ciemat.es (O.P.J.); rocio.millan@ciemat.es (R.M.)

* Correspondence: rafael.rivilla@uam.es

Received: 25 March 2019; Accepted: 11 June 2019; Published: 14 June 2019

Abstract: Diesel is a complex pollutant composed of a mixture of aliphatic and aromatic hydrocarbons. Because of this complexity, diesel bioremediation requires multiple microorganisms, which harbor the catabolic pathways to degrade the mixture. By enrichment cultivation of rhizospheric soil from a diesel-polluted site, we have isolated a bacterial consortium that can grow aerobically with diesel and different alkanes and polycyclic aromatic hydrocarbons (PAHs) as the sole carbon and energy source. Microbiome diversity analyses based on 16S rRNA gene showed that the diesel-degrading consortium consists of 76 amplicon sequence variants (ASVs) and it is dominated by *Pseudomonas*, *Aquabacterium*, *Chryseobacterium*, and *Sphingomonadaceae*. Changes in microbiome composition were observed when growing on specific hydrocarbons, reflecting that different populations degrade different hydrocarbons. Shotgun metagenome sequence analysis of the consortium growing on diesel has identified redundant genes encoding enzymes implicated in the initial oxidation of alkanes (AlkB, LadA, CYP450) and a variety of hydroxylating and ring-cleavage dioxygenases involved in aromatic and polyaromatic hydrocarbon degradation. The phylogenetic assignment of these enzymes to specific genera allowed us to model the role of specific populations in the diesel-degrading consortium. Rhizoremediation of diesel-polluted soil microcosms using the consortium, resulted in an important enhancement in the reduction of total petroleum hydrocarbons (TPHs), making it suited for rhizoremediation applications.

Keywords: rhizoremediation; diesel; bacteria; consortium; metagenomics; PAHs; TPH

1. Introduction

Soil pollution by petroleum hydrocarbons, including diesel fuel, is produced by spills and leakages and is a major environmental concern due to the large number of hazardous and toxic constituents [1,2] that lead to reduced germination rates of plant seeds and a decrease in the diversity of the associated soil biota [3,4]. Diesel is a complex mixture of alkanes and polycyclic aromatic hydrocarbons (PAHs) which varies widely depending on the geographical source of the crude oil fraction used during petroleum separation. Furthermore, diesel pollution is usually associated with the presence of heavy metals [5], which also poses an environmental concern due to its toxic effects and produces an acute inhibition of the diesel biodegradation process by microorganisms [6].

Many microorganisms can aerobically degrade alkanes, using them as carbon and energy source [7,8]. Four different pathways for aerobic biodegradation of alkanes have been uncovered to date [7]. Two of these well-studied pathways are initiated by a terminal or subterminal oxidation of a methyl or methylene group, mediated by alkane monooxygenase enzymes, and resulting in the production of primary or secondary alcohols respectively [8–10], which are further oxidized by alcohol and aldehyde dehydrogenases into fatty acids before they can enter the beta-oxidation [11]. Alternatively, fatty acids resulting in the alkane terminal oxidation can be further oxidized at the terminal omega methyl group via biterminal oxidation. This process results in the formation of omega-hydroxy fatty acids that are then converted by alcohol and aldehyde dehydrogenases into dicarboxylic acids, which are also funneled into the beta-oxidation [12]. Initial terminal or subterminal oxidation of alkanes is carried out by alkane 1-monooxygenases (AlkB) or long-chain alkane monooxygenases (LadA), which have been extensively characterized [13–16]. On the other hand, the omega-hydroxy fatty acid formation via the biterminal oxidation pathway is primarily attributed to cytochrome P450 of the PYC153 family [12,17], which can also hydroxylate alkanes on terminal positions to primary alcohols.

Bacteria can aerobically metabolize PAHs via different well-established pathways [18–20]. Metabolism of low molecular weight PAHs, such as naphthalene, anthracene, phenanthrene, and fluorene is usually initiated by the addition of molecular oxygen into the aromatic nucleus mediated by ring-hydroxylating dioxygenases [21,22]. The dihydrodiols formed in this initial step follow a dehydrogenation and then a meta-cleavage mediated by extradiol dioxygenases to give the ring cleavage products, which are further converted into central aromatic intermediates via subsequent series of enzymatic reactions. Many of these PAHs can also be degraded by co-metabolism in the environment [23].

Aged hydrocarbon-polluted soils are characterized by the presence of recalcitrant TPHs, such as branched aliphatic, PAHs and substituted aromatic hydrocarbons, which are usually associated to the organic and clay soil fractions, limiting the access to microorganisms and therefore reducing their biodegradation ability [24–26]. In fact, an important factor affecting the bioremediation of hydrocarbon-polluted soils is the presence of appropriate microorganisms [27], along with the soil physicochemical characteristics and other environmental conditions required to support this biota [28]. For bioremediation purposes, the use of indigenous bacterial consortia isolated from the polluted sites, rather than using allochthonous strains, might be advantageous. The effective cooperation of several specialized local microorganisms already adapted to the polluted site in terms of complementary substrate specificity can result in the mineralization of complex hydrocarbon mixtures [29]. Furthermore, as several studies indicate, microbial structure and function are influenced by total petroleum hydrocarbons (TPHs) [30,31]. Therefore, choosing an indigenous population might overcome the problems of bacterial composition shift when introduced into a new environment, which could be replaced by indigenous non-degrading populations that are, however, more competitive. To this purpose, metagenomic approaches have been used to analyze the bacterial composition and its changes [32], and to identify key genes encoding enzymes involved in the pollutant degradation process [33,34]. Rhizoremediation, a type of bioremediation that involves the use of plants to stimulate the activity of petroleum-degrading microorganisms, has been reported to be a cost-effective method for the removal of petroleum hydrocarbons from soil [35–38]. The combined use of indigenous microorganisms and plants to stimulate their degradation abilities could enhance the removal of hydrocarbons in soil-polluted environments.

Although most of the studies to date have focused on evaluating the degradation abilities of specific bacterial strains or synthetic bacterial consortia [39], in this work, we report the isolation and characterization of an indigenous soil bacterial consortium that can grow aerobically with diesel oil and other aliphatic and aromatic constituents of diesel as sole carbon and energy source. Metagenomic analysis of the diesel-degrading microbiota allowed us to identify active populations in the degradation of both, aliphatic and aromatic polycyclic hydrocarbons by the assignation of specific key coding DNA sequences (CDSs) to certain genera. Finally, we have tested this consortium in real diesel-polluted soil microcosms to address its potential for rhizoremediation.

2. Materials and Methods

2.1. Isolation of the Bacterial Consortium and Growth Conditions

Standard successive enrichment culture procedures were used to isolate the diesel-degrading bacterial consortium. Briefly, samples were collected from the rhizosphere of two plant species: *Tamarix gallica* and *Pistacia lentiscus*, planted in an aged diesel-polluted soil. Pollution came from ship fuel tanks spills in San Fernando (Cádiz, Spain. 36.497624 N, 6.191080 W). 2 g of diesel-polluted rhizospheric soil was added to 500 mL of sterile liquid minimum salt medium (MM) [40], supplemented with 1 mL/L of phosphate-buffered mineral medium salts (PAS) [41] and 0.005% yeast extract, and grown at 28 °C with shaking (140 rpm). One mL/L of diesel oil (from the ship fuel tank) was added as the sole carbon and energy source. After five subcultures, 20 mL aliquots after 48 h of growth were centrifuged for 10 min at 4000× *g*. The pellet was then resuspended in 0.75 mL of MM+PAS and mixed with 0.25 mL of glycerol 80% and deep-frozen at −80 °C. The isolated consortium was routinely grown in a liquid culture of MM+PAS supplemented with 1 mL/L of diesel oil as the sole carbon and energy source and 0.005% yeast extract, at 28 °C with shaking (140 rpm).

The culture growth on different aliphatic and aromatic compounds as sole carbon and energy sources was evaluated as above, but *n*-hexane, *n*-heptadecane, *n*-tetracosane, naphthalene, and phenanthrene (1 mL or 1 g/L) were added as the sole carbon and energy source. In the case of hexane, sterile filter paper was soaked in hexane, added to the flask caps and sealed to prevent its evaporation.

2.2. DNA Extraction, Sequencing, and Assembly

DNA extraction from the bacterial consortium after 48 h of growth (Supplementary Figure S1), on diesel oil, hexane, pentadecane, heptadecane, tetracosane, naphthalene, or phenanthrene as sole carbon and energy source was performed using the Realpure Genomic DNA Extraction Kit (Durviz, Spain). Illumina sequencing of 16S rRNA amplicons in all samples and whole-metagenome shotgun of the consortium growing with diesel as sole carbon source was carried out by Parque Científico de Madrid (Spain). Briefly, the 16S rRNA genes in each sample were sequenced by means of amplification of the V3-V4 16S rRNA region with the primers 16SV3-V4-CS1 (5′-CCT ACG GGN GGC WGC AG-30) and 16SV3-V4-CS2 (5′-GAC TAC HVG GGT ATC TAA TCC-3′), position 341 to 785 in *Escherichia coli*, prior to libraries preparation with Illumina MiSeq v3 reagent kit according to supplier specifications, and sequenced by Illumina MiSeq paired 300-bp platform. Whole-metagenome of the diesel-growing bacterial consortium was sequenced using Illumina TruSeq preparation kit, a mean library size of 483 bp and illumina MiSeq paired 300-pb.

Raw reads were trimmed and quality-filtered using Trimmomatic v0.36 software [42] to remove those with less than 50 nts in the case of microbiomes or 100 nts in the metagenome, resulting in a read recovery rate ranging from 95.9% to 97.5% in the microbiomes and 97.2% in the metagenome reads. Trimmed reads from the metagenome sequencing were assembled using SPAdes v3.12 software [43], metaSPAdes option, and default settings. Assembly quality was evaluated using QUAST v4.4 [44]. The resulting contigs were annotated using the RAST pipeline [45].

2.3. Diversity Analysis of the 16S rRNA Gene and Coding DNA Sequences (CDSs)

Microbiome 16S rRNA gene diversity was assessed with QIIME v2-2019.4 [46]. Briefly, cleaned and trimmed paired reads (described above) were filtered and denoised using DADA2 [47]. For chimera identification, 200,000 training sequences were used. Identified amplicon sequence variants (ASVs) were aligned using MAFFT [48] and further processed to construct a phylogeny with fasttree2 [49]. Rarefaction curves and Shannon Index were estimated using the plugin q2-diversity running 10 iterations, and 1000 sequence steps up to the maximum number of sequences per sample. Taxonomy was assigned to ASVs using the q2-feature-classifier [50], classify-sklearn naïve Bayes taxonomy classifier against the SILVA v132 99% 16S sequence database [51]. A specific classifier for the amplified 16S region was trained using the primers specified above and a maximum fragment size of 300 nts.

To assess the diversity of coding DNA sequences (CDSs), after whole-metagenome assembly and annotation (as specified above), CDSs were searched against the NCBI nucleotide (nt) database (October 2018) using blastn from BLAST v2.2.31+ software [52]. For each query, the first hit with a minimum of 75% sequence identity and 50% coverage was used for genus assignation.

2.4. Identification of CDSs Involved in Alkanes and Aromatic Hydrocarbons Metabolism

Coding DNA sequences of alkane 1-monooxygenase (AlkB), long-chain alkane monooxygenase (LadA), cytochrome P450 alkane hydrolase (CYP153 family) and extra and intradiol ring-cleavage and ring-hydroxylating dioxygenases (catechol 2,3-dioxygenase, biphenyl-2,3-dioxygenase, 3-carboxyethylcatechol 2,3-dioxygenase, 3-hydroxyantranilate 3,4-dioxygenase, 3-*O*-methylgallate 3,4-dioxygenase, 2,3-dihydryphenylpropionate 1,2-dioxygenase, 4,5-DOPA dioxygenase, 2-aminophenyl-1,6-dioxygenase, protocatechuate 4,5-dioxygenase, 3,4-dihydroxyphenylacetate 2,3-dioxtygenase, catechol 1,2-dioxygenase, protocatechuate 3,4-dioxygenase, gentisate 1,2-dioxygenase, homogentisase 1,2-dioxygenase, anthranilate 1,2-dioxygenase, benzoate 1,2-dioxygenase, naphthalene 1,2-dioxygenase, 2-halobenzoate 1,2-dioxygenase, biphenyl 2,3-dioxygenase, 3-phenylpropanoate dioxygenase and p-cumate 2,3-dioxygenase) were identified in the diesel oil-degrading bacterial metagenome by means of annotations and validated by blast searches against the nucleotide (nt) NCBI database (October 2018). Results were further filtered by 75% sequence identity, 50% coverage and a minimum of 1×10^{-10} expected value. For queries without significant hits against nt NCBI database, protein searches against non-redundant (nr) NCBI database (October 2018) were used instead.

2.5. Bioremediation Treatments in Microcosms

To evaluate the bioremediation feasibility of the diesel-degrading bacterial consortium, four-month microcosms systems with two different treatments were evaluated. The microcosms and the treatments are detailed below:

(a) Soil homogenization processing. Diesel-polluted bulk soil from ship fuel tank spills was collected in San Fernando (Cádiz, Spain). The soil was homogenized by a first sieving process with a < 4 mm net, followed by manual homogenization. The soil was then automatically quartered with 2, 4 and 8 divisions. Finally, 200 g of this sieved, homogenized and quartered soil was included in an automatic tumbler for 12 h to ensure homogeneity before placing it into pots. The initial diesel concentration of the pot's soil was 2974 ± 143 mg·kg^{-1}. The soil had a water holding capacity of 32.25 mL·100 g^{-1}, a pH of 8.165, an electrical conductivity of 203 µL·cm^{-1}, 512.1 mg·L^{-1} of nitrogen, 4.32 mg·kg^{-1} of phosphate and 23.16 mg·kg^{-1} of easily oxidizable carbon (EOC).

(b) Treatment 1. Pots with 200 g of the homogenized soil previously described were surface-inoculated once, at the beginning of the experiment, with 1 mL of washed diesel-degrading bacterial consortium after 48 h of growth and concentrated to a final $DO_{600} = 0.6$. Four replicates of the treatment 1 together with other four control replicates without the bacterial inoculum were placed.

(c) Treatment 2. Five one-week old alfalfa (*Medicago sativa*) were transplanted into each pot, consisting in 200 g of the homogenized soil previously described. Alfalfa seeds were surface sterilized with 70% ethanol for 3 min and 5% NaClO for 10 min, washed 10 times with sterile distilled water and pre-germinated in 1% (w/v) sterilized agar-water plates at 28 °C before transplant. 1 mL of the bacterial consortium specified above was inoculated per pot to the stem base of the plants (0.2 mL per plant). Four replicates of the treatment 2 together with other four control replicates without the bacterial inoculum were placed.

(d) Microcosms conditions. The two treatments together with the controls were kept for four months in culture chambers with a photoperiod of 16/8 h light/dark and 25/18 °C and maintaining an 80% soil humidity with Fahraeus Plant (FP) medium [53] when needed. The experiment started when the bacterial inoculum was added.

2.6. Total Petroleum Hydrocarbon and PAHs Characterization

Total petroleum hydrocarbons (TPHs) and other hydrocarbon fractions in soils, were analyzed by gas chromatography (GC) according to the procedure previously described [54]. Briefly, 1 g of duplicates dry soil samples were microwave-extracted by a mixture of hexane/acetone (1:1) and extracts with petroleum hydrocarbons were subsequently fractioned by a solid phase extraction (SPE) procedure. Aliphatic and aromatic fractions were finally analyzed by GC with a flame ionization detector (GC-FID). Sample analyses for PAH determination in the diesel fuel used for enrichment cultures were performed on an Agilent series 1200 high-performance liquid chromatograph (HPLC) coupled to an Agilent 1100 fluorescent detector (FD, Waldbronn, Germany). Diesel was weighted to obtain more precise PAHs measurements. Particular conditions were previously optimized [55].

2.7. Sequence Deposition

Raw reads of the microbiomes 16S rRNA gene amplicons and the whole-metagenome shotgun sequence of the diesel-degrading consortium have been deposited in the NCBI Sequence Read Archive (SRA) and are available under the BioProject accession number PRJNA525339 and SRAs SRR8663212-SRR8663218.

3. Results and Discussion

3.1. Diesel Characterization

The initial characterization of the aliphatic and aromatic hydrocarbon fractions in the diesel oil from ship fuel tanks and the aged diesel-polluted soil used in this study by means of gas chromatography [54], shows a prevalence of middle-chain to long-chain aliphatic hydrocarbons (C_{12} to C_{35}) and C_{16}–C_{35} aromatic hydrocarbons (Table 1). The aged soil, compared with the diesel oil from the tanks, is enriched in aliphatic >C_{21}–C_{35} and >C_{35} and aromatic >EC_{21}–EC_{35} and >EC_{35} fractions while a reduction in aliphatic >C_{10}–C_{12} in the diesel-polluted soil is observed. This was expected as short-chain alkanes are more volatile and prone to bioremediation than long-chain alkanes and PAHs [56].

Table 1. Aliphatic and aromatic hydrocarbon fraction composition of the diesel oil and aged diesel-polluted soil used in this study.

TPH Fraction	Diesel Oil (mg·mL^{-1})	Soil (µg·g^{-1})
	Aliphatic hydrocarbons	
>C_{10}–C_{12}	82 ± 1	3.5 ± 0.5
>C_{12}–C_{16}	257 ± 7	151 ± 4
>C_{16}–C_{21}	283 ± 8	563 ± 28
>C_{21}–C_{35}	55 ± 4	1086 ± 73
>C_{35}	0.05 ± 0.001	116 ± 16
	Aromatic hydrocarbons	
>EC_{10}–C_{12}	17 ± 1	11 ± 5
>EC_{12}–C_{16}	13 ± 1	8 ± 1
>EC_{16}–C_{21}	57 ± 3	484 ± 48
>EC_{21}–C_{35}	2 ± 0.1	530 ± 70
>EC_{35}	0.1 ± 0.004	22 ± 4
TPHs	764 ± 7	2974 ± 143

Among the aromatic hydrocarbon fraction of the diesel oil analyzed by HPLC/FD, the composition of the diesel oil is mainly supported by 2-methylnaphthalene (4000 µg·g^{-1}), 1-methylnaphthalene (1300 µg·g^{-1}), and naphthalene (870 µg·g^{-1}). Other constituents are phenanthrene (720 µg·g^{-1}), fluorene (230 µg·g^{-1}), acenaphthene (90 µg·g^{-1}), pyrene (40 µg·g^{-1}), anthracene (30 µg·g^{-1}), chrysene (27 µg·g^{-1}), benzo(b)fluoranthene (2.6 µg·g^{-1}), benzo(k)fluoranthene (2 µg·g^{-1}), and benzo(a)pyrene

(2 μg·g^{-1}). Trace aromatic hydrocarbons are fluoranthene (<1 μg·g^{-1}), benzo(a)anthracene (<1 μg·g^{-1}), dibenzo(ah)anthracene (<1 μg·g^{-1}), and benzo(ghi)perylene (<1 μg·g^{-1}).

3.2. Bacterial Diversity in the Diesel-Degrading Consortium

Sequencing of the 16S rRNA gene in the diesel-degrading consortium resulted in a total of 47,306 sequences assigned to 76 different amplicon sequence variants (ASV). The rarefaction curve obtained (Figure 1a) shows a clear community coverage, as saturation of observed ASVs is achieved before 40,000 sequences and the presence of other taxa in the consortium is unlikely. The relative abundance of genera assigned to these sequences shows dominance of *Pseudomonas*, *Aquabacterium*, and *Chryseobacterium*, with relative abundances of 27.01%, 22.36%, and 15.34%, respectively (Figure 1b). Other genera with a representative abundance in the diesel-degrading consortium are *Sphingobium*, *Novosphingobium*, *Dokdonella*, *Parvibaculum*, and *Achromobacter* (5.2%, 3.65%, 3.29%, 3.24%, and 2.45%% of relative abundance, respectively). This abundance is detailed in Supplementary Table S1. On the other hand, the metagenome shotgun sequencing of the diesel-degrading consortium resulted in 140 Mbps, distributed in 114,357 contigs (Supplementary Table S2), 18,473 of them > 1 Kbp. After annotation, 120,867 CDSs were identified and roughly 65% of them could be assigned to the genus level (78,110). The relative abundance of these CDSs in the diesel-degrading consortium shows a major difference of populations, as shown in Supplementary Table S1. Although CDSs of *Pseudomonas* remain as the most abundant (15.53%), CDSs of *Aquabacterium* and *Chryseobacterium* are scarce with relative abundances of 0.04% and 2.82%, respectively, while CDSs from *Achromobacter* and *Commamonadaceae* bacteria are increased (11.07% and 15.08% in CDSs abundance, while 2.45% and 2.64% in 16S rRNA abundance, Supplementary Table S1). Interestingly, other genera that have little representation in 16S rRNA sequences appear in the CDSs genus assignation, such as *Cupriavidus* (6.99% in CDSs while 1.15% in 16S rRNA). However, these differences between 16S rRNA and CDSs are diminished at the class level (Supplementary Table S1), which might suggest an unreliable genus assignation of CDSs, lack of representative sequences for all genera in the NCBI nt database, a primer bias of 16S rRNA sequence or failed prediction of ORFs in small contigs. This result was not unexpected, as it has been previously reported [33]. The taxa identified in the microbiome of the diesel-degrading consortium are in agreement with previous works, where it has been shown what *Pseudomonas* is one of the most abundant genera on hydrocarbon-polluted soils [34,57,58] and *Aquabacterium* and *Chryseobacterium* are also common members in hydrocarbon-degrading bacterial communities [58,59]. Although *Pseudomonas* also rules the degradation of PAHs in sediments [60], other genera present in the diesel-degrading consortium belonging to the *Sphingomonadaceae* family have also been previously reported to be responsible for the degradation of different PAHs [61,62].

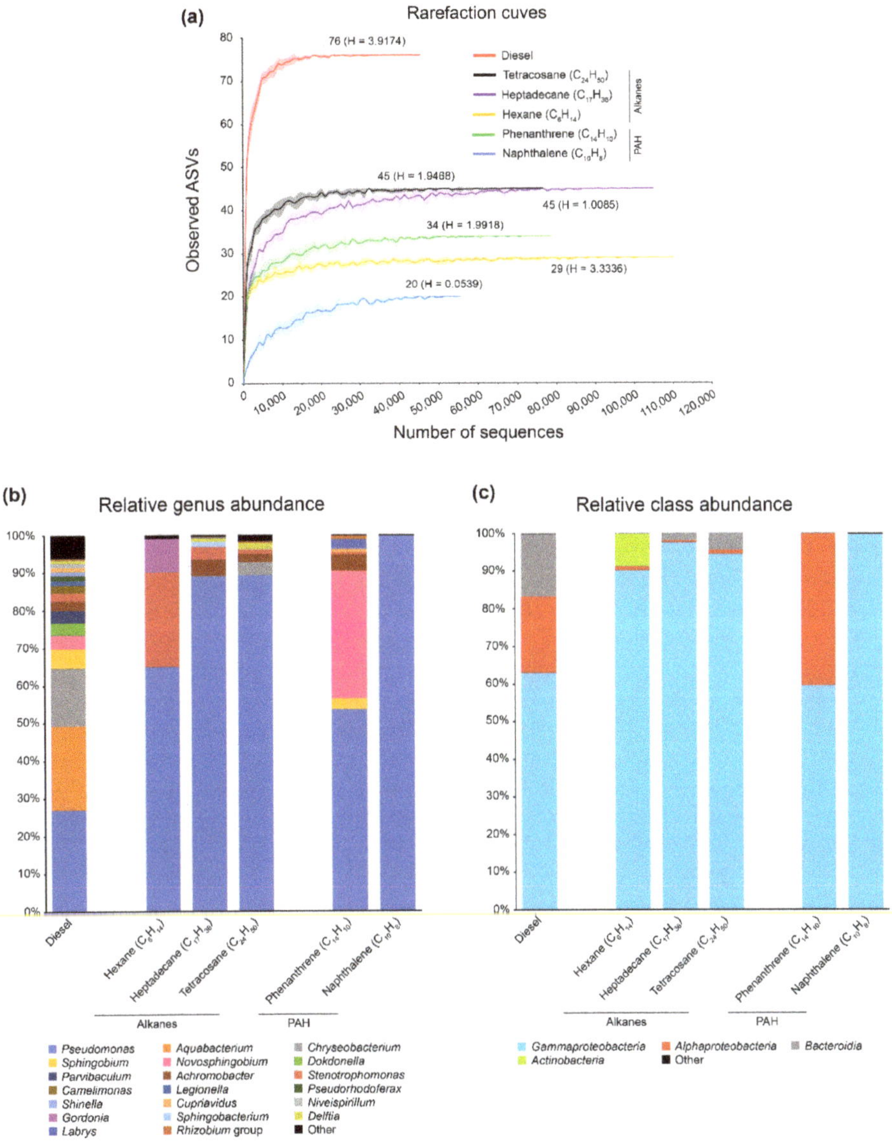

Figure 1. Diversity and taxonomic composition of the diesel-degrading consortium growing on diesel and different alkanes and polycyclic aromatic hydrocarbons (PAHs) as sole carbon and energy source. (**a**) Rarefaction curves of observed amplicon sequence variants (ASVs) over the number of sequences sampled. Lines represent mean values while colored shadows represent standard deviation over 10 iterations. All curves show complete community coverages. Number of ASVs in each microbiome and Shannon diversity index (H) are indicated above/below each curve. (**b**) Relative genus abundance or (**c**) class abundance based on 16S rRNA. Only taxa with a minimum relative abundance of 1% across samples are represented. For detailed abundance distribution, see Supplementary Table S1.

3.3. Substrate-Specific Diversity

To address the changes in populations of the diesel-degrading consortium that might be occurring due to specific constituents of diesel, microbiome analyses were performed with the consortium growing on three different *n*-alkanes (hexane, heptadecane, and tetracosane) and two PAHs (phenanthrene and naphthalene) as sole carbon and energy source. It is important to note that growth patterns and yield were different on different hydrocarbons (Supplementary Figure S1). The yield was very low in the case of phenanthrene. When growing on this hydrocarbon, OD_{600} at sampling time (48 h) was only 0.06, compared to 0.25–0.5 for the other hydrocarbons. Therefore, phenanthrene is a poor carbon and energy source for bacteria present in the consortium. Regarding the growth pattern, polyphasic curves were obtained for growth on hexane, heptadecane, tetracosane, and phenanthrene, indicating probably a succession in the bacterial populations present in the consortium. Taken together, it is likely that depending on the carbon and energy source, different populations are thriving at different times, and therefore, the detected microbiota reflects a snapshot at the sampling time. As expected, the number of ASVs varies greatly depending on the specific substrate, as shown in Figure 1a. While the consortium growing in tetracosane and heptadecane presents the highest number of ASVs (45), followed by phenanthrene (34), hexane (29), and naphthalene (20), the Shannon diversity index is considerably higher in the consortium growing with hexane (H = 3.3) and lower in the consortium growing with naphthalene (H = 0.05) (Figure 1a). Rarefaction curves, in all cases show nearly complete community coverage. Regarding bacterial abundance, all alkanes are dominated by *Pseudomonas*. In the case of heptadecane and tetracosane, *Pseudomonas* represent the ~89% of the bacterial community and little changes are observed in the remaining genera, none of them representing more than 5% of relative abundance (Figure 1b, Supplementary Table S1). The similarity of the bacterial populations in both, middle and long-chain alkanes, suggest that the same populations are involved in the degradation of both hydrocarbons. On the other hand, diversity of the consortium growing with hexane, shows a dominancy of *Pseudomonas* (64.92%), *Stenotrophomonas* (25.23%), and *Gordonia* (8.7%). *Gordonia* is known to degrade short-chain gaseous alkanes, such as propane [10,63], which could explain the abundance of this genera in the consortium growing with hexane, although there are also reports of *Pseudomonas* strains that are also able to degrade short-chain alkanes [64]. In the case of *Stenotrophomonas*, it has been suggested that the high metabolic versatility of this genus [65] might contribute to its ubiquity in different bacterial populations, including hydrocarbon and PAH-degrading communities [33,60,66] by cross-feeding on secondary metabolites.

Regarding the relative abundance of the consortium growing with two different PAHs as sole carbon and energy source, naphthalene-degrading diversity is almost exclusive to *Pseudomonas*, representing the 99.72% of the bacterial community (Figure 1b). On the other hand, the bacterial populations that thrive in the phenanthrene culture are mainly distributed between *Pseudomonas* (53.6%) and *Novosphingobium* (33.69%), which is in agreement with previous reports [60–62].

3.4. Identification of Alkane-Degrading CDSs

In order to identify putative active populations in the degradation of alkanes, the metagenome CDSs of the diesel-degrading consortium were screened to find alkane 1-monooxygenases (AlkB), cytochrome P450 alkane hydroxylases from the CYP153 family and long-chain alkane monooxygenases (LadA), whose role in *n*-alkane degradation have been extensively studied [7,16,17,67,68]. The results are summarized in Figure 2 (for details see Supplementary Table S3).

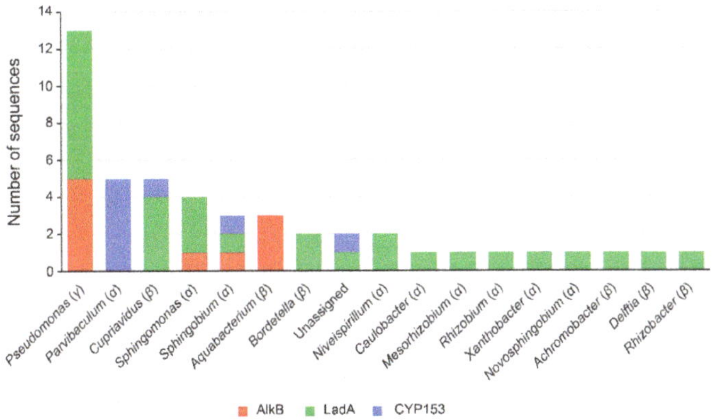

Figure 2. Number and taxonomic assignation at the genus level of the enzymes belonging to alkyl group hydroxylases; AlkB (alkane 1-monooxygenase), LadA (long-chain alkane monooxygenase), and CYP153 (cytochrome P450 family CYP153). Class adscription of the genera depicted is indicated under parenthesis. For additional information, see Supplementary Table S3.

AlkB is a non-heme iron integral membrane protein that is responsible for the initial hydroxylation of a diverse range of *n*-alkanes [8,13,67]. Ten putative AlkB have been identified in the metagenome of the diesel-growing bacterial consortium. Half of these AlkB have been classified as belonging to the *Pseudomonas* genus (5), three were assigned to *Aquabacterium*, and the remaining ones were classified as belonging to *Sphingomonas* and *Sphingobium* (Figure 2). On the other hand, LadA, a flavoprotein monooxygenase that inserts an oxygen atom into long-chain alkanes [16], was putatively found 29 times in the metagenome and was mainly assigned to *Pseudomonas* (8), *Cupriavidus* (4) and *Sphingomonas* (3) among others (Figure 2). Finally, CYP153 family of cytochrome P450 have been reported to display hydroxylating activity toward alkanes [17,68]. The metagenome of the diesel-growing consortium contains eight of these enzymes, which have been classified as belonging to *Parvibaculum* (5), *Sphingobium* (1), and *Cupriavidus* (1), while the remaining one could not be assigned to any genera.

These results are in agreement with the relative abundance of *Pseudomonas*, *Aquabacterium*, and *Sphingobium*, and other *Sphingomonadaceae* genera in the diesel-growing consortium and suggest that different genera are active in the degradation of the alkane constituents of diesel. The fact that the consortium growing on heptadecane and tetracosane is dominated by *Pseudomonas* could indicate that the rest of AlkB, LadA, and CYP153-containing bacteria plays a predominant role in the degradation of other alkanes or are specific for a certain alkane length or pathway. For instance, CYP153 coding sequences, which have been primarily assigned to *Parvivaculum* suggest that biterminal oxidation of alkanes is specific to this genus, although it could also hydroxylate alkanes on terminal positions. On the other hand, the diversity of ASVs found in the consortium growing with the long-chain alkane tetracosane (Figure 1) is congruent with the number of LadA enzymes found in the metagenome (29), although AlkB could also be involved on long-chain alkane degradation [15]. It is important to note that also genes participating in the early oxidation of alkanes or those belonging to low abundant bacteria could be missing from the analysis given that metagenome analysis was performed after 48 h of the diesel-degrading consortium growth.

3.5. Identification of PAH-Degrading CDSs and Central Aromatic Metabolism CDSs

Among PAHs present in diesel oil, naphthalene and its methyl derivatives are the most abundant (see above). Naphthalene biodegradation is initiated by the ring-hydroxylating naphthalene 1,2-dioxygenase (NahA) enzyme, whose implication in a wide range of different PAHs degradative reactions have been uncovered [69,70], including hydroxylation of anthracene, phenanthrene, and fluorene, and monooxygenation of acenaphthene among others [22,69,71]. Initial oxidation of naphthalene and other PAHs is followed by subsequent reactions until funneled into central aromatic degradation pathways, usually with catechol, gentisate, or protocatechuate as intermediaries depending on the specific PAH (for a review see [20]). The screening of the diesel-degrading consortium metagenome revealed the presence of 83 putative ring-hydroxylating dioxygenases, nine of which were annotated as naphthalene 1,2-dioxygenases (Figure 3a). Most of these NahA were assigned to *Sphingomonas* (5), *Sphingobium* (2), and *Bordetella* (2) (Figure 3b, Supplementary Table S3), which is in agreement with the fraction of *Alphaproteobacteria* present in the microbiome of the consortium growing with phenanthrene as the sole carbon and energy source (Figure 1). Unexpectedly, none of these NahA were assigned to *Pseudomonas* or even to *Gammaproteobacteria*, class that dominates both PAH-degrading microbiomes (59.59% and 99.75% in phenanthrene or naphthalene, respectively). The number of other ring-hydroxylating and ring-cleavage intra and estradiol dioxygenases is also scarce among *Gammaproteobacteria*, which suggest that the involvement of *Pseudomonas* in the degradation of PAHs in the diesel-degrading consortium might be attributed to the use of products of PAHs degradation rather than being involved on its initial oxidation. However, other causes cannot be ruled out, including low representation of enzymes not appearing at the metagenome depth this study was carried out.

Regarding central metabolism of aromatic compounds, catechol 1,2-dioxygenase, catechol 2,3-dioxygenase, gentisate 1,2-dioxygenase, and homogentisate 1,2-dioxygenase were among the most abundant CDSs found in the diesel-degrading metagenome (16, 17, 19 and 19 CDSs respectively, Figure 3a). Protocatechuate 3,4-dioxygenase, protocatechuate 4,5-dioxygenases, and benzoate 1,2-dioxygenase were also found in the metagenome (7, 8, and 10 CDSs, respectively). These results agree with the degradation pathways of the catabolic products of naphthalene, anthracene, phenanthrene, and fluorene, among others, via ortho or meta cleavage [20]. According to the taxonomic assignation of these central aromatic degradation enzymes, most of them belong to genera such as *Pseudomonas*, *Bordetella*, *Achromobacter*, *Sphingomonas*, *Sphingobium*, and *Cupriavidus*, among others (Figure 3b), which could explain their presence in the diesel-degrading consortium.

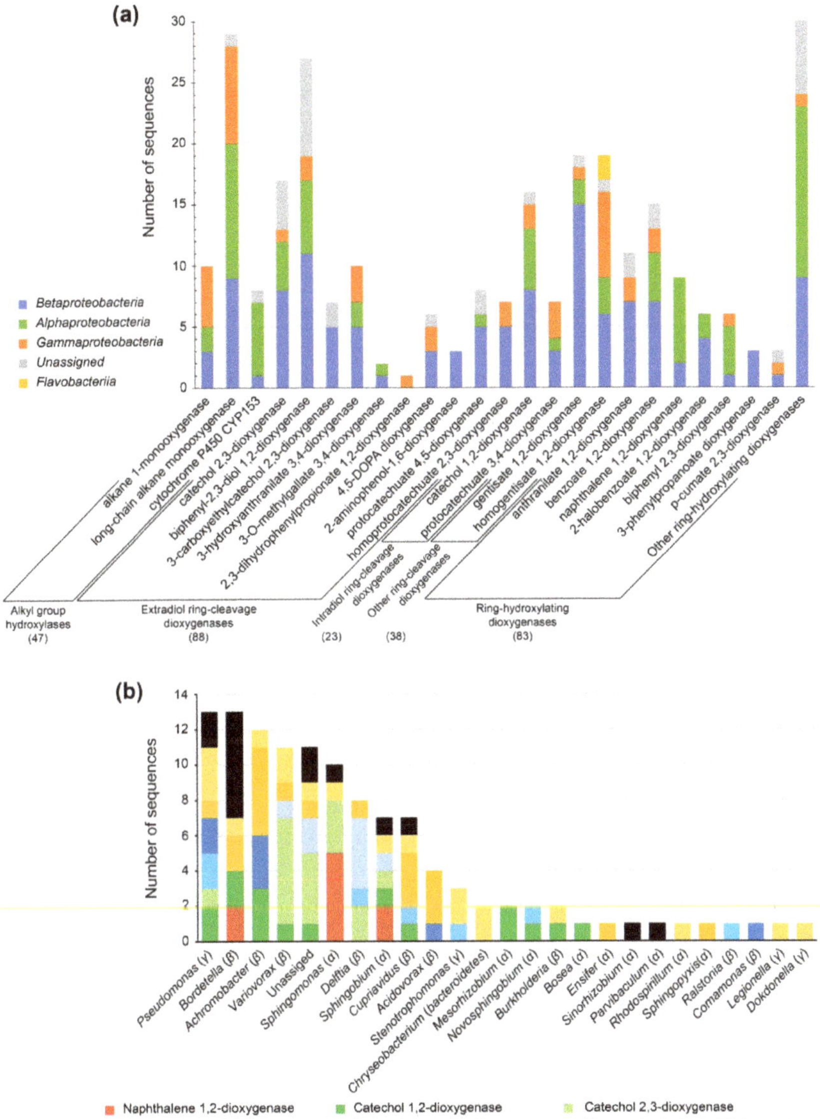

Figure 3. (**a**) Number and taxonomic assignation at the class level of the enzymes belonging to alkanes, PAHs and aromatic metabolism pathways. Total sequences of each main protein groups are indicated in parenthesis. (**b**) Classification at the genus level of naphthalene 1,2-dioxygenase and other central aromatic biodegradation pathways involving catechol, protocatechuate, gentisate, and benzoate. For additional information, see Supplementary Table S3.

3.6. Metabolic Roles of Specific Populations in the Diesel-Degrading Consortium

The identification and taxonomic assignment of enzymes involved in the initial hydroxylation of alkanes and PAHs within the diesel-degrading consortium metagenome, along with the characterization of the enzymes responsible of central aromatic degradation pathways, provides a profound understanding of the different roles of the main bacterial populations that thrive in the consortium with regard of their relative abundance. These results are summarized in Figure 4. The initial hydroxylation of alkanes is carried out by different bacteria, including *Pseudomonas*, *Aquabacterium*, *Sphingomonadaceae* family bacteria, and *Achromobacter*. Among these, alkane group hydroxylases (AlkB, LadA, and CYP153 cytochrome P450 family) are more abundant in *Pseudomonas* and *Sphingomonadaceae* family, containing 13 and 8 of these enzymes, respectively (Figure 4), which is consistent with their relative abundance in the consortium and with previous reports [34,57–59]. These genera could be responsible for the initial terminal or subterminal oxidation or biterminal oxidation of different chain-length alkanes. Conversely, the initial hydroxylation of PAHs based on the presence of naphthalene 1,2-dioxygenases found in the metagenome of the diesel-degrading consortium is primarily attributed to bacteria of the *Sphingomonadaceae* family. High redundancy of central aromatic degradation pathways is observed among the different genera present in the consortium, which could explain the diversity found in the diesel-degrading consortium and the population shift towards *Pseudomonas* when growing in alkanes as sole carbon and energy source (Figure 1). Although none of the naphthalene 1,2-dioxygenases found in the metagenome have been classified as belonging to *Pseudomonas*, the fact that this genus dominates the naphthalene-growing population (Figure 3b), might be also related with the functional redundancy of central aromatic pathways this genus exhibits (Figure 4). Nonetheless, genes from key species playing an important role in the early oxidation of alkanes and PAHs might be missing, since 48 h growth culture of the consortium culture was used to perform the analyses. The different growth pattern of the consortium on different carbon substrates (Supplementary Figure S1) shows that different populations could evolve on time. Further analyses to see the evolution of the community over time and the genes present in different growth stages could provide deeper insights into the biodegradation process.

Interestingly, two of the most abundant genera within the diesel-degrading consortium do not harbor many of the CDSs for these pathways. This is the case of *Aquabacterium* and *Chryseobacterium* (22.36% and 15.34% 16S rRNA relative abundance, respectively). The differences in relative abundance observed between 16S rRNA and CDSs in these genera might suggest a misclassification of CDSs. In the specific case of *Burkholderiales* order, the CDSs abundance of *Commamonadaceae* family and *Cupriavidus* genus (15.08% and 6.99%, respectively) is similar to *Aquabacterium* 16S rRNA (22.36%), another *Burkholderiales* order genus. This finding could explain the relatively low representation of *Aquabacterium* CDSs and its presence in the consortium. Nevertheless, this is not the case of *Chryseobacterium*, a *Flavovacteriia* class whose representation in CDSs is less than 3%. It is unclear if unclassified coding sequences could belong to this genus or the role it might have in the diesel-degrading consortium, even though *Chryseobacterium* have been previously identified in diesel fuel degrading consortia [66]. Other possibilities, such as the bacterial shift towards more metabolically versatile members in late states of the biodegradation process, which do not participate in the initial oxidation of diesel constituents, could also explain the presence of these genera.

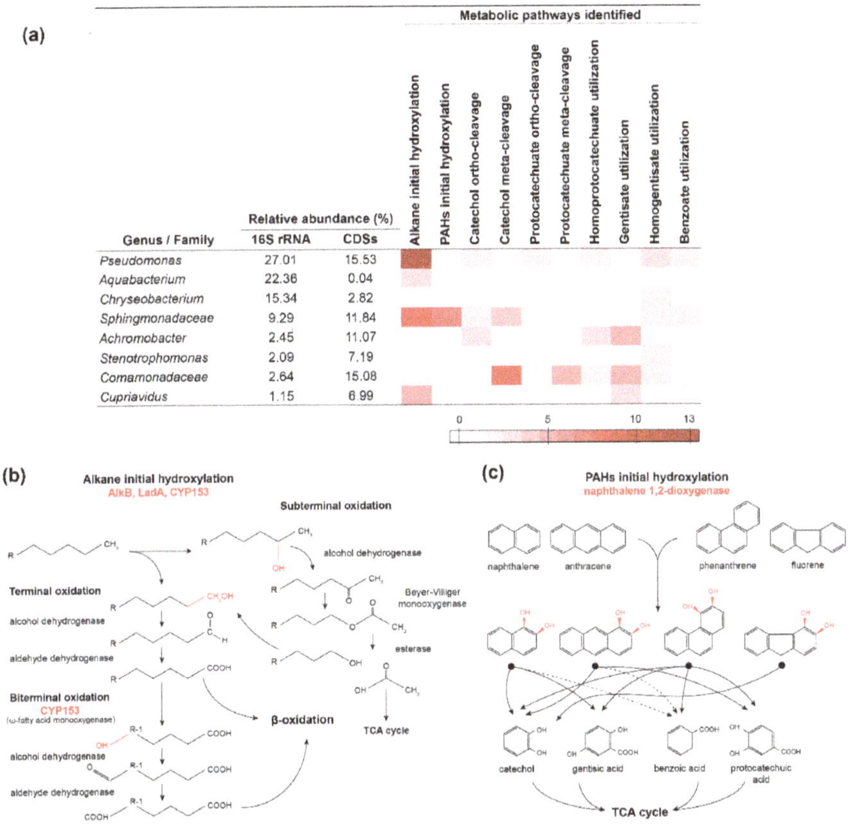

Figure 4. (**a**) Summary of alkanes, PAHs, and central aromatic biodegradation pathways found in the most abundant genera/families within the metagenome of the diesel-degrading consortium. Red scale bar represents the number of enzymes for each pathway identified. (**b**) Schematic view of the initial terminal, subterminal, and biterminal alkane aerobic oxidation mediated by AlkB, LadA, and CYP153 enzymes. Adapted from [7,8]. (**c**) Schematic view of initial naphthalene 1,2-dioxygenase hydroxylating reactions in PAHs components of diesel and aromatic central metabolites generated by further oxidation reactions of these *cis*-diol intermediates. Dotted lines indicate metabolic products depending on specific degradation pathways. Enzymes and chemical reactions catalyzed by these enzymes are indicated in red typing.

3.7. Rhizoremediation Assays in Diesel-Polluted Soil Microcosms

Rhizoremediation with indigenous hydrocarbon-degrading microorganisms have been proposed as one of the most effective techniques in restoring diesel-polluted soils [35,36,38], which could be enhanced by stimulation of the catalytic activities of microorganisms by plant roots [37,72] and can also be combined with other techniques such as chemical oxidation [73]. In order to test whether the bacterial consortium isolated in this study could be suited for rhizoremediation of the original diesel-polluted soil from where it was isolated, soil microcosms assays inoculated with the consortium were evaluated. Additionally, alfalfa (*Medicago sativa*) plants were used to address a possible stimulating effect. The results show a clear soil TPHs reduction after 4 months (Figure 5). In the control, untreated pot, TPHs reduced from the original 2974 mg·kg^{-1} to 2588 mg·kg^{-1}. In the soil treatment with the consortium resulted in a further reduction of 8.35%. This reduction was 12.36% with alfalfa plants without inoculum, probably due to the stimulation of indigenous populations already present in the soil. However, the combined effect of the consortium with alfalfa plants resulted in a 27.91% decrement in TPHs, when compared with the original soil, which tripled that of the consortium alone. The aromatic fraction in all cases was the most degraded, showing a 44.14% reduction in the consortium with plants treatment while the aliphatic fraction accounted for a 21.42% reduction (Figure 5).

(a)

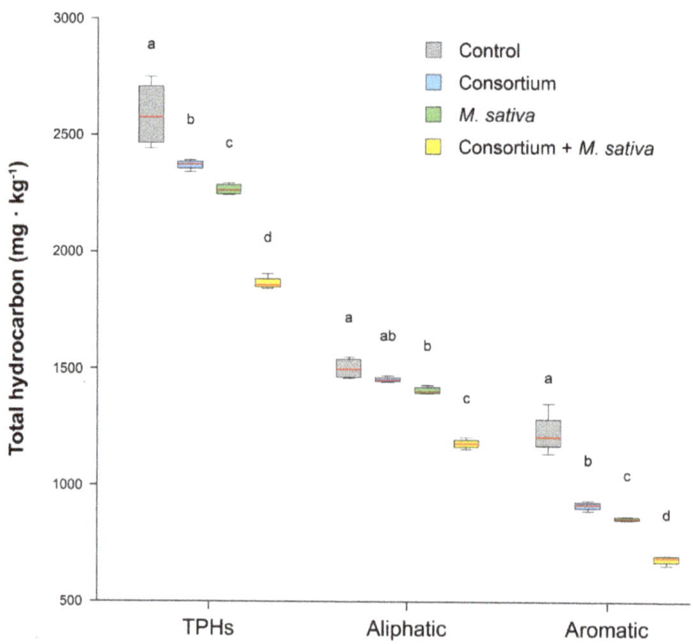

(b)

	Control	Consortium	M. sativa	Consortoum M. sativa	Control	Consortium	M. sativa	Consortoum M. sativa	Control	Consortium	M. sativa	Consortoum M. sativa
Mean	2,588.25	2,372.25	2,268.25	1,865.75	1,502.25	1,454.75	1,407.75	1,180.50	1,226.75	917.75	860.50	685.25
%	100	91.65	87.64	72.09	100	96.84	93.71	78.58	100	74.81	70.14	55.86
% Reduct.	0.00	8.35	12.36	27.91	0.00	3.16	6.29	21.42	0.00	25.19	29.86	44.14

Figure 5. (**a**) Gas chromatography with a flame ionized detector (GC-FID) results of total hydrocarbons remaining after four-month treatments with the diesel-degrading consortium, alfalfa (*M. sativa*) plants without inoculum, and combined consortium and alfalfa plants in four-month microcosms assays. Each box indicates Q1 and Q3, while the red line indicates the median values out of four replicates. Error bars indicate maximum and minimum values. Different letters indicate statistically significant differences ($p < 0.05$) using two-way ANOVA with Tukey's multiple comparison test corrections. (**b**) Mean values in each treatment in mg·kg^{-1} out of four replicates and percentages compared to the control.

Regarding the specific aliphatic and aromatic chain content, the consortium combined with alfalfa plants showed a major reduction of short and long-chain alkanes (>C_{10}–C_{12}, >C_{21}–C_{35} and >C_{35}), showing a 65.96%, 27.31%, and 31.36% hydrocarbon reduction respectively (Supplementary Figure S2). These results are consistent with the number of long-chain alkane monooxygenases found in the metagenome and the fact that the consortium is able to grow in the presence of the short-chain alkane hexane. In the case of aromatic hydrocarbons, the diesel-degrading consortium combined with alfalfa plants resulted in a major decrease in most cases (Supplementary Figure S2). This reduction is more substantial in the case of the >EC_{10}–EC_{16} fraction (41.67% and 57.69% for >EC_{10}–C_{12} and

$>EC_{12}$–C_{16}, respectively) which is compatible with naphthalene (C_{10}), fluorene (C_{13}), anthracene (C_{14}), and phenanthrene (C_{14}), among others. The observed effect caused by alfalfa plants in both, aliphatic and aromatic hydrocarbons was not unexpected, as it has been previously reported [36,38,74] and might be attributed to induction of microbial biodegradation pathways by plant metabolites [75] and other stimulating effects [37,76].

The microcosms results show that the bacterial consortium isolated in this study could serve as an inoculum for effective rhizoremediation of diesel-polluted soils. However, further analyses are required to evaluate its potential in other polluted sites, whose hydrocarbon composition might vary, and to evaluate the plant factors affecting the rhizoremediation process. Additional analyses at different times of the bioremediation process could also provide powerful insights into the community evolution and to identify key bacterial roles of the consortium.

4. Conclusions

Complex pollutants such as diesel require multiple microorganisms for their degradation. We have shown here that an effective autochthonous bacterial consortium can be constructed by successive enrichment cultivation of soils from contaminated sites. By metagenomic analysis of the consortium, growing on diesel and on specific aliphatic and polyaromatic hydrocarbons, we have been able to determine the bacterial genera composition of the consortium, the genes and enzymes implicated in diesel degradation and the specific degradative roles of the major populations within the consortium. The functional redundancy observed in the metagenome might be related to the plasticity that allows the populations to adapt to changes in the environment, and therefore conferring robustness to the degrading hydrocarbon system.

Supplementary Materials: The following are available online at http://www.mdpi.com/2073-4425/10/6/456/s1. Table S1: Relative abundance of the diesel-degrading consortium based on 16S rRNA gene and CDSs from the metagenome annotations. Table S2: Statistics of the 16S rRNA and whole-metagenome shotgun sequencing and processing of reads. Table S3: CDSs for alkanes and PAHs initial hydroxylation and others involved in central aromatic metabolism. Figure S1: Growth curves of the diesel-degrading consortium in each of the substrates used in this study. Figure S2: Detailed GC-FID results of aliphatic and aromatic hydrocarbons fractions remaining after microcosms treatments.

Author Contributions: Conceptualization, D.G.-S., M.R.-N., M.M., R.M., and R.R.; methodology, D.G.-S., M.G., O.P.J., and M.R.-N.; software, D.G.-S. and M.R.-N.; validation, R.R., M.M., R.M., and M.R.-N.; formal analysis, D.G.-S., M.G., O.P.J., M.R.-N., M.M., R.M., and R.R.; investigation, D.G.-S., M.G., and O.P.J.; resources, D.G.-S., M.G., O.P.J., M.R.-N., R.M., M.M., and R.R.; data curation, D.G.-S. and M.R.-N.; writing—original draft preparation, D.G.-S. and R.R.; writing—review and editing, D.G.-S., M.R.-N., M.M., and R.R.; visualization, D.G.-S., and M.G.; supervision, D.G.-S., M.R.-N., R.M., M.M., and R.R.; project administration, M.M., R.M., and R.R.; funding acquisition, M.M., R.M., and R.R.

Funding: This research was funded by GREENER-H2020 (EU), grant number 826312, Ministerio de Ciencia, Innovación y Universidades/FEDER EU, grant number RTI2018-0933991-B-I00 and the BIOXISOIL project, grant number LIFE11/ENV/ES/000505. D.G.-S. was granted by the MECD FPU fellowship program, grant number FPU14/03965.

Acknowledgments: We thank Mª José Sierra Herráiz for her advice on the selection of alfalfa seeds and their optimal growth conditions in the culture chamber.

Conflicts of Interest: The authors declare no conflict of interest.

References

1. Risher, J.; Rhodes, S. *Toxicological Profile for Fuel Oils*; U.S. Department of Health and Human Services, Public Health Service, Agency for Toxic Substances and Disease Registry: Atlanta, GA, USA, 1995.
2. Nessel, C.S. A comprehensive evaluation of the carcinogenic potential of middle distillate fuels. *Drug Chem. Toxicol.* **1999**, *22*, 165–180. [CrossRef] [PubMed]
3. Hentati, O.; Lachhab, R.; Ayadi, M.; Ksibi, M. Toxicity assessment for petroleum-contaminated soil using terrestrial invertebrates and plant bioassays. *Environ. Monit. Assess.* **2013**, *185*, 2989–2998. [CrossRef] [PubMed]

4. Ramadass, K.; Megharaj, M.; Venkateswarlu, K.; Naidu, R. Ecological implications of motor oil pollution: Earthworm survival and soil health. *Soil Biol. Biochem.* **2015**, *85*, 72–81. [CrossRef]
5. Pulles, T.; van der Gon, H.D.; Appelman, W.; Verheul, M. Emission factors for heavy metals from diesel and petrol used in European vehicles. *Atmos. Environ.* **2012**, *61*, 641–651. [CrossRef]
6. Riis, V.; Babel, W.; Pucci, O.H. Influence of heavy metals on the microbial degradation of diesel fuel. *Chemosphere* **2002**, *49*, 559–568. [CrossRef]
7. Ji, Y.; Mao, G.; Wang, Y.; Bartlam, M. Structural insights into diversity and n-alkane biodegradation mechanisms of alkane hydroxylases. *Front. Microbiol.* **2013**, *4*, 58. [CrossRef] [PubMed]
8. Van Beilen, J.; Li, Z.; Duetz, W.; Smits, T.; Witholt, B. Diversity of alkane hydroxylase systems in the environment. *Oil Gas Sci. Technol.* **2003**, *58*, 427–440. [CrossRef]
9. Feng, L.; Wang, W.; Cheng, J.; Ren, Y.; Zhao, G.; Gao, C.; Tang, Y.; Liu, X.; Han, W.; Peng, X.; et al. Genome and proteome of long-chain alkane degrading *Geobacillus thermodenitrificans* NG80-2 isolated from a deep-subsurface oil reservoir. *Proc. Natl. Acad. Sci. USA* **2007**, *104*, 5602–5607. [CrossRef]
10. Kotani, T.; Yamamoto, T.; Yurimoto, H.; Sakai, Y.; Kato, N. Propane monooxygenase and NAD+-dependent secondary alcohol dehydrogenase in propane metabolism by *Gordonia* sp. strain TY-5. *J. Bacteriol.* **2003**, *185*, 7120–7128. [CrossRef]
11. Throne-Holst, M.; Wentzel, A.; Ellingsen, T.E.; Kotlar, H.K.; Zotchev, S.B. Identification of novel genes involved in long-chain n-alkane degradation by *Acinetobacter* sp. strain DSM 17874. *Appl. Environ. Microbiol.* **2007**, *73*, 3327–3332. [CrossRef]
12. Coon, M.J. Omega oxygenases: Nonheme-iron enzymes and P450 cytochromes. *Biochem. Biophys. Res. Commun.* **2005**, *338*, 378–385. [CrossRef] [PubMed]
13. Van Beilen, J.B.; Wubbolts, M.G.; Witholt, B. Genetics of alkane oxidation by *Pseudomonas oleovorans*. *Biodegradation* **1994**, *5*, 161–174. [CrossRef] [PubMed]
14. Hamamura, N.; Yeager, C.M.; Arp, D.J. Two Distinct Monooxygenases for Alkane Oxidation in *Nocardioides* sp. Strain CF8. *Appl. Environ. Microbiol.* **2001**, *67*, 4992–4998. [CrossRef] [PubMed]
15. Bihari, Z.; Szvetnik, A.; Szabó, Z.; Blastyák, A.; Zombori, Z.; Balázs, M.; Kiss, I. Functional analysis of long-chain n-alkane degradation by *Dietzia* spp. *FEMS Microbiol. Lett.* **2011**, *316*, 100–107. [CrossRef] [PubMed]
16. Li, L.; Liu, X.; Yang, W.; Xu, F.; Wang, W.; Feng, L.; Bartlam, M.; Wang, L.; Rao, Z. Crystal structure of long-chain alkane monooxygenase (LadA) in complex with coenzyme FMN: Unveiling the long-chain alkane hydroxylase. *J. Mol. Biol.* **2008**, *376*, 453–465. [CrossRef] [PubMed]
17. Scheps, D.; Malca, S.H.; Hoffmann, H.; Nestl, B.M.; Hauer, B. Regioselective omega-hydroxylation of medium-chain n-alkanes and primary alcohols by CYP153 enzymes from *Mycobacterium marinum* and *Polaromonas* sp. strain JS666. *Org. Biomol. Chem.* **2011**, *9*, 6727–6733. [CrossRef]
18. Cerniglia, C.E. Biodegradation of polycyclic aromatic hydrocarbons. *Curr. Opin. Biotechnol.* **1993**, *4*, 331–338. [CrossRef]
19. Samanta, S.K.; Singh, O.V.; Jain, R.K. Polycyclic aromatic hydrocarbons: Environmental pollution and bioremediation. *Trends Biotechnol.* **2002**, *20*, 243–248. [CrossRef]
20. Mallick, S.; Chakraborty, J.; Dutta, T.K. Role of oxygenases in guiding diverse metabolic pathways in the bacterial degradation of low-molecular-weight polycyclic aromatic hydrocarbons: A review. *Crit. Rev. Microbiol.* **2011**, *37*, 64–90. [CrossRef]
21. Ensley, B.; Gibson, D. Naphthalene dioxygenase: Purification and properties of a terminal oxygenase component. *J. Bacteriol.* **1983**, *155*, 505–511.
22. Jerina, D.M.; Selander, H.; Yagi, H.; Wells, M.C.; Davey, J.F.; Mahadevan, V.; Gibson, D.T. Dihydrodiols from anthracene and phenanthrene. *J. Am. Chem. Soc.* **1976**, *98*, 5988–5996. [CrossRef]
23. Bouchez, M.; Blanchet, D.; Vandecasteele, J. Degradation of polycyclic aromatic hydrocarbons by pure strains and by defined strain associations: Inhibition phenomena and cometabolism. *Appl. Microbiol. Biotechnol.* **1995**, *43*, 156–164. [CrossRef] [PubMed]
24. Cunliffe, M.; Kertesz, M.A. Effect of *Sphingobium yanoikuyae* B1 inoculation on bacterial community dynamics and polycyclic aromatic hydrocarbon degradation in aged and freshly PAH-contaminated soils. *Environ. Pollut.* **2006**, *144*, 228–237. [CrossRef] [PubMed]

25. Dandie, C.E.; Weber, J.; Aleer, S.; Adetutu, E.M.; Ball, A.S.; Juhasz, A.L. Assessment of five bioaccessibility assays for predicting the efficacy of petroleum hydrocarbon biodegradation in aged contaminated soils. *Chemosphere* **2010**, *81*, 1061–1068. [CrossRef] [PubMed]
26. Ranc, B.; Faure, P.; Croze, V.; Simonnot, M.O. Selection of oxidant doses for in situ chemical oxidation of soils contaminated by polycyclic aromatic hydrocarbons (PAHs): A review. *J. Hazard. Mater.* **2016**, *312*, 280–297. [CrossRef] [PubMed]
27. Solano-Serena, F.; Marchal, R.; Vandecasteele, J. Biodegradation of gasoline in the environment: From overall assessment to the case of recalcitrant hydrocarbons. *Oil Gas Sci. Technol.* **2001**, *56*, 479–498. [CrossRef]
28. Gallego, J.L.R.; Sierra, C.; Permanyer, A.; Peláez, A.I.; Menéndez-Vega, D.; Sánchez, J. Full-scale remediation of a jet fuel-contaminated soil: Assessment of biodegradation, volatilization, and bioavailability. *Water Air Soil Pollut.* **2011**, *217*, 197–211. [CrossRef]
29. Richard, J.; Vogel, T. Characterization of a soil bacterial consortium capable of degrading diesel fuel. *Int. Biodeterior. Biodegrad.* **1999**, *44*, 93–100. [CrossRef]
30. Dos Santos, H.F.; Cury, J.C.; do Carmo, F.L.; dos Santos, A.L.; Tiedje, J.; van Elsas, J.D.; Rosado, A.S.; Peixoto, R.S. Mangrove bacterial diversity and the impact of oil contamination revealed by pyrosequencing: Bacterial proxies for oil pollution. *PLoS ONE* **2011**, *6*, e16943. [CrossRef]
31. Margesin, R.; Hammerle, M.; Tscherko, D. Microbial activity and community composition during bioremediation of diesel-oil-contaminated soil: Effects of hydrocarbon concentration, fertilizers, and incubation time. *Microb. Ecol.* **2007**, *53*, 259–269. [CrossRef]
32. Sutton, N.B.; Maphosa, F.; Morillo, J.A.; Al-Soud, W.A.; Langenhoff, A.A.; Grotenhuis, T.; Rijnaarts, H.H.; Smidt, H. Impact of long term diesel contamination on soil microbial community structure. *Appl. Environ. Microbiol.* **2013**, *79*, 619–630. [CrossRef] [PubMed]
33. Garrido-Sanz, D.; Manzano, J.; Martin, M.; Redondo-Nieto, M.; Rivilla, R. Metagenomic Analysis of a Biphenyl-Degrading Soil Bacterial Consortium Reveals the Metabolic Roles of Specific Populations. *Front. Microbiol.* **2018**, *9*, 232. [CrossRef] [PubMed]
34. Yergeau, E.; Sanschagrin, S.; Beaumier, D.; Greer, C.W. Metagenomic analysis of the bioremediation of diesel-contaminated Canadian high arctic soils. *PLoS ONE* **2012**, *7*, e30058. [CrossRef] [PubMed]
35. Miya, R.K.; Firestone, M.K. Enhanced phenanthrene biodegradation in soil by slender oat root exudates and root debris. *J. Environ. Qual.* **2001**, *30*, 1911–1918. [CrossRef] [PubMed]
36. Chaineau, C.; Morel, J.; Oudot, J. Biodegradation of fuel oil hydrocarbons in the rhizosphere of maize. *J. Environ. Qual.* **2000**, *29*, 569–578. [CrossRef]
37. Kuiper, I.; Lagendijk, E.L.; Bloemberg, G.V.; Lugtenberg, B.J. Rhizoremediation: A beneficial plant-microbe interaction. *Mol. Plant-Microbe Interact.* **2004**, *17*, 6–15. [CrossRef]
38. Liste, H.-H.; Felgentreu, D. Crop growth, culturable bacteria, and degradation of petrol hydrocarbons (PHCs) in a long-term contaminated field soil. *Appl. Soil Ecol.* **2006**, *31*, 43–52. [CrossRef]
39. Wu, M.; Li, W.; Dick, W.A.; Ye, X.; Chen, K.; Kost, D.; Chen, L. Bioremediation of hydrocarbon degradation in a petroleum-contaminated soil and microbial population and activity determination. *Chemosphere* **2017**, *169*, 124–130. [CrossRef]
40. Brazil, G.M.; Kenefick, L.; Callanan, M.; Haro, A.; de Lorenzo, V.; Dowling, D.N.; O'Gara, F. Construction of a rhizosphere pseudomonad with potential to degrade polychlorinated biphenyls and detection of *bph* gene expression in the rhizosphere. *Appl. Environ. Microbiol.* **1995**, *61*, 1946–1952.
41. Bedard, D.L.; Unterman, R.; Bopp, L.H.; Brennan, M.J.; Haberl, M.L.; Johnson, C. Rapid assay for screening and characterizing microorganisms for the ability to degrade polychlorinated biphenyls. *Appl. Environ. Microbiol.* **1986**, *51*, 761–768.
42. Bolger, A.M.; Lohse, M.; Usadel, B. Trimmomatic: A flexible trimmer for Illumina sequence data. *Bioinformatics* **2014**, *30*, 2114–2120. [CrossRef] [PubMed]
43. Bankevich, A.; Nurk, S.; Antipov, D.; Gurevich, A.A.; Dvorkin, M.; Kulikov, A.S.; Lesin, V.M.; Nikolenko, S.I.; Pham, S.; Prjibelski, A.D.; et al. SPAdes: A new genome assembly algorithm and its applications to single-cell sequencing. *J. Comput. Biol. J. Comput. Mol. Cell Biol.* **2012**, *19*, 455–477. [CrossRef] [PubMed]
44. Gurevich, A.; Saveliev, V.; Vyahhi, N.; Tesler, G. QUAST: Quality assessment tool for genome assemblies. *Bioinformatics* **2013**, *29*, 1072–1075. [CrossRef] [PubMed]

45. Aziz, R.K.; Bartels, D.; Best, A.A.; DeJongh, M.; Disz, T.; Edwards, R.A.; Formsma, K.; Gerdes, S.; Glass, E.M.; Kubal, M.; et al. The RAST Server: Rapid annotations using subsystems technology. *BMC Genom.* **2008**, *9*, 75. [CrossRef] [PubMed]
46. Bolyen, E.; Rideout, J.R.; Dillon, M.R.; Bokulich, N.A.; Abnet, C.; Al-Ghalith, G.A.; Alexander, H.; Alm, E.J.; Arumugam, M.; Asnicar, F. QIIME 2: Reproducible, interactive, scalable, and extensible microbiome data science. *PeerJ Preprints* **2018**, *6*, e27295v2.
47. Callahan, B.J.; McMurdie, P.J.; Rosen, M.J.; Han, A.W.; Johnson, A.J.; Holmes, S.P. DADA2: High-resolution sample inference from Illumina amplicon data. *Nat. Methods* **2016**, *13*, 581–583. [CrossRef] [PubMed]
48. Katoh, K.; Misawa, K.; Kuma, K.; Miyata, T. MAFFT: A novel method for rapid multiple sequence alignment based on fast Fourier transform. *Nucleic Acids Res.* **2002**, *30*, 3059–3066. [CrossRef]
49. Price, M.N.; Dehal, P.S.; Arkin, A.P. FastTree 2—Approximately maximum-likelihood trees for large alignments. *PLoS ONE* **2010**, *5*, e9490. [CrossRef]
50. Bokulich, N.A.; Kaehler, B.D.; Rideout, J.R.; Dillon, M.; Bolyen, E.; Knight, R.; Huttley, G.A.; Caporaso, J.G. Optimizing taxonomic classification of marker-gene amplicon sequences with QIIME 2's q2-feature-classifier plugin. *Microbiome* **2018**, *6*, 90. [CrossRef] [PubMed]
51. Quast, C.; Pruesse, E.; Yilmaz, P.; Gerken, J.; Schweer, T.; Yarza, P.; Peplies, J.; Glockner, F.O. The SILVA ribosomal RNA gene database project: Improved data processing and web-based tools. *Nucleic Acids Res.* **2013**, *41*, D590–D596. [CrossRef]
52. Camacho, C.; Coulouris, G.; Avagyan, V.; Ma, N.; Papadopoulos, J.; Bealer, K.; Madden, T.L. BLAST+: Architecture and applications. *BMC Bioinform.* **2009**, *10*, 421. [CrossRef] [PubMed]
53. Fåhraeus, G. The infection of clover root hairs by nodule bacteria studied by a simple glass slide technique. *Microbiology* **1957**, *16*, 374–381. [CrossRef] [PubMed]
54. Jiménez, O.P.; Pastor, R.M.P.; Segovia, O.E. An analytical method for quantifying petroleum hydrocarbon fractions in soils, and its associated uncertainties. *Anal. Methods* **2014**, *6*, 5527–5536. [CrossRef]
55. García-Alonso, S.; Pérez-Pastor, R.; García-Frutos, F. An evaluation of analytical quality for selected PAH measurements in a fuel-contaminated soil. *Accredit. Qual. Assur.* **2011**, *16*, 369–377. [CrossRef]
56. Leahy, J.G.; Colwell, R.R. Microbial degradation of hydrocarbons in the environment. *Microbiol. Rev.* **1990**, *54*, 305–315.
57. Whyte, L.G.; Schultz, A.; Beilen, J.B.; Luz, A.P.; Pellizari, V.; Labbe, D.; Greer, C.W. Prevalence of alkane monooxygenase genes in Arctic and Antarctic hydrocarbon-contaminated and pristine soils. *FEMS Microbiol. Ecol.* **2002**, *41*, 141–150. [CrossRef]
58. Gontikaki, E.; Potts, L.D.; Anderson, J.A.; Witte, U. Hydrocarbon-degrading bacteria in deep-water subarctic sediments (Faroe-Shetland Channel). *J. Appl. Microbiol.* **2018**, *125*, 1040–1053. [CrossRef]
59. Viggor, S.; Joesaar, M.; Vedler, E.; Kiiker, R.; Parnpuu, L.; Heinaru, A. Occurrence of diverse alkane hydroxylase alkB genes in indigenous oil-degrading bacteria of Baltic Sea surface water. *Mar. Pollut. Bull.* **2015**, *101*, 507–516. [CrossRef]
60. Wald, J.; Hroudova, M.; Jansa, J.; Vrchotova, B.; Macek, T.; Uhlik, O. Pseudomonads Rule Degradation of Polyaromatic Hydrocarbons in Aerated Sediment. *Front. Microbiol.* **2015**, *6*, 1268. [CrossRef]
61. Rentz, J.A.; Alvarez, P.J.; Schnoor, J.L. Benzo[a]pyrene degradation by *Sphingomonas yanoikuyae* JAR02. *Environ. Pollut.* **2008**, *151*, 669–677. [CrossRef]
62. Van Herwijnen, R.; Wattiau, P.; Bastiaens, L.; Daal, L.; Jonker, L.; Springael, D.; Govers, H.A.; Parsons, J.R. Elucidation of the metabolic pathway of fluorene and cometabolic pathways of phenanthrene, fluoranthene, anthracene and dibenzothiophene by *Sphingomonas* sp. LB126. *Res. Microbiol.* **2003**, *154*, 199–206. [CrossRef]
63. Kotani, T.; Yurimoto, H.; Kato, N.; Sakai, Y. Novel acetone metabolism in a propane-utilizing bacterium, *Gordonia* sp. strain TY-5. *J. Bacteriol.* **2007**, *189*, 886–893. [CrossRef] [PubMed]
64. Takahashi, J.; ICHIKAWA, Y.; Sagae, H.; Komura, I.; Kanou, H.; Yamada, K. Isolation and identification of n-butane-assimilating bacterium. *Agric. Biol. Chem.* **1980**, *44*, 1835–1840. [CrossRef]
65. Hauben, L.; Vauterin, L.; Moore, E.R.; Hoste, B.; Swings, J. Genomic diversity of the genus *Stenotrophomonas*. *Int. J. Syst. Bacteriol.* **1999**, *49 Pt 4*, 1749–1760. [CrossRef]
66. Zanaroli, G.; Di Toro, S.; Todaro, D.; Varese, G.C.; Bertolotto, A.; Fava, F. Characterization of two diesel fuel degrading microbial consortia enriched from a non acclimated, complex source of microorganisms. *Microb. Cell Fact.* **2010**, *9*, 10. [CrossRef] [PubMed]

67. Kok, M.; Oldenhuis, R.; van der Linden, M.P.; Raatjes, P.; Kingma, J.; van Lelyveld, P.H.; Witholt, B. The *Pseudomonas oleovorans* alkane hydroxylase gene. Sequence and expression. *J. Biol. Chem.* **1989**, *264*, 5435–5441. [PubMed]
68. Maier, T.; Forster, H.H.; Asperger, O.; Hahn, U. Molecular characterization of the 56-kDa CYP153 from *Acinetobacter* sp. EB104. *Biochem. Biophys. Res. Commun.* **2001**, *286*, 652–658. [CrossRef] [PubMed]
69. Resnick, S.; Lee, K.; Gibson, D. Diverse reactions catalyzed by naphthalene dioxygenase from *Pseudomonas* sp strain NCIB 9816. *J. Ind. Microbiol.* **1996**, *17*, 438–457.
70. Ferraro, D.J.; Okerlund, A.; Brown, E.; Ramaswamy, S. One enzyme, many reactions: Structural basis for the various reactions catalyzed by naphthalene 1,2-dioxygenase. *IUCrJ* **2017**, *4*, 648–656. [CrossRef]
71. Selifonov, S.A.; Grifoll, M.; Eaton, R.W.; Chapman, P.J. Oxidation of naphthenoaromatic and methyl-substituted aromatic compounds by naphthalene 1,2-dioxygenase. *Appl. Environ. Microbiol.* **1996**, *62*, 507–514.
72. Pilon-Smits, E. Phytoremediation. *Annu. Rev. Plant Biol.* **2005**, *56*, 15–39. [CrossRef] [PubMed]
73. Del Reino, S.; Rodríguez-Rastrero, M.; Escolano, O.; Welte, L.; Bueno, J.; Fernández, J.; Schmid, T.; Millán, R. In Situ Chemical Oxidation Based on Hydrogen Peroxide: Optimization of Its Application to an Hydrocarbon Polluted Site. In *Environment, Energy and Climate Change I*; Springer: Berlin, Germany, 2014; pp. 207–228.
74. Huang, X.-D.; El-Alawi, Y.; Gurska, J.; Glick, B.R.; Greenberg, B.M. A multi-process phytoremediation system for decontamination of persistent total petroleum hydrocarbons (TPHs) from soils. *Microchem. J.* **2005**, *81*, 139–147. [CrossRef]
75. Singer, A.C.; Crowley, D.E.; Thompson, I.P. Secondary plant metabolites in phytoremediation and biotransformation. *Trends Biotechnol.* **2003**, *21*, 123–130. [CrossRef]
76. Nichols, T.; Wolf, D.; Rogers, H.; Beyrouty, C.; Reynolds, C. Rhizosphere microbial populations in contaminated soils. *Water Air Soil Pollut.* **1997**, *95*, 165–178. [CrossRef]

© 2019 by the authors. Licensee MDPI, Basel, Switzerland. This article is an open access article distributed under the terms and conditions of the Creative Commons Attribution (CC BY) license (http://creativecommons.org/licenses/by/4.0/).

MDPI
St. Alban-Anlage 66
4052 Basel
Switzerland
Tel. +41 61 683 77 34
Fax +41 61 302 89 18
www.mdpi.com

Genes Editorial Office
E-mail: genes@mdpi.com
www.mdpi.com/journal/genes

www.ingramcontent.com/pod-product-compliance
Lightning Source LLC
Chambersburg PA
CBHW040825250926
51850CB00007BA/1265
9783039362363